FOOD CHEMISTRY

Related AVI Books

ADVANCED SUGAR CHEMISTRY
  Shallenberger
AGRICULTURAL AND FOOD CHEMISTRY: PAST, PRESENT, FUTURE
  Teranishi
EXPERIMENTAL FOOD CHEMISTRY
  Mondy
FLAVOR TECHNOLOGY
  Heath
FOOD ANALYSIS LABORATORY EXPERIMENTS
  2nd Edition   *Meloan and Pomeranz*
FOOD ANALYSIS: THEORY AND PRACTICE
  Revised Edition   *Pomeranz and Nelson*
FOOD AND THE CONSUMER
  Revised Edition   *Kramer*
FOOD FOR THOUGHT
  2nd Edition   *Labuza and Sloan*
FOOD OILS AND THEIR USES
  Weiss
FOOD, PEOPLE AND NUTRITION
  Eckstein
FOOD SCIENCE
  3rd Edition   *Potter*
FUNDAMENTALS OF DAIRY CHEMISTRY
  2nd Edition   *Webb, Johnson and Alford*
FUNDAMENTALS OF FOOD CHEMISTRY
  American Edition   *Heimann*
HANDBOOK OF SUGARS
  Pancoast and Junk
PRESCOTT & DUNN'S INDUSTRIAL MICROBIOLOGY
  4th Edition   *Reed*
PRINCIPLES OF FOOD CHEMISTRY
  Revised Edition   *de Man*
SOURCE BOOK OF ENZYMOLOGY
  Schwimmer
SOURCE BOOK OF FLAVORS
  Heath
VITAMIN C IN HEALTH AND DISEASE
  Basu and Schorah

FOOD CHEMISTRY

Edited by   LILLIAN HOAGLAND MEYER

Professor and Head
Department of Chemistry
Western Michigan University
Kalamazoo, Michigan

THE AVI PUBLISHING COMPANY, INC.

WESTPORT, CONNECTICUT

Published by arrangement with
VAN NOSTRAND REINHOLD COMPANY

Fourth Printing
1982

Library of Congress Catalog Card Number: 60-11080
ISBN-0-87055-171-X

Printed in the United States of America

Preface

THE RAPID GROWTH of the food industry into big business and the changes in the number of items on the grocers' shelves, the many ready-to-eat products, the new controls on food additives, and the attempts to standardize some food articles all serve to emphasize the growing importance of the chemistry of foods. For centuries the production of food followed traditional procedures, and new ideas or methods were largely the result of empirical trial and error or accidental discoveries. Now a complete revolution in growing, processing, and marketing of food is here. The accumulation of scientific knowledge about food composition has been, in the main, slow. Dependent as the science of food is on biology, bacteriology, and mycology as well as chemistry, it was necessary for all sciences to develop to the point where elucidation of the complex mixtures encountered in foods became accurate and meaningful. Fortunately, research of significance to food chemistry is now appearing not only in journals devoted exclusively to food problems but also in those in the fields of biology, chemistry, chemical engineering, and even physics.

It is the author's belief that an attempt to consolidate the fundamentals of food chemistry with recent advances in the food industry is now appropriate. Although there have been splendid books available for many years in the area of food analysis and more lately in food technology, we have not had one that unifies these areas with basic chemistry in a simple fashion. The present book is designed to meet this need. It is intended primarily as a text in food chemistry for undergraduate students in home economics, food technology, and chemistry. However it should be useful also as a reference book for students and research workers in these fields.

Thus the aim of the book is to provide a unified picture of foods from a chemical standpoint. The primary emphasis is on the composition of foods and the changes that occur when they are subjected to processing. The author has tried to keep in mind undergraduates who have had a course in organic chemistry; however, an effort has been made to provide necessary background chemistry and sufficient explanations for students to proceed without additional references. Some brief descriptive orientation is given on the structure of plants and animals, but this is kept quite simple and is used to promote greater unity of presentation. Occasionally, material on food habits is included to add interest and to round out the close relationship of food chemistry to everyday life. The chapter on food additives and the appendix on permissible food additives are special features of this text that provide up-to-date information on this vital topic. At the end of each chapter are bibliographical references to aid the reader who wishes to investigate further the respective areas of food chemistry.

Thanks are due to many people for the successful completion of this book. Calvin A. VanderWerf and Betty Watts were especially helpful in their critical evaluation of the manuscript and Betty Taylor and Curtis Meyer for their suggestions, criticisms, and continued interest.

LILLIAN HOAGLAND MEYER

Kalamazoo, Michigan
June, 1960

Contents

CHAPTER ONE. DEVELOPMENT OF FOOD CHEMISTRY *1*

Individual Variability *3*
Individual Uniformity *4*
Methods of Sampling *4*
Moisture in Foods *5*
    HYDROGEN BONDING *6*
    BOUND WATER *8*
    DETERMINATION OF MOISTURE *8*
References *11*

CHAPTER TWO. FATS AND OTHER LIPIDS *12*

Occurrence in Foods and Composition *12*
Edible Fats and Oils *17*
    FATTY ACIDS *18*
    IDENTIFICATION OF NATURAL FATS AND OILS *21*
        Physical Properties *23*
            Melting Point *23*
            Softening Point *24*
            Slipping Point *25*
            Shot Melting Point *25*
            Specific Gravity *25*
            Refractive Index *25*
            Smoke, Flash, and Fire Points *26*
            Turbidity Point *26*
        Chemical Properties *27*

Flavor Changes in Fats and Oils *32*
RANCIDITY *32*
   Tests for Rancidity *38*
REVERSION *39*
The Technology of Edible Fats and Oils *42*
RENDERING *42*
PRESSING *43*
SOLVENT EXTRACTION *43*
REFINING *45*
   Steam Refining *46*
   Alkali Refining *46*
   Bleaching *46*
   Steam Deodorization *46*
HYDROGENATION *48*
TECHNOLOGY OF INDIVIDUAL FAT PRODUCTS *51*
   Butter *51*
   Oleo Oil and Oleostearin *52*
   Lard *52*
      Interesterification in Lard *53*
   Salad, Cooking, and Frying Oils *54*
   Shortenings *55*
   Margarine *57*
THE SHORTENING VALUE OF DIFFERENT FATS *60*
Importance in Diet *62*
References *63*

CHAPTER THREE. CARBOHYDRATES *65*

Monosaccharides *66*
Disaccharides *68*
Oligosaccharides *71*
Polysaccharides *72*
CELLULOSE *73*
STARCH *75*
   Enzymes and Starch *76*
      Alpha Amylase *76*
      Beta Amylase *77*
   Technology of Starch *77*
FRUCTANS *79*
MANNANS AND GALACTANS *79*
CHITIN *80*
HYALURONIC ACID AND THE CHONDROITIN SULFATES *80*

HEMICELLULOSE 83
PECTIC SUBSTANCES 87
 Protopectin 90
 Pectin Gels 92
  Jelly Grade 93
  Setting Time 93
  Gel Formation with Low Ester Pectins 94
  Pectins in Alcoholic Fermentation 95
GUMS AND MUCILAGES 95
 Gum Arabic 95
 Seaweed Polysaccharides 96
  Agar 96
  Alginic Acid 97
  Carageenan 97
 Locust Bean Gum 97
Identification 97
COLOR REACTION 97
REACTION PRODUCTS OF MONO- AND DISACCHARIDES
 WITH HYDRAZINES 100
FERMENTATION WITH YEAST 100
SEPARATION BY CHROMATOGRAPHY 101
OPTICAL ACTIVITY 102
OTHER QUANTITATIVE DETERMINATIONS 103
Changes of Carbohydrates on Cooking 104
(1) SOLUBILITY 104
(2) HYDROLYSIS 104
(3) GELATINIZATION OF STARCH 105
Crude Fiber 107
Browning Reactions 108
References 112

CHAPTER FOUR. PROTEINS IN FOODS 114

Proteins in Man's Diet 115
Chemical and Physical Properties 116
MOLECULAR WEIGHTS AND HOMOGENEITY 117
 Electrophoresis 117
 Sedimentation 120
 Osmotic Pressure 121
 Light Scattering 121
 Analysis of Unusual Component 121

CHEMICAL PROPERTIES OF PROTEINS   122
  Amphoterism of Proteins   122
  Binding of Ions   122
  Hydration of Proteins   122
  Precipitation with Antibodies   123
NATIVE AND DENATURED PROTEINS   124
GEL FORMATION   125
  Theories of Gel Formation   126
    (1) Adsorption of Solvent   126
    (2) Three-dimensional Network   126
    (3) Particle Orientation   127
  Gelatin   128
  Denatured Protein Gels   128
Determination of Protein in Foods   129
  KJELDAHL METHOD   131
  DUMAS METHOD   133
  AMINO ACIDS   133
Heat Treatment   136
Pure Proteins from Some Foods   138
  PLANT PROTEINS   138
  MILK PROTEINS   140
    Casein   140
    Whey Proteins   142
    Colostrum   142
  EGG PROTEINS   143
    Egg White Proteins   143
    Egg Yolk   146
References   147

CHAPTER FIVE. THE FLAVOR AND AROMA OF FOOD   148

The Sensation of Flavor   148
  TASTE   148
    Influence of Chemical Constitution   150
    Influence of Other Factors   151
  ODOR   152
    Testing Sense of Smell or Aroma   154
  FEELING   154
  BLENDS   155
Control of Flavor and Aroma in Processed Food   156
  MEASUREMENT OF FLAVOR   157
  FLAVOR INTENSIFIER: MONOSODIUM GLUTAMATE   160

FLAVORING EXTRACTS   *161*
SYNTHETIC FLAVORING SUBSTANCES   *162*
Recent Developments in Flavoring Research   *162*
References   *169*

CHAPTER SIX. MEAT AND MEAT PRODUCTS   *171*

Animal Structure   *171*
MUSCLE   *171*
CONNECTIVE TISSUE   *174*
   Collagen   *175*
   Elastin   *178*
   Adipose Tissue   *179*
   Cartilage   *180*
   Bone   *181*
Postmortem Changes   *181*
RIPENING ON AGING   *186*
Color of Meat   *188*
Cured and Smoked Meats   *196*
SMOKING MEATS   *199*
Changes in Meat on Cooking   *201*
DENATURATION OF THE PROTEIN   *202*
THE HYDROLYSIS OF COLLAGEN   *203*
COLOR CHANGE   *204*
DRIP FORMATION   *204*
MEAT AROMA   *205*
DISPERSION OF FAT   *205*
DECREASE IN VITAMINS   *205*
SURFACE REDDENING   *209*
OVERCOOKING   *210*
Tenderness   *210*
CONNECTIVE TISSUE   *211*
HYDRATION   *212*
NATURE OF PROTEIN MOLECULES   *213*
FAT   *213*
Juiciness   *213*
Phosphates in Canned Meat   *214*
References   *214*

CHAPTER SEVEN. VEGETABLES AND FRUITS   *218*

Vegetables   *219*
Structure of Fruits and Vegetables   *220*

Texture of Fruits and Vegetables *225*
Pigments in Fruits and Vegetables *226*
   THE CAROTENOIDS *228*
   CHLOROPHYLLS *235*
The Flavonoids *239*
   THE ANTHOCYANINS *241*
     In Cookery and Processing *244*
   THE ANTHOXANTHINS AND FLAVONES *247*
     Changes in Cooking and Processing *249*
   TANNINS *250*
     Colors in Foods from Tannins *252*
     Astringency from Tannins *253*
The Browning Reaction *254*
   ENZYMATIC BROWNING *255*
   NONENZYMATIC BROWNING *259*
Pectic Substances *261*
Changes on Cooking and Processing *265*
   CHANGES IN PECTIC SUBSTANCES *265*
   CHANGES IN CELLULOSE *269*
   CHANGES IN STARCH GRANULES *269*
   CHANGES IN INTERCELLULAR AIR *274*
   PRODUCTION OF VOLATILE ACIDS *274*
   VOLATILE SULFUR COMPOUNDS *278*
Post Harvest Changes in Fruit *280*
References *288*

CHAPTER EIGHT. MILK AND MILK PRODUCTS *293*

Formation of Milk *294*
Composition of Milk *295*
Chemical Analysis of Milk *299*
Checks for Purity *302*
Special Milks *303*
Butter *305*
Cheese *306*
   COMPOSITION *307*
   UNRIPENED CHEESE *307*
   RIPENED CHEESE *310*
   RIPENING *311*
   PROCESS CHEESE *314*
References *315*

CHAPTER NINE. CEREALS AND THEIR USE *316*

The Cereals *316*
  STRUCTURE OF THE GRAIN *317*
    The Primary Unit: The Cell *317*
    Parts of the Seed *317*
      Composition of Seed Parts *318*
  STORAGE OF GRAIN *318*
Wheat and Wheat Flour *319*
  MILLING *320*
    Types of Flours Produced *320*
    Relation to Dough *321*
  FLOUR PROTEINS *321*
    Gluten *322*
      Strength of Gluten *324*
  LIPIDS AND CARBOHYDRATES *324*
  VITAMINS *325*
  ENRICHMENT OF FLOUR *326*
Wheats and Their Baking Qualities *329*
  DOUGHS *329*
    Gas Production in Yeast-Leavened Doughs *329*
      "Starters" and Cultures *329*
      The Nutritional Needs of Yeast *331*
      Amylase Content and Activity *332*
      Pre-ferments *333*
    Gas Retention in Doughs *335*
      Mechanical Action *335*
      Proteolytic Enzymes *335*
      Oxidation *336*
      Other Factors *337*
  BATTERS *337*
    Gas Production in Batters *337*
    Gas Retention in Batters *339*
    Fat *341*
    Sugar *341*
    pH *343*
    Baking Temperature and Time *345*
    Moisture *345*
  STALING *346*
  SCORING BAKED PRODUCTS *347*
  PREVENTION OF MOLD *347*

References *349*

CHAPTER TEN. FOOD ADDITIVES *351*
Legal Safeguards *351*
Why Use Food Additives? *353*
Methods of Demonstrating Safety *354*
Color Additives Legislation *355*
References *355*

APPENDIX ON FOOD ADDITIVES *357*

INDEX *365*

CHAPTER ONE

Development of Food Chemistry

THE ORIGINS OF SCIENCE go back to prehistory when man could not write a record of his successes and his failures, his trials and experiments, and his traditions. We assume that as man emerged as a thinking animal he began to explore the world around him, to wonder at its operation, and finally to try to control it. Consequently, it seems that as the centuries roll by man's understanding and control of his environment, although still incomplete, continue to improve. In the early days every science had a tenuous start—even one as dependent on the development of mathematics and laboratory procedures as chemistry. At this time man had learned only a few chemical skills, some of which were in food chemistry. Probably they were accidently discovered, then carefully passed on from one generation to another so that by the time writing was invented man could not only make pottery and soap, cook food, and smelt a few metals but also make wine and vinegar. These processes were hedged around with superstitions, but the vintner could control the process; although he had never heard of a yeast or an *Acetobacter,* he knew how to handle the fermenting fruit.

In the seventeenth century, as laboratory procedures and an appreciation of the scientific method developed and as scientists began to exchange information, chemistry began to emerge as a true science. A considerable accumulation of information and a beginning study of foods reflected the general interest in the composition of the entire world. Some of the early experiments attempted to separate food components, to study the pigments that give color to the world, or to find the "life-giving" compound. As time passed these studies were successful to a limited extent. In the eighteenth century simple compounds were isolated, and during the first half of the nineteenth century the old idea that the living and the nonliving world were completely different was abandoned. The terms "inorganic" and "organic" reflect this early belief in two types of chemistry. When some compounds such as sodium chloride were found in the "inorganic" earth as well as in the blood and urine of a living animal and when "organic" compounds were shown to obey the same laws as "inorganic," the idea was discarded. However, a complete understanding of the composition and the changes in composition of such complex mixtures as living cells was impossible at this stage.

The nineteenth century saw the development of organic chemistry, analytical chemistry and physical chemistry—all essential to the growth of food chemistry and our understanding of it. The pace at which discoveries were made and at which advances occurred increased throughout the century. The field of carbohydrates began to fit together, proteins were recognized, and many other compounds of importance in food chemistry were studied.

Biology also saw rapid growth during the nineteenth century; understanding of this field was also required before much progress could be made in food chemistry. The Cell Theory and the Theory of Evolution were fundamental to much insight into either animal or plant materials. These theories and the bulk of knowledge supporting them gave scientists an appreciation of the complex nature of cells as well as the common patterns that cells may have through their common heritage and the different patterns that separate families, species, and even varieties.

In the twentieth century, the passage of the Pure Food and Drug Act by the United States Congress in 1906 had a great influence on the channels that research in food chemistry took. Harvey Wiley, shocked at the food which could be legally sold in the United States, campaigned for years for the passage of a bill to restrict filth, decay, and adulteration in food. After the passage of the bill, there came a long fight for enforcement. As always in the United States, a company or individual must be proved guilty; consequently under the Pure Food and Drug Act proving that a food did not come up to standard was necessary. The act was originally administered by the Department of Agriculture; chemists in this department worked out many tests to detect adulteration and to set up standards. Eventually the methods developed for foods and other products were published as the Official Methods of Analysis of the Association of Official Agricultural Chemists. The eighth edition, published in 1955, runs to 1,008 pages.

A phenomenal growth in our knowledge of biochemistry has occurred in recent years. Metabolic pathways in plants and animals are being clarified, the structure of large molecules is yielding to study, applications of physical-chemical principles become easier and more meaningful every day. These developments result in a better understanding of the biological materials which we use as food. It is not yet possible to separate completely all of the compounds which make up a cell or to describe even sketchily all of the reactions that occur in a cell, but the body of knowledge has now become large and is rapidly increasing. Probably only a rather small per cent of the reactions that occur when a food is cooked or processed in some way are known, but more are being studied every day.

Through the centuries that man has known fire, cooking procedures have developed empirically. Many of these processes were the result of trial and error, while some must have been discovered accidentally. When a man stumbled on a process, cheese making for example, he handed it down to his descendants. Sometimes it underwent modification, but often the same rules have been followed for generation after generation. These procedures have seldom been the result of scientific experimentation where a series of controlled batches are used and where variations are planned and at least partially understood. Empirical processes usually contain a number of procedures which are essential for success, but sometimes there are procedures which have little or no bearing on the problem. The salt added in salt-rising bread was thought for generations to be important in keeping the "starter" fresh and sweet. Actually, it had little effect on the bacteria and yeasts present.

The food industry developed from small operations, sometimes built around one small kitchen with one stove, or a little butcher or grocery shop. Many industrial processes also developed by trial and error methods. Today, with nationwide industries and with the growth of a consumer market which expects and demands a uniform product, it is necessary to control processes carefully. This control demands understanding of the process. Some industries have been very active in attempting to develop a scientific basis for their procedures. There has been, for example, a number of studies on orange juice concentration. While these researches are directed to solving a problem of the orange juice industry, they have contributed to our knowledge of some basic problems in food chemistry. Browning of orange juice during processing is one of the problems which plagues the processors; the solution of this problem may have a bearing on browning reactions in other foods.

INDIVIDUAL VARIABILITY

In biochemical research it has been recognized for years that an adequate sample must include a fairly large number of individual subjects since variability between members of the same plant or animal species is common. Thus we can expect every tomato on a single vine to contain ascorbic acid, but we cannot expect every one to have the identical concentration of ascorbic acid. Added to this is the variability which is introduced by variety and growing conditions. In animals we also find a range of values rather than exactly the same composition. When we analyze a compound, we find that each sample has exactly the same composition; but when we analyze a plant or animal, we find that the values fall into a range.[3]

An appreciation of variability is essential for anyone concerned with food chemistry, since it requires that an adequate sample include a number of individuals and also that samples be adequately mixed. If broccoli, for instance, is analyzed, a number of samples from different plants, grown in different locations to the same degree of maturity should be used. The buds and stem must be chopped finely, well mixed, and representative samples withdrawn. The importance of adequate sampling cannot be overemphasized. Too often conclusions have been drawn from small samples. These conclusions are often not valid.

INDIVIDUAL UNIFORMITY

On the other hand, both plant and animal tissues have many things in common. Since all living things appear to have sprung from the same, though distant, ancestor, they possess some compounds and some metabolic pathways in common. With mutations and the occurrence of variations or even the development of new species, some differences in chemistry must necessarily be present. Often the chemical variation is significant but relatively small. Thus every living cell contains a number of proteins. These proteins are not identical, but they have many properties and some aspects of their structure in common. Another example of uniformity is the metabolic pathway of glucose or glycogen to pyruvic acid. This series of reactions is used by all plant and animal cells which have been studied thus far.

An appreciation of uniformity is also important in food chemistry since the accumulated knowledge is still scanty. We often infer that if a reaction explains an observation in one food, it will apply to another. Thus the study of the browning of orange juice may tell us something about other browning reactions. It is, however, necessary to maintain a critical as well as a skeptical attitude and to recognize that although transfer of information from one food to another may be valid, it must be verified.

METHODS OF SAMPLING

When samples of foods are taken for analysis, they are usually fragmented, minced, or macerated in some manner. Dry samples are sometimes ground in a ball mill, sometimes by some other device; wet samples such as plant and animal tissue are usually ground or macerated. Loss of juice through the squeezing of the grinder knives must be avoided or the juice squeezed out must be returned and mixed with the sample. It is essential that adequate mixing of the sample occur so that the composition does not vary from one part to another. With liquids and pulverized materials, it is fairly easy to pour the sample from one container to another or to mix by stirring. With solid material such as a frozen food, it is necessary to take drillings or chucks from different parts of the package, and after proper treatment such as thawing, mix them thoroughly. When the package or batch is exceedingly large, such as a boxcar, it is necessary to take a number of samplings. Various devices have been developed for taking cores of samples at different places in the batch, which are then carefully mixed and representative samples removed. Often, *quartering* is used. The sample is spread on a sheet of paper in a cone and flattened to a circle. If the sample is very large, a blanket or sheet must be used. The circle is divided into quarters and two opposite segments, quarters 1 and 3, removed. These in turn are quartered, but segments 2 and 4 are saved this time. The process is repeated until the sample is reduced to a size suitable for analysis. The sample is mixed and often pulverized further between each quartering.

MOISTURE IN FOODS

The most abundant compound, and the one which is almost always present in foods, is water. Occasionally a food such as an oil will be dry; but even crystallized substances which are relatively pure, such as sugar and salt, contain small amounts of water adsorbed on the surfaces of the crystals. Cellular material, whether plant or animal, contains an abundance of water. In leafy green vegetables there is 90 or more per cent water, while even in cooked meat where some water has been driven off the amount is between 50 and 65 per cent.

In plants and animals water is present in the circulating fluid—the sap or the blood and lymph—between the cells as intercellular fluid and within the cells. If an animal is cut, fluid drips out; even when the stalk of a plant is cut, fluid can sometimes be seen to drip or ooze out of the vessels which carry the sap. Within the cells water is plentiful in the cytoplasm. In some types of plant cells special cell sap vacuoles contain solutions of compounds dissolved in water.

The water which is present in foods may be held (1) as free liquid in which substances are dissolved or dispersed, (2) as hydrates, (3) as imbibed water in gels, or (4) by adsorption on the surfaces of solids. Examples of the first type are found in cytoplasm, intercellular fluid, and any of the circulating fluids of tissues. In the second type, hydrates form either when hydrogen bonds are established between water molecules and ions or molecules which contain oxygen or nitrogen or when the unshared electrons of the oxygen are coordinated with an ion. Starches, proteins, and many other organic compounds important in foods, as well as salts, form hydrates. Imbibed water may not be different from water held as a hydrate. Some substances pick up water and swell when they come in contact with water. They are said to "imbibe" water; this may be accomplished by hydrogen bonding. The fourth type of water is held on all surfaces exposed to air in which water vapor is present. "Dry" cocoa holds water and air on the surface of the particles. Solids which are very finely divided have a very large surface area and consequently have a high adsorptive capacity.

Hydrogen Bonding

Water molecules are able to attach themselves to other molecules by means of a hydrogen bond. Water molecules are dipoles in which the hydrogen atoms are slightly positive and the oxygen atom slightly negative. Within a water molecule the hydrogen atoms are bonded to the oxygen by a covalent pair of electrons, but the angle between these atoms is 105°. Thus the four unshared electrons of the oxygen atom are on one side and create a slight negativity. The oxygen atom holds the electrons closely and draws them slightly away from the proton. As a result, the oxygen is slightly negative in comparison to the hydrogens which are positive.

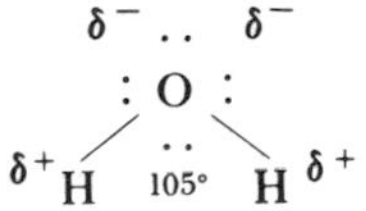

$$\delta^- \overset{..}{} \delta^-$$
$$: O :$$
$$\delta^+ H \quad 105° \quad H \; \delta^+$$

The hydrogen bond or bridge is able to form whenever a slightly positive hydrogen approaches an atom such as oxygen which tends to be slightly negative. The small size of the hydrogen atom allows it to come very close to this atom and establish a weak bond or bridge to it.

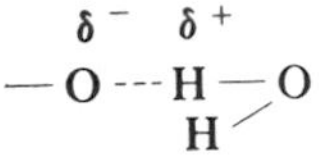

$$\delta^- \quad \delta^+$$
$$- O \cdots H - O$$
$$H$$

In pure water two molecules associate at some temperatures when one hydrogen is attracted to the slightly negative oxygen of another molecule of water and association occurs through the hydrogen bond. The strength of the bond between the two molecules is not nearly as great as the strength of the covalent bonds holding the two hydrogens and oxygen together, but it is much greater than the van der Waal's forces which hold molecules close together in liquids and even closer in solids.

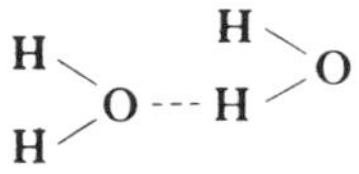

$$\begin{array}{c} H \\ \diagdown \\ O \text{---} H \\ H \diagup \end{array} \begin{array}{c} H \\ \diagdown \\ O \\ \diagup \end{array}$$

Energy in the form of heat must be supplied in order to disrupt hydrogen bonds. The relatively high boiling point of water, a very low molecular weight compound, is explained on the basis of the heat required to break hydrogen bonds.

Not only can hydrogen bonding hold one water molecule to another, it can also cause association or hydrate formation between water and compounds which have polar oxygens. A compound such as methanol, CH_3OH, or a carbohydrate which has a hydroxyl group will bind water through hydrogen bonding. Indeed, pure methanol will associate with itself through hydrogen bonding. In large complex molecules which are coiled and folded water will penetrate within the molecule, forming hydrates through hydrogen bonding and causing a change in the size and shape of the molecule.

The nitrogen atom is frequently linked to another nitrogen or an oxygen atom through a hydrogen bond. Like oxygen, nitrogen tends to be slightly negative, and it will have an attraction for the slightly positive hydrogen of a molecule such as water and bond it to the nitrogen. An amino group in a compound such as a protein will associate with the hydrogen of a hydroxyl group in, say, methanol. The strength of the hydrogen bond is not as strong in this case as it is with oxygen since nitrogen is not as negative as oxygen and the hydrogen is not held as closely.

Hydrates form with those metallic ions which tend to form complexes. The unshared electrons of the oxygen atom will fill out the shells of the

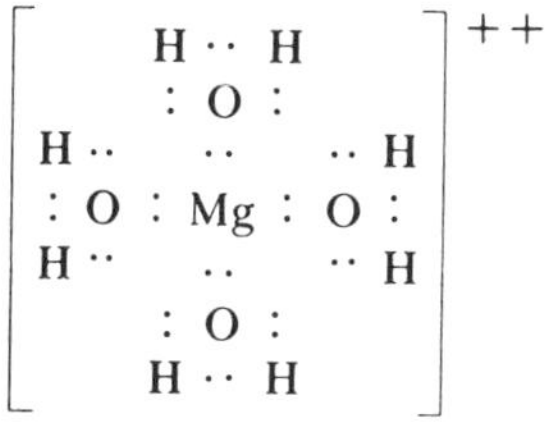

$$\begin{bmatrix} & H \cdot\cdot H & \\ & : O : & \\ H \cdot\cdot & \cdot\cdot & \cdot\cdot H \\ : O : & Mg : & O : \\ H \cdot\cdot & \cdot\cdot & \cdot\cdot H \\ & : O : & \\ & H \cdot\cdot H & \end{bmatrix}^{++}$$

Hydrated Magnesium Ion

ion and hold water to it. Sodium, magnesium, calcium, and many other elements exist as hydrated ions in solution. The hydration of the hydrogen ion has been emphasized in recent years by use of a special name, the "hydronium ion."

Bound Water

Many years ago it was discovered that some of the water in cellular material can be readily removed by pressing or heating, while some cannot. Gortner attempted to discover why pine needles did not freeze in the winter. He found that needles gathered at that time of year have little water which can be squeezed out, although in the summer needles this is not true. In summer less of the water is "bound." In all cellular material there is some bound water.

In working with water in food it is important to remember that some of the water may be bound and very difficult to separate. Often it is extremely difficult to separate the water without decomposing other molecules present in the sample.

Determination of Moisture

In some foods the determination of moisture is relatively simple. The sample is weighed and heated in an oven to constant weight. The difference in weight is the water which has evaporated. The sample is usually weighed into a flat bottomed, shallow dish made of aluminum or similar material which will not react with the food nor pick up water readily. The oven must be thermostatically controlled and is usually set at 100°C or 105°C. A thin layer of sand, pumice, or asbestos is often added to the bottom of the dish to support the food particles and accelerate drying.

Many foods decompose to some degree if they are heated to 100°C. This is true, for example, of all foods which contain fructose. It is necessary to dry them in a vacuum oven where the temperature is maintained at a lower figure and the pressure is reduced to facilitate loss of moisture. An alternative method with foods sensitive to heat is use of a vacuum desiccator with sulfuric acid as the drying agent. The samples are again dried to constant weight.

Those foods which contain volatile compounds other than water must be treated by another method. None of the weight-loss methods are adequate to differentiate between loss of water and loss of some other volatile substance. The immiscible solvent distillation method can be used for this purpose. The sample is placed in a flask which is connected with a reflux condenser equipped with a distillate trap. (See Figure 1.1.) The sample is covered with a suitable solvent and the trap filled with the solvent. The

FIGURE 1.1 APPARATUS FOR THE DETERMINATION OF MOISTURE IN FOODS. An immiscible liquid such as toluene boils off with the water. Water is trapped and measured in side vessel.

solvent must be immiscible with water so that as they distill separation of the two liquids can occur. Toluene is most commonly used, although xylene and heptane are sometimes employed. The flask is heated and the vapors of water and solvent are condensed by the condenser and drop into the trap. The lighter solvent flows over into the flask, but the water is captured. If the trap is calibrated, the amount of water distilled out of the sample can be read directly.[1,2]

Nuclear magnetic resonance has been developed into a rapid and simple method for the determination of moisture in samples, particularly solids. The equipment is rather expensive, but it is possible for an untrained person to make numerous determinations in a very short time. The instrument can be calibrated to read per cent moisture directly. Nuclear-magnetic-resonance measurements depend on the magnetic behavior of the nuclei of atoms. All nuclei are positively charged owing to their load of protons and many spin either clockwise or counterclockwise. The rotation of a charged body creates a magnetic field. Only those nuclei which have an even number of protons and neutrons ($_6C^{12}$, $_8O^{16}$, and $_{16}S^{32}$) do not appear to have an angular momentum or create this tiny magnetic field. If spinning nuclei are placed in the field of a magnet, the magnetic field exerts a torque on them and tends to align them with the field. They absorb radio-frequency

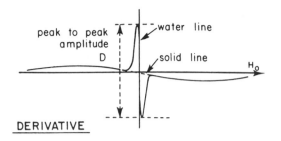

FIGURE 1.2 NUCLEAR MAG-
NETIC RESONANCE DERIVATIVE
OF ABSORPTION CURVE OF MA-
TERIAL CONTAINING SOLID AND
WATER. The contribution of
protons in the solid ("solid
line") is small compared to
those in water ("water line").
*Courtesy of H. Rubin,
Ridgefield Instrument Group,
Schlumberger Corp.*

energy and precess at a definite frequency like tiny gyroscopes. The frequency of the energy which can be absorbed is characteristic of each isotope. Thus in a magnetic field of 1700 gauss, hydrogen will absorb energy at 7.25 megacycles and change to another magnetic energy level. The amount of energy absorbed will be proportional to the quantity of hydrogen present in the sample.

The apparatus is designed to measure the absorbence of energy when the sample is placed inside a coil supplied with radio-frequency current. The coil is mounted between a large permanent magnet. The field strength of the magnet is slowly varied; and when it crosses the NMR value for the

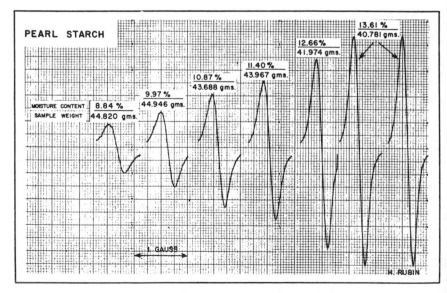

FIGURE 1.3. NUCLEAR MAGNETIC RESONANCE. Curves of moisture in pearl starch.
Courtesy of H. Rubin, Ridgefield Instrument Group, Schlumberger Corp.

particular isotope, the nuclei absorb energy and shift magnetic energy levels. The amount of energy absorbed is recorded.

Many compounds other than water contain hydrogen, but the nature of the compounds and particularly the physical state of the compounds have an effect on the width of the line and shape of the curve produced. In liquids the motion of molecules is so rapid that the effect of the magnetic field created by neighboring nuclei is smoothed out. In many samples water is the only liquid present.

REFERENCES

1. ANDREW, E. R., "Nuclear Magnetic Resonance," Cambridge University Press, New York, N. Y., 1958.
2. GUTOWSKY, H. S., "Analytical Application of Nuclear Magnetic Resonance," in Berl, W. G., ed., "Physical Methods in Chemical Analysis," Vol. III, Academic Press, Inc., New York, N. Y., 1950.
3. WILLIAMS, R. J., "Biochemical Individuality," John Wiley & Sons, Inc., New York, N. Y., 1956.

CHAPTER TWO

Fats and
Other Lipids

LIPIDS ARE ONE OF THE LARGE GROUPS of organic compounds which are of great importance in the food we eat because they are readily digested and utilized in the body. They are widely distributed and almost every natural food has considerable quantities of them. Fruits and vegetables are not ordinarily thought of as sources of lipids, but we find that most of them contain between 0.1 and 1 per cent total lipid, usually listed as fat.

A few fruits and vegetables are rich sources of these compounds. Thus an avocado contains an average of 20 per cent lipid while a ripe olive has about 19 per cent. Whole-grain cereals have from 1 per cent for whole barley on the dry basis to 7.4 per cent for dry oatmeal, while nuts are very rich in lipid. Pecans average 73 per cent and English walnuts about 64 per cent. But the natural foods which contribute the largest amounts of these compounds to our diet are the animal products—meats and fowl, milk and milk products, and eggs.

OCCURRENCE IN FOODS AND COMPOSITION

The American diet is unusually rich in fats and other lipids. During recent years the percentage of calories derived from fat has increased markedly in the average diet. Many nutritionists and physicians are alarmed at this high consumption and consider it unwise. Table 2.1 shows the consumption of fats in the United States in recent years. The problem of diet fat is briefly discussed on page 62.

The lipid fraction of a food is usually separated from the other compounds present in the food by extraction with a solvent such as petroleum ether, ethyl ether, chloroform, or benzene, and is reported either as "ether-soluble fraction" or "crude fat." Actually the fraction contains not only true fats but also waxes, complex lipids such as phospholipids, derived lipids such as the sterols, and many pigments, hormones, and volatile oils.

TABLE 2.1. CONSUMPTION OF FOOD FATS AND OILS

Source	Pounds per Person		Percentage Distribution	
	1957–59	1976	1957–59	1976
Visible fats				
Butter (fat content)	6.6	4.1	5.7	3.2
Lard (direct use)	9.3	4.4	8.1	3.4
Margarine (fat content)	7.2	9.0	6.3	7.0
Shortening	11.4	17.0	9.9	13.3
Other edible fats and oils	10.8	18.6	9.4	14.5
Total visible	45.3	53.2	39.4	41.6
Invisible fats				
Dairy products (excluding butter)	19.0	15.7	16.5	12.3
Eggs	4.6	4.1	4.0	3.2
Meat, poultry and fish	38.0	46.0	33.0	36.0
Dry beans, peas, nuts, soya, flour and cocoa	5.3	6.1	4.6	4.7
All fruits and vegetables	1.0	1.1	0.9	0.9
Grain products	1.7	1.7	1.5	1.3
Total invisible	69.7	74.7	60.6	58.4
Total fats and oils	115.0	127.9	100.0	100.0

Occasionally this crude fat fraction is further separated. The sample is boiled with an alkali such as sodium hydroxide and by saponification the fats, waxes, compound lipids, and free fatty acids then form soaps. These disperse in the water layer and other products such as glycerol and phosphates dissolve in water, but the sterols, pigments, hydrocarbons, etc. are insoluble. The "saponifiable fraction" thus includes all lipids except the sterols, pigments, and hydrocarbons. Saponification of the "saponifiable fraction" (fats, waxes, and phosphatides) is shown diagrammatically below:

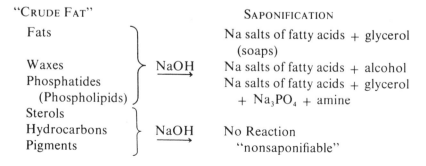

The fat content reported in foods is often this "crude fat" and represents total lipid content rather than true fat.

Little information on the distribution of lipids in the crude fat fraction of most foods is generally available. In animal tissues true fat is found primarily in the adipose tissues, while the active tissues, those which use considerable oxygen and produce much carbon dioxide—the muscles, nervous tissue, and glands—contain relatively little fat in the lipid fraction and much more of the complex lipids and sterols.

The fats and oils of commerce are for the most part mixtures of fats, the triglycerides of fatty acids.

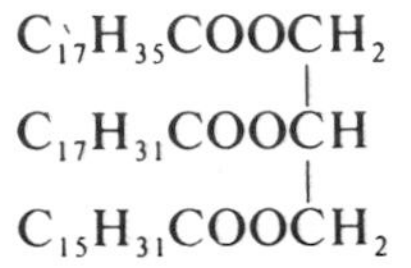

$$C_{17}H_{35}COOCH_2$$
$$C_{17}H_{31}COOCH$$
$$C_{15}H_{31}COOCH_2$$

A Fat

Some of the purified oils and fats contain only very small amounts of other compounds, and these, for the most part, have little or no effect on flavor, color, or development of rancidity. In the crude oils, particularly those from seeds, there are considerable quantities of *lecithins,*

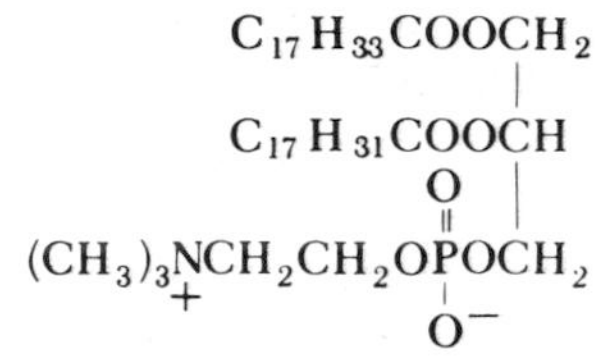

$$C_{17}H_{33}COOCH_2$$
$$C_{17}H_{31}COOCH$$
$$\underset{+}{(CH_3)_3}NCH_2CH_2O\overset{O}{\overset{\|}{P}}OCH_2$$
$$O^-$$

A Lecithin

cephalins, and other lipids which have not been completely identified.

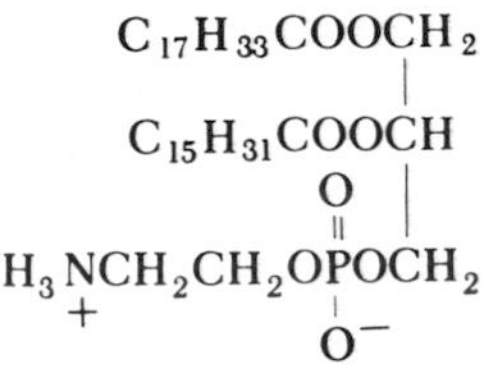

$$C_{17}H_{33}COOCH_2$$
$$C_{15}H_{31}COOCH$$
$$\underset{+}{H_3}NCH_2CH_2O\overset{O}{\overset{\|}{P}}OCH_2$$
$$O^-$$

A Cephalin

In freshly extracted corn or soybean oil, the phosphatides present may be as high as 2 or 3 per cent, but on purification most of these compounds

are removed. *Sterols* are

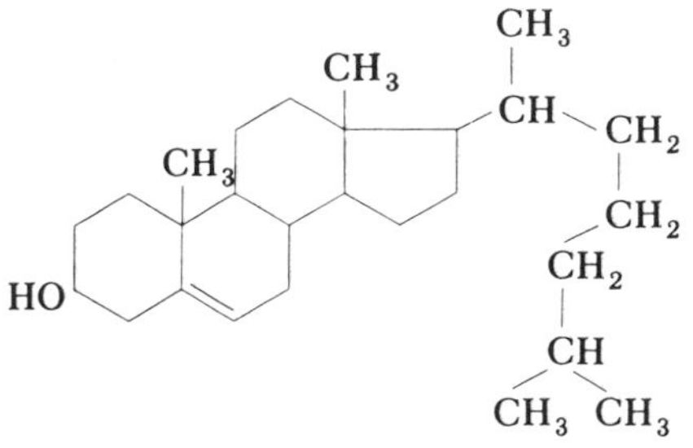

Cholesterol, A Common Animal Sterol

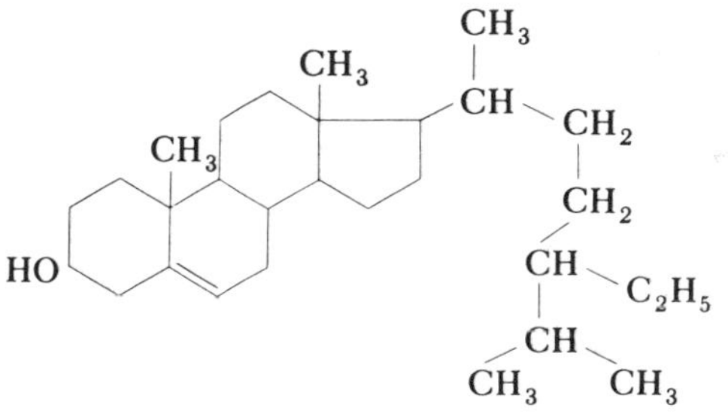

Sitosterol, A Common Plant Sterol

also present in crude oils to an appreciable extent. They vary from halibut liver oil with 7.6 per cent sterol to beef tallow with 0.08 per cent, with fourteen oils in the range from 0.1 to 1 per cent. The sterols occur in crude fats both as free sterols and as esters of fatty acids. They are partially removed during the alkaline refining of the oil, but even in refined oil there is still some sterol. The phytosterol content of common vegetable oils is given in Table 2.2. "Phytosterol" includes all plant sterols with sitosterol the most common. They constitute the major group of compounds in the non-saponifiable fraction of fats and oils.

Hydrocarbons are also present in some natural fats and oils. Squalene, $C_{30}H_{50}$, is a highly unsaturated hydrocarbon with the formula

$$[(CH_3)_2C\!=\!CH(CH_2)_2C(CH_3)\!=\!CH(CH_2)_2C(CH_3)\!=\!CHCH_2\!-\!]_2$$

and occurs not only in fish liver oils but also in olive, rice bran, and many others. Gadusene, $C_{18}H_{32}$, is named for the cod, Gadus, and was first isolated from that oil. Pristane, $C_{18}H_{28}$, zamene, $C_{19}H_{38}$, and cetorhinene are also in fish oil. The formula for zamene has recently been determined by Christensen and Sorensen[4] to be

$$CH_2{=}C(CH_3)(CH_2)_3CH(CH_3)(CH_2)_3CH(CH_3)(CH_2)_3CH(CH_3)_2$$

2, 6, 10, 14 pentadec-l-ene

Waxes and some free fatty alcohols occur in very small amounts in fats and oils. The wax in corn oil has been reported to be composed of a mix-

TABLE 2.2. PHYTOSTEROLS IN VEGETABLE OILS*

	mg/100 g
Castor Bean	500
Cocoa Butter	170
Cocoa Butter	200
Coconut, Hydrog.	79.8
Coconut	60
Coconut	80
Corn	580
Corn	1000
Corn Germs	800-1000
Cottonseed	311.2
Cottonseed	260
Crisco	232
Olive	133.7
Olive	210
Peanut	247.9
Peanut	200
Peanut	190
Sesame	594.4
Sesame	520
Soybean	90
Soybean	380
Soybean	195
Soybean	230
Wheat Germ	3600-6700

*From Lange, W., *J. Am. Oil Chemist's Soc.,* **27,** 414-422 (1950).

ture of myricyl isobehenate and lignocerate, while another sample was reported to contain the esters of cetyl alcohol rather than myricyl.

$C_{21}H_{43}COOC_{14}H_{29}$ $C_{23}H_{47}COOC_{30}H_{61}$ $C_{23}H_{47}COOC_{16}H_{33}$
Myricyl Isobehenate   Myricyl Lignocerate   Cetyl Lignocerate

The coloring matter of fats and oils is composed of numerous pigments.

An orange-red or yellow pigmentation is usually caused by the presence of carotenoids which are soluble in oils. Carotenoids are highly unsaturated hydrocarbons, and when oils are hydrogenated, hydrogenation of the pigments also occurs with a reduction in color. Carotenoids are unstable at high temperatures, and if a fat or oil is treated with steam there is some loss of color. They cannot be removed by oxidation since it brings the fat or oil at or near to rancidity. They are, however, readily adsorbed on some adsorbants such as fuller's earth.

The darkening which some oils show on limited oxidation is caused by the oxidation of tocopherols, the vitamin E substances present in some. Tocopherol content of some fats is shown in Table 2.3. Occasionally when the oil is made from a green plant product, some of the chlorophyll will be carried along and color the oil. This is undesirable and very difficult to remove. Brown pigments are usually only present in oils which are prepared from rotten or damaged plant products and are believed to arise from the disintegration of protein and carbohydrate molecules.

We see, therefore, that most foods and food products contain not only true fats, but a great many related compounds. The classification "crude fat" or even "fat" in food composition tables usually refers to the sum of these compounds. Even purified fats and oils contain small amounts of compounds other than simple fats, the triglycerides of fatty acids. The other compounds present are (1) the complex lipids—lecithins, cephalins, other phosphatides, and glycolipids; (2) sterols, both free and combined with fatty acids as waxes; (3) free fatty acids; (4) waxes; (5) pigments which are lipid soluble; and (6) hydrocarbons.

EDIBLE FATS AND OILS

Much of our discussion will be confined to the prepared edible fats and oils which are sold in a fairly pure state. A great body of research has been built up around these foods because of efforts to differentiate one from the other so that a cheap oil is not sold for a high-priced one and also because the manufacturers of shortenings have engaged in considerable investigation. For many generations lard was the animal fat of choice in preparing doughs and batters since it has sufficient plasticity at room temperature so that it will cream with sugar and mix with egg or egg yolk. But today in the United States the use of lard is small compared to the use of "shortenings." Sometimes the shortenings are hydrogenated vegetable oils; sometimes they are purified and standardized animal fats. The market is highly competitive and during the years of development of these products much has been learned about fats.

Fatty Acids

Natural fats are mixtures of mixed glycerides in which the three fatty acids esterifying glycerol differ from each other. Little or none of the simple glycerides are present. Since the solubilities of these mixed glycerides are very similar, it is extremely difficult to fractionate them and to succeed in describing them in terms of the molecules present. After hydrolysis it is possible to separate the fatty acids; the available analyses of natural

TABLE 2.3. TOCOPHEROL CONTENT OF FATS*

	mg/100 g
Cocoa	12.5
Cocoa	2.8
Coconut	5
Coconut	8.3
Corn, Mazola	119, 102
Crude	110
Refined	104
Cottonseed, Refined	83-110
Crude	110
Olive	3-30
Peanut, Refined	26-51
Crude	40-52
Safflower, Crude	80
Sesame, Refined	18-65
Soybean, Refined	99-175
Wheat Germ	140-520
Influence of Processing on Cottonseed Oil	
Crude	102.8
Water Washed	102.0
Alk. Refined, Bleached, Filtered	100.7
Refined, Bleached, Filtered, Deodor.	95.9
Refined, Bleached, Filtered, Hydrog.	98.3
Refined, Bleached, Filtered, Hydrog., Deodor.	97.6

*From Lange, W., J. Am. Oil Chemist's Soc., 27, 414-422 (1950).

fats are usually based on an analysis of the fatty acids rather than the actual mixed glycerides which occur in the natural product. Percentages of fatty acids by weight in some common fats are shown in Table 2.4.

Sometimes groups of fatty acids are separated. Molecular weight and the presence or absence of unsaturation affects (1) solubility in water, (2) volatility with steam, and (3) solubility of salts in water and alcohol. Thus no fatty acid above lauric acid, $C_{11}H_{23}COOH$, is soluble in water even at 100°C. Butyric (C_4), caproic (C_6), caprylic (C_8), and capric (C_{10}) are volatile with steam, while lauric (C_{12}) and myristic (C_{14}) are slightly volatile. The lead salts of the low molecular weight acids and unsaturated acids are more soluble in ethyl alcohol than are the high molecular weight, saturated acids.

TABLE 2.4. PERCENTAGE OF FATTY ACIDS BY WEIGHT IN SOME EDIBLE FATS AND OILS*

Fatty Acids	Corn, %	Cotton Seed, %	Olive, %	Leaf Lard, %	Beef Tallow, %	Mutton Tallow, %	Butter, %
Myristic							
$C_{13}H_{27}COOH$		1	1	1	2	2	10
Palmitic							
$C_{15}H_{31}COOH$	6	21	9	28	32	34	30
Stearic							
$C_{17}H_{35}COOH$	2	2	1	8	15	19	11
Arachidic							
$C_{19}H_{39}COOH$	1	1	1	—	—	—	—
Oleic							
$C_{17}H_{33}COOH$	37	25	80	56	49	43	30
Linoleic							
$C_{17}H_{31}COOH$	54	50	8	5	2	2	3
Butyric							
C_3H_7COOH	—	—	—	—	—	—	3
Caproic							
$C_5H_{11}COOH$	—	—	—	—	—	—	2
Total	100	100	100	98	100	100	99

*Adapted from Bailey, A. E., "The Chemistry and Technology of Food and Food Products," edited by Jacobs, M. B., 1, 581–582, Interscience Publishers, Inc., New York, N. Y., 1944.

The groups of fatty acids are further separated into individual compounds by esterification to form either the methyl or ethyl esters followed by fractionation. They have been fractionated by very careful distillation at extremely low pressures, by crystallization from solvents at low temperatures, by counter-current distribution in which they are partitioned between two solvents, and by gas chromatography. Adsorption spectra with fine separation of wave lengths is used to identify and even determine quantities of fatty acids.

Gas chromatography is a technique in which the methyl or ethyl esters of the fatty acids are passed over a column composed of a solid wet with some liquid in which these esters will dissolve. A gas, usually helium,

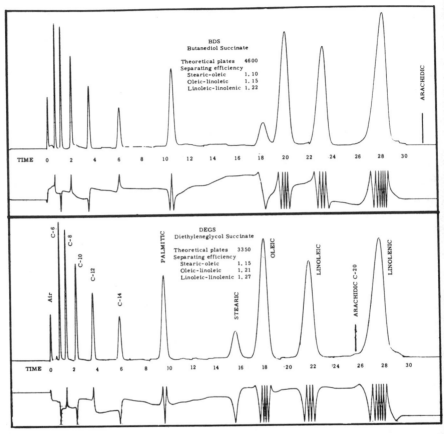

FIGURE 2.1. SEPARATION OF METHYL ESTERS OF FATTY ACIDS OF LINSEED OIL BY
GAS CHROMATOGRAPHY.

Courtesy of Wilkens Instrument and Research, Inc.

then flows over the column and removes fractions. The most volatile will
come off the column first.

The infra-red spectra of fatty acids and their esters have now been well
established, and study of the spectrum of a fat or oil gives much information about the composition. High resolution of the lines in the infra-red
portion of the spectrum is necessary in order that the presence of double
bonds and cis or trans isomerism may be demonstrated.[22]

The *total* unsaturated fatty acids are determined by methanolysis with
methyl alcohol-sodium hydroxide followed by oxidation with standard
potassium permanganate. The method has now been extended to include

the saturated fatty acids. After oxidation the acid oxidation products are removed by washing with alkali. The saturated methyl esters are removed, dried, and weighed. Esters of fatty acids containing 16 or more carbon atoms with only small amounts of C_{14} are included among these saturated methyl esters.[15]

$$R_{sat}.COOCH_2$$
$$|$$
$$RCOOCH \quad + \quad 3CH_3OH \quad \longrightarrow$$
$$|$$
$$R_1CH=CH(CH_2)_nCOOCH_2$$

$$HOCH_2$$
$$|$$
$$R_{sat}.COOCH_3 \qquad HOCH$$
$$+ \qquad + \qquad |$$
$$RCOOCH_3 \qquad HOCH_2$$
$$+$$
$$R_1CH=CH(CH_2)_nCOOCH_3 \quad \xrightarrow{KMnO_4} \quad R_1COOH$$
$$+$$
$$HOOC(CH_2)_nCOOCH_3$$

The fatty acids which are commonly present in natural fats are restricted to a surprisingly small number of possible compounds. The common fatty acids are given in Table 2.5. Most of the acids are straight-chain acids and, except for the low molecular weight acids, fatty acids are exclusively composed of acids with an even number of carbon atoms. The unsaturated acids have the possibility of either cis or trans isomerism, but in nature only the cis isomer occurs. However, since formation of the trans isomers occurs when a fat is heated to a high temperature, is hydrogenated, or comes in contact with a number of catalysts, many fats and oils in commerce do contain trans isomers. In the unsaturated acids, 18 is a common number of carbon atoms and scarcely any fat lacks oleic acid. A common position for the double bond is between the ninth and tenth carbons as in oleic, linoleic, and linolenic acids. When there are two or more double bonds, they are separated by a CH_2 group. Conjugation does occur in a few fatty acids, but this is rare and occurs primarily in those fats which are used as drying oils.

Identification of Natural Fats and Oils

We will consider in this section the role of physical and chemical properties.

TABLE 2.5. MELTING POINTS OF COMMON FATTY ACIDS (MOST STABLE FORM)

No. of Carbon Atoms	Fatty Acids		Melting Point in °C
	Saturated		
4	Butyric	C_3H_7COOH	−7.9
6	Caproic	$C_5H_{11}COOH$	−3.4
8	Caprylic	$C_7H_{15}COOH$	16.7
10	Capric	$C_9H_{19}COOH$	31.6
12	Lauric	$C_{11}H_{23}COOH$	44.2
14	Myristic	$C_{13}H_{27}COOH$	54.4
16	Palmitic	$C_{15}H_{31}COOH$	62.9
18	Stearic	$C_{17}H_{35}COOH$	69.6
20	Arachidic	$C_{19}H_{39}COOH$	75.3
22	Behenic	$C_{21}H_{43}COOH$	79.9
24	Lignoceric	$C_{23}H_{47}COOH$	84.1
	Unsaturated		
16	Palmitoleic 9-hexadecenoic	$CH_3(CH_2)_5CH{=}CH(CH_2)_7COOH$	0.5
18	Oleic cis-9-octadecenoic	$CH_3(CH_2)_7CH{=}CH(CH_2)_7COOH$	16.3
18	Elaidic trans-9-octadecenoic	$CH_3(CH_2)_7CH{=}CH(CH_2)_7COOH$	43.7
18	Linoleic cis-cis-9-12 octadecadienoic	$CH_3(CH_2)_4CH{=}CHCH_2CH{=}CH(CH_2)_7COOH$	−5.0
18	Linolenic cis-cis-cis-9-12-15 octadecatrienoic	$CH_3CH_2CH{=}CHCH_2CH{=}CHCH_2CH{=}CH(CH_2)_7COOH$	−11.0
20	Arachidonic cis-cis-cis-cis-5-8-11-14 eicosatetraenoic	$CH_3(CH_2)_4CH{=}(CHCH_2CH)_3{=}CH(CH_2)_3COOH$	−49.5
22	Erucic cis-13 cocosenoic	$CH_3(CH_2)_7CH{=}CH(CH_2)_{11}COOH$	33.7

Physical Properties. The physical properties of the natural fats and oils are often used to identify them. Usually more than one property is measured so that the identification can be made with some assurance since natural fats and oils vary somewhat in their properties. Their composition is not constant but varies slightly with climate, soil, and variety for vegetable oils, and with nutrition, season, and breed for animal oils.

Melting Point. Fats do not melt sharply but soften over a range of temperatures, and it is therefore impossible to apply the melting point technique, used in the identification of pure organic compounds, to them.

If a fat, a fatty acid, or some esters of fatty acids are heated very slowly, they will melt, exist as a liquid as the temperature rises, and then solidify again. A second melting will then occur at a higher temperature. If the material is then chilled rapidly, it will melt at a lower temperature when it is warmed again. This behavior has been known for many years, but it is only in recent times with the use of modern techniques that the explanation has been possible. Polymorphism, the occurrence of more than one crystalline form, explains this phenomenon.

Polymorphism is found in many long-chain carbon compounds. The number of crystalline forms possible for each fatty acid or each ester is still a matter of debate since many of the forms show melting points so close together that it is difficult to separate them. However, for some compounds several crystalline forms have been established. These are designated alpha, beta, and gamma; the melting points of triglycerides are given in Table 2.6. In compounds with several crystalline forms, one form is the most stable and tends to be established.

Polymorphism is important to the understanding of the melting-point behavior of fats, fatty acids, and their esters. Furthermore, polymorphism plays a significant role in any operation where fats are solidified. In this latter instance environmental conditions must be controlled so that the product will be uniform—for example, when chocolates are dipped, the formation of the high-melting forms is produced by controlling the temperature of solidification.

When melting point is used to identify a fat, all aspects of the procedure must be carefully controlled. A number of procedures have been developed and a few will be described briefly.

Solid fats are plastic over a fairly wide range of temperatures. By plastic we mean that they are soft and can be deformed, but do not have the ability to flow. The spreading quality of butter is the result of its plastic nature. When solid fats are examined microscopically, we see that they are composed of a mass of tiny crystals in a matrix of liquid fat. The crystals are not enmeshed but are able to slide by one another and consequently give

TABLE 2.6. MELTING POINTS OF DIFFERENT FORMS OF SIMPLE TRIGLYCERIDES IN °C*

Triglyceride	γ-form	α-form	β-form
Tricaprin	−15	18.0	31.5
Trilaurin	15	35.0	46.4
Trimyristin	33	46.5	57.0
Tripalmitin	45	56.0	65.5
Tristearin	54.5	65.0	71.5
Triolein	−32	−12	5.0
Trielaidin	15.5	37.0	41.5
Trierucin	6	25	32.5
Trilinolein	—	−43	−13.1

*From Bailey, A. E., "Industrial Oil and Fat Products," 60, Interscience Publishers, Inc., New York, N. Y., 1945.

the mixed fat its plastic nature. A fat composed of only one kind of molecules does not possess this property of plasticity since it is, at any given temperature, composed either entirely of crystals or is liquid.

As a fat is warmed, the number of crystals distributed through the liquid fat diminishes and the amount of liquid increases so that the fat softens. If the number of crystals exceeds a critical amount, the fat will be hard and brittle and will lose plasticity. On the other hand, if the amount of liquid exceeds a critical level, the fat will flow.

Natural fats are complex mixtures of glycerides each with its own characteristic melting point. As the temperature of the fat is raised, the melting point of first one and then another of these glycerides is exceeded. Eventually a temperature is reached at which all of the glycerides have melted and the fat is liquid. The temperature at which this occurs is not sharply defined. The problem is further complicated by the fact that glycerides are soluble in one another. A particular type of molecule may *dissolve* in the liquid portion of the fat at a temperature considerably below its melting point.

Glycerides also have a tendency to supercool, i.e., to remain as liquids at a temperature below their melting point as they are cooled down from a higher temperature. The consistency of a fat at say 30°C may be different if it is heated slowly from room temperature than if it has been quickly cooled from a higher temperature. After standing, it will again come to equilibrium and the effects of supercooling will disappear.

The *Softening Point* of a fat is sometimes determined as a means of identification, but it cannot be applied to all fats. Capillary tubes are filled with oil and packed in ice over night so that the oil can solidify and come to equilibrium. The capillary tubes are clamped to a thermometer and submerged in a beaker of water. The temperature is slowly raised and the temperature at which the column of fat rises in the capillary tube is called the softening point. The method gives reproducible results with some fats, rather poor results with others, while on lard compounds, for example, it cannot be used.

The *Slipping Point* is another empirical method used to identify some natural fats and fat compounds. Small brass cylinders, filled with the solid fat, are suspended in a bath close to the thermometer. As the bath is stirred, the temperature is slowly raised. The point at which the fat rises in the cylinder, or slips, is recorded as the slip point. The slip point is related to the air or water beaten into the fat during its manufacture as well as the composition of the fat. The slip point cannot, therefore, be repeated on a particular sample with reproducible results.

The *Shot Melting Point* is the temperature at which a small lead shot will fall through a sample. This method has some usefulness.

Why natural fats and oils differ in their melting behavior has been the subject of a great deal of study. How variations in the composition of a fat and oil influence the melting has been studied in many free fatty acids and pure glycerides. Likewise, analysis of the fatty acids present in fats and a comparison with the melting point gives some information. In general fats which contain relatively large amounts of unsaturated fatty acids have relatively low melting points and are usually oils at room temperature, while those with relatively large amounts of saturated fatty acids have higher melting points. When the melting points of pure, simple triglycerides are determined, it is found that lengthening the carbon chain of the fatty acids increases the melting point. For example, the melting point of trimyristin (alpha form) is 46.5°C, tripalmitin, 56.0°C, and tristearin, 65.0°C. Thus fats which are mixtures of glycerides of long chain, saturated fatty acids have higher melting points than those which have either numerous unsaturated fatty acids or short chain ones, or both (see Table 2.6).

Specific Gravity. The specific gravity of oils and fats is determined by the usual methods. The temperature is carefully controlled since significant changes in these compounds occur in short ranges of temperature. The specific gravity of a fat or oil is usually measured at 25°C, but it may be necessary to use temperatures of 40°C or even 60°C for high-melting fats. Variations in the specific gravity from one oil or fat to another are not great. In general, either unsaturation of the fatty acid chains or increase in chain length of the fatty acid residues tends to increase the specific gravity.

Refractive Index. The index of refraction is the degree of deflection of a beam of light that occurs when it passes from one transparent medium to another. The refractive indices of fats and oils are often measured both because they can be rapidly and accurately determined and because they are useful in identification of these substances and the testing of their purity. An Abbé Refractometer with temperature control is used and the measurement is usually at 25°C. With high melting fats 40°C or even 60°C can be used, but temperature must be controlled and noted. The index of refraction decreases as the temperature rises; however, it increases with increase in the length of the carbon chains and also with the number of double bonds present.

Smoke, Flash, and Fire Points. The smoke point is the temperature at which a fat or oil gives off a thin bluish smoke. It is measured by a standard method in an open dish specified by the American Society for Testing Materials so that the evolution of smoke can be readily seen and reproduced. The flash point is the temperature at which the mixtures of vapor with air will ignite; the fire point is the temperature at which the substance will sustain continued combustion.

For a given sample of oil or fat, the temperature is progressively higher for the smoke point, flash point, and fire point. Table 2.7 gives data for a few oils with flash points measured in an open cup. The temperatures vary with the amount of free fatty acids present in an oil or fat, decreasing with increased free fatty acids. Since the amount of free fatty acids changes with variations in refining, the history of the oil or fat is important. The smoke point of a fat used for deep fat frying decreases with use of the fat. Fats and oils with low molecular weight fatty acids have low smoke, flash, and fire points. The number of double bonds present has little effect on the temperature required. Smoke, flash, and fire points are particularly useful in connection with fats used for any kind of frying.

Turbidity Point. The turbidity point of an oil is determined by cooling a mixture of it and a solvent in which it has a limited solubility. The mixture is warmed until complete solution occurs and then slowly cooled until the oil begins to separate and turbidity occurs. The temperature at which turbidity first is detectable is known as the turbidity point. The first solvent employed was glacial acetic acid in the Valenta test; but since it is difficult to keep the acid pure and since moisture has a marked effect on the test, other solvents have been substituted. In the Crismer test the solvent is methyl alcohol while in the Fryer and Weston modification of the Crismer test it is an equal mixture of 92 per cent ethyl alcohol and amyl alcohol. Data for some oils are given in Table 2.8.

The turbidity point determined for any one oil does show a range of values. It is particularly sensitive to the presence of free fatty acids and a

TABLE 2.7. SMOKE, FLASH, AND FIRE POINTS OF OILS*

Oil	Smoke Points °F	°C	Flash Points (Open Cup) °F	°C	Fire Points °F	°C
Castor, refined	392	200	568	298	635	335
Castor, dehydrated	348	176	570	299	638	337
Corn, crude	352	178	562	294	655	346
Corn, refined	440	227	618	326	678	359
Linseed, raw	325	163	540	287	667	353
Linseed, refined	320	160	588	309	680	360
Olive, virgin	391	199	610	321	682	361
Soybean, expeller, crude	357	181	564	296	664	351
Soybean, extracted, crude	410	210	603	317	670	354
Soybean, refined	492	256	618	326	673	356
Perilla, raw	321	161	575	302	678	359
Perilla, refined	352	178	608	320	685	363
Perilla, refined	408	209	615	324	685	363

*From Detwiler, S. B. and Markley, K. S., "Smoke, Flash, and Fire Points of Soybean and Other Vegetable Oils," *Oil and Soap*, **17**, 39 (1940).

correction factor must be introduced for these acids. Nevertheless, different oils show a wide enough range of values so that the test has value in the differentiation of some oils and in the detection of adulteration.

Chemical Properties. A number of chemical tests have been evolved during the years of study of oils and fats which are based on the partial determination of the chemical composition of the oil or fat. These tests serve both to identify the fat and to detect the presence of adulteration. All oils and fats show some range of values; therefore sometimes more than one test is necessary. A few of the most commonly used tests are given below. The Reichert Meissl Number is a measure of the amount of water-soluble volatile fatty acids; the Polenske Number measures the amount of volatile insoluble fatty acids; the Saponification Number, the amount of potassium hydroxide required to saponify the fat; and the Iodine Number, the amount of unsaturation present. These chemical tests, then, differentiate fats and oils on the basis of the chemical composition of the various triglycerides present in the mixture.

The *Reichert Meissl Number* is defined as the number of milliliters of $0.1N$ alkali (such as potassium hydroxide) required to neutralize the volatile *water-soluble* fatty acids in a 5 g sample of fat. The volatile acids will be those in the range of molecular weights from butyric (C_4) to myristic (C_{14}) acid. The Reichert Meissl test determines the amount of butyric and caproic acids which are readily soluble in water and the caprylic and capric acids which are slightly soluble. The *Polenske Number* is the number of

TABLE 2.8. CRISMER TESTS FOR VARIOUS OILS (FRYER AND WESTON)*

Fat or Oil	Acidity (As Oleic) Per Cent	Observed Crismer Value	Correction Factor for Acidity	Corrected Value
Perilla	5.5	49.0	2.05	60.3
Linseed	2.0	58.3	2.05	62.4
Tung	0.9	74.0	2.05	75.8
Soybean	1.2	65.0	2.05	67.0
Niger	2.2	57.5	2.05	60.0
Sunflower	2.2	59.5	2.05	64.0
Corn	2.8	62.5	2.03	68.2
Cottonseed	0.1	65.0	2.03	65.2
Sesame	4.0	60.5	2.03	68.1
Rape	0.6	82.3	1.61	83.3
Almond	0.9	68.2	2.07	70.1
Peanut	1.1	72.0	2.07	74.3
Olive	0.7	67.8	2.07	69.2
Olive	1.8	65.5	2.07	69.2
Olive	3.6	61.5	2.07	69.0
Cacao	2.9	71.0	1.72	76.0
Chinese Vegetable Tallow	6.9	54.0	1.72	65.9
Palm	0.1	68.0	1.72	68.2
Lard	0.9	70.8	2.13	72.7
Tallow	0.1	72.5	2.13	72.7
Butter fat	1.9	43.0	1.54	46.0
Coconut	0.0	34.0	2.01	34.0
Coconut	1.6	30.0	2.01	33.2
Palm Kernel	0.0	40.0	2.01	40.0

*From Jamieson, G. S., "Vegetable Fats and Oils," 2nd ed., 36, Reinhold Publishing Corp., New York, N. Y., 1943.

milliliters of $0.1N$ alkali necessary to neutralize the volatile, *water-insoluble* fatty acids which are present in a 5 g sample. These two determinations are readily run on the same sample of fat and the Kirschner Value, described below, also can be carried out on the same sample.

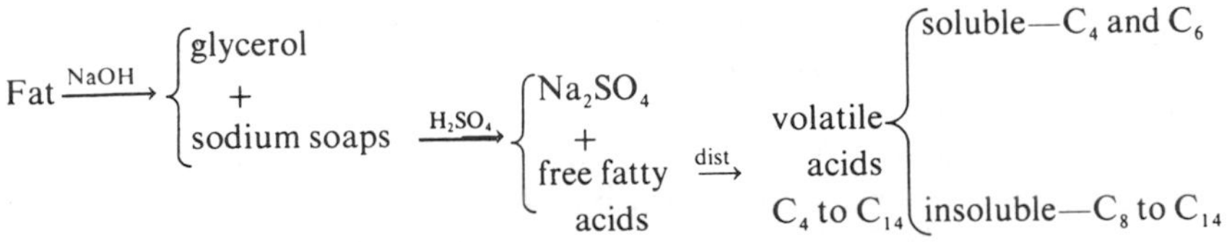

$$\text{Fat} \xrightarrow{\text{NaOH}} \begin{cases} \text{glycerol} \\ + \\ \text{sodium soaps} \end{cases} \xrightarrow{H_2SO_4} \begin{cases} Na_2SO_4 \\ + \\ \text{free fatty} \\ \text{acids} \end{cases} \xrightarrow{\text{dist}} \begin{array}{l} \text{volatile} \\ \text{acids} \\ C_4 \text{ to } C_{14} \end{array} \begin{cases} \text{soluble}-C_4 \text{ and } C_6 \\ \\ \text{insoluble}-C_8 \text{ to } C_{14} \end{cases}$$

A 5 g sample of fat is introduced into a flask and treated either with alcoholic sodium hydroxide or sodium hydroxide in glycerol. If alcohol is used, it must be removed by evaporation before the fatty acids are neutralized. Saponification occurs rapidly under the conditions of the experiment;

the soaps are neutralized with sulfuric acid, and the mixture distilled. On condensation of the distillate the insoluble fatty acids precipitate. The distillate is filtered, an aliquot of the filtrate is titrated with standard potassium or sodium hydroxide and the Reichert Meissl Number calculated. The condenser, receiver, and filter paper are washed with water, and the insoluble fatty acids are dissolved in succeeding portions of ethyl alcohol. This solution is titrated with standard alkali and the Polenske Number calculated.

All of these determinations are particularly valuable in differentiating butter from coconut oil and in detecting adulteration of butter or the substitution of fat mixtures with the physical constants of butter for it. Only a small number of other oils contain appreciable amounts of volatile fatty acids. The Reichert Meissl Number is particularly valuable in detecting adulteration in butter. Although the Reichert Meissl Number varies for butter with season, nutrition, and time in the lactational cycle of the cow, it is usually between 24 and 34, higher than other edible oils. The distillation procedure as set up in the Reichert Meissl and Polenske methods does not remove all of the volatile acids from the saponification mixture; but if the procedure is followed accurately and the size of the sample restricted to 5 g, reproducible results and valuable information can be obtained.

The chief volatile, water-soluble fatty acid in butter is butyric acid. This is determined by the *Kirschner Value* which measures the potential amount of soluble silver salts in the Reichert Meissl distillate. Silver butyrate is soluble in water while the silver salts of the other volatile, water-soluble fatty acids are relatively insoluble. The neutralized distillate from the Reichert Meissl determination is treated with silver sulfate and filtered. The filtrate is acidified with sulfuric acid and distilled. The distillate is carefully collected and titrated with $0.1N$ alkali, either sodium, potassium, or barium hydroxide. Kirshner Value is calculated from the following equation:

$$\text{Kirschner Value} = \frac{A \times 121(100 + B)}{10,000}$$

where:

A = corrected Kirschner titration, and
B = milliliters of alkali to neutralize the 100 ml distillate from Reichert Meissl.

The equation results from the fact that the quantity of distillate collected from the 5 g sample in the Reichert Meissl (first) distillation is 110 ml while 100 ml is titrated, and likewise in the Kirschner (second) distillation. Butter gives a Kirschner Value of 19 to 26, coconut oil approximately 1.9, palm kernel around 1, and other fats and oils between 0.1 to 0.2. Butter is therefore remarkable for the large amount of butyric acid present in it, and it can be identified on this basis.

The *Saponification Number* (the Koettstorfer Number) is defined as the number of milligrams of potassium hydroxide required to saponify 1 g of fat or oil. When potassium hydroxide reacts with a triglyceride, three moles of potassium hydroxide react with one mole of fat.

$$
\begin{array}{l}
R_1COOCH_2 \\
\quad | \\
R_2COOCH \; + \; 3\,KOH \\
\quad | \\
R_3COOCH_2
\end{array}
\rightarrow
\begin{array}{l}
R_1COOK \quad HOCH_2 \\
\quad + \qquad\quad | \\
R_2COOK \; + \; HOCH \\
\quad + \qquad\quad | \\
R_3COOK \quad HOCH_2
\end{array}
$$

If the triglyceride contains low molecular weight fatty acids, the number of molecules present in a 1 g sample of the fat will be greater than if the fatty acids have long carbon chains and high molecular weights. The fat with the low molecular weight fatty acids will consequently have a high Saponification Number. We find that butter with its unusually high percentage of butyric acid has the highest Saponification Number.

The method depends on the saponification of a weighed sample of fat of about 5 g with an excess of standard alcoholic potassium hydroxide. The potassium hydroxide is carefully measured into the flask with a buret or pipet. The mixture is boiled under reflux condensation until saponification is complete, and the remaining potassium hydroxide is determined by back titration with $0.5N$ hydrochloric acid.

$$
\text{Total KOH (meq)}
\begin{cases}
\nearrow & \text{Part used in saponification} \\
\longleftarrow & \text{(Total meq–remaining meq)} \\
\searrow & \text{Remaining meq by titration}
\end{cases}
$$

The Saponification Number can then be determined by subtracting the number of milliequivalents of hydrochloric acid used for back titration from the number of milliequivalents of alcoholic potassium hydroxide introduced, multiplying by 56.1 mg (milliequivalent weight of KOH) and dividing by the weight of the sample in grams.

$$
\text{Saponification Number} = \frac{[(\text{ml} \times N\,\text{KOH}) - (\text{ml} \times N\,\text{HCl})] \times 56.1}{\text{g sample}}
$$

The *Hehner Value* measures the amount of fatty acids which are *insoluble* in water. Thus fats and oils which have high Reichert Meissl Numbers will have low Hehner Values. Since most fatty acids present in natural fats are not soluble in water, most fats have relatively high Hehner Values. The value can be determined either by weighing a sample of fat and saponifying it with alcoholic potassium hydroxide or by using the solution from the Saponification Number titration. The alcohol is evaporated, the soaps dissolved in hot water, and treated with concentrated hydrochloric acid. Free fatty acids are formed; and when the mixture is cooled, the insoluble fatty acids form a cake on the top of the solution. This is filtered on a weighed filter paper, dried, and weighed.

The *Iodine Number* is the number of grams of iodine or iodine compound absorbed by 100 g of fat. The double bonds present in the unsaturated fatty acids react readily with iodine or certain iodine compounds to form an addition compound even while the fatty acid is combined with glycerol in the fat. The Iodine Number is therefore a measure of the extent of unsaturation of the fatty acids present in a fat. While oleic acid contains one double bond in its 18 carbon chain, linolenic acid contains three double bonds in its 18 carbon chain. Thus a molecule of fat containing one oleic acid can absorb or react with only one third as much iodine (or IBr or ICl) as a molecule of fat containing one linolenic acid residue. The fatty acid assortment present in natural fats is fairly characteristic of the fat. While there will be variation in each vegetable oil with climate, soil, and variety and in each animal fat with nutrition and breed, the variations are small compared to the variations between fats. The Iodine Number is therefore of great value in identifying fats and oils

The Iodine Number is determined by dissolving a weighed sample of fat (0.1 g to 0.5 g) in chloroform or carbon tetrachloride and adding an excess of halogen. After standing in the dark for a controlled period of time, the excess, unreacted iodine is measured by thiosulfate titration. In the Hanus method the standard iodine solution is made up in glacial acetic acid and contains not only iodine but iodine bromide which accelerates the reaction. The Wijs method uses an iodine solution made up in glacial acetic acid but containing iodine chloride as accelerator. The excess iodine reacts with sodium thiosulfate according to the following equation:

$$2\,Na_2S_2O_3 + I_2 \rightarrow 2\,NaI + Na_2S_4O_6$$

The end point is determined by the disappearance of the blue starch–iodine color.

The *Acetyl Value* is a measure of the amount of hydroxy fatty acids present in a fat. Castor oil is rich in ricinoleic acid, $C_{17}H_{32}(OH)COOH$, and the determination is of value for this and a few other oils. Most edible oils and fats contain only very small amounts or no hydroxy fatty acids.

By the use of the physical constants and the chemical methods it is possible to differentiate and identify natural oils and fats. Some of these data are given in Table 2.9. Some variation occurs in all of the determinations because of the slight variations in composition of these natural fats. When it is said that a fat is characteristic of the species of plant or animal from which it comes, the statement is not true to an absolute degree. Nevertheless, by means of a few tests, an oil or fat can be identified and the presence of adulteration can be detected.

FLAVOR CHANGES IN FATS AND OILS

When fats and oils are stored, they undergo flavor changes which markedly influence their market value. Consequently, a great deal of research has been devoted to an attempt to discover why some fats undergo changes more rapidly than others, what the changes are, what causes the changes and how they can be controlled. The flavor of rancidity is well known to everyone; and the problem of rancidity, although by no means completely understood, has been the subject of extensive research. Some oils and fats develop off-flavors before the onset of rancidity. This change is called reversion, and it, too, is important because it makes the fat or oil undesirable for food products.

Rancidity

It has been known for many years that fats and oils slowly take up oxygen for a period of time before it is possible to detect the flavor of the products of rancidity. This period is called the induction period, and it is followed by a second period in which the uptake is much more rapid. Rapid oxidation often continues for an extended period of time after which the rate falls off. The length of each period is markedly affected by many factors for each fat, and the course of the oxidation can apparently take a number of paths. Temperature, moisture, the amount of air in contact with the fat, light, particularly that in the ultraviolet or near ultraviolet, as well as the presence or absence of antioxidants and prooxidants influence the reaction.

Long ago farm women learned to store their lard in crocks with as small a surface exposed to the air as possible, and in a cool place. Darkness and coolness are not always possible when fats are shipped and stored during the ordinary course of commerce. Vegetable fats, particularly those from seeds, show a marked resistance to the onset of rancidity. Some seeds, if they are not bruised or crushed, can be stored for years without any change in the fats. But in general animal fats deteriorate fairly rapidly. Many years ago it was shown that the resistance of vegetable fats to oxidation rested

TABLE 2.9. CONSTANTS OF FATS, OILS, AND WAXES*

Name	Specific Gravity 15°/15°C	Solidification Point, °C	Acid Value	Saponification Value	Iodine Value	Reichert Meissl Value
Almond	0.914 to 0.921	−15 to −20	0.5 to 3.5	183.3 to 207.6	93 to 103.4	0.5
Beef Tallow	0.895	31 to 38	0.25	196 to 200	35.4 to 42.3	0.25
Black Walnut	0.918 to 0.921	turbid −12	8.6 to 9.0	190.1 to 191.5	141 to 142.7	—
Butter Fat	0.907 to 0.912 $\frac{40°}{15°}$	20 to 23	0.45 to 35.4	210 to 230	26 to 38	17.0 to 34.5
Castor	0.960 to 0.967	turbid −12	0.12 to 0.3	175 to 183	84	1.4
Chicken Fat	0.924	21 to 27	1.2	193 to 204.6	66 to 71.5	1.8
Coconut	0.926	14 to 22	2.5 to 10	253.4 to 262	6.2 to 10	6.6 to 7.5
Cocoa (Cacao) Butter	0.964 to 0.974	21.5 to 23	1.1 to 1.9	192.8 to 195	32.8 to 41.7	0.3 to 1
Cod Liver	0.922 to 0.931	−3	5.6	171 to 189	137 to 166	0.2
Corn	0.921 to 0.928	−10 to −20	1.37 to 202	187 to 193	111 to 128	4.3
Cottonseed	0.917 to 0.918 $\frac{25°}{25°}$	+12 to −13	0.6 to 0.9	194 to 196	103 to 111.3	0.95
Lard Oil	0.934 to 0.938	27.1 to 29.9	0.5 to 0.8	195 to 203	47 to 66.5	0.5 to 0.8
Linseed	0.930 to 0.938	−19 to −27	1 to 3.5	188 to 195	175 to 202	0.95
Mutton Tallow	0.937 to 0.953	32 to 41	1.7 to 14	195 to 196	48 to 61	—
Olive	0.914 to 0.918	turbid + 2, ppt − 6	0.3 to 1.0	185 to 196	79 to 88	0.6 to 1.5
Palm	0.924 to 0.858$^{100°}$	35 to 42	10	200 to 205	49.2 to 58.9	0.9 to 1.9
Peanut	0.917 to 0.926	3	0.8	186 to 194	88 to 98	0.4
Perilla	0.930 to 0.937	—	—	188 to 194	185 to 206	—
Poppy Seed	0.924 to 0.926	−16 to −18	2.5	193 to 195	128 to 141	0.6
Pumpkin Seed	0.923 to 0.925	−15	—	188 to 193	121 to 130	4.45
Rape Seed	0.913 to 0.917	−10	0.36 to 1.0	168 to 179	94 to 105	0.0 to 7.9
Sesame	0.919 $\frac{25°}{25°}$	−4 to −6	9.8	188 to 193	103 to 117	1.1 to 1.2
Soya, Soy or Soja Bean	0.924 to 0.927	−10 to −16	0.3 to 1.8	189 to 193.5	122 to 134	0.5 to 2.8
Sunflower	0.924 to 0.926	−17	11.2	188 to 193	129 to 136	0.5
Walnut	0.925 to 0.927	−15 to −27	2.5	190.1 to 197	139 to 150	0.92
Whale	0.917 to 0.924	−2 to 0	1.9	160 to 202	90 to 146	14

*Adapted from Lange, N. A., "Handbook of Chemistry," 14th ed., 678, Handbook Publishers Inc., Sandusky, Ohio, 1944.

on the presence of antioxidants which occur naturally in the tissues and which are present in the oil when it is pressed. Some natural fats contain prooxidants which accelerate the onset of rancidity although most of the prooxidants, particularly the metals and their compounds, are picked up during processing from the metallic equipment.

The uptake of oxygen and the onset of rancidity seems to be related to the unsaturation of the fat, although this has been exceedingly difficult to show by direct comparison of natural fats. Since natural fats vary to a great degree in the occurrence of antioxidants, it is not surprising that contradictory results have been obtained. Studies of the autooxidation of simple esters of fatty acids have been more revealing. In one study, the relative rates of autooxidation under standard conditions were followed by measur-

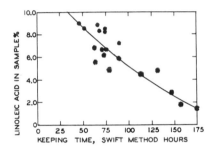

FIGURE 2.2. RELATIONSHIP BETWEEN LINOLEIC ACID CONTENT OF ALL-HYDROGENATED COTTON SEED OIL SHORTENING AND THE STABILITY OF THE SHORTENING. Reproduced from Bailey, A. E., "Industrial Oil and Fat Products," Interscience Publishers, Inc., New York, N. Y., 1945.

ing peroxide formation when oxygen was bubbled through the methyl ester: oleate, 1; linoleate, 12; linolenate, 25. Many other studies have been made, and it is generally agreed that the site of oxidation is the double bond.

The oxidation is not, however, a simple oxidation of the double bond. Many attempts to analyze the volatile compounds which can be smelled in a rancid fat show that a very complex mixture of compounds is formed. Heptyl aldehyde has been isolated most commonly and probably in largest amount. But heptyl aldehyde alone does not have the same quality of odor which is readily detected in a rancid fat. The compounds isolated have been relatively short chain compounds and include aldehydes, acids, hydroxy acids, ketones, and keto acids.

The products formed during the induction period have for many years been called peroxides and tests for the most part have centered around the ability of these compounds to release iodine from potassium iodide. This reaction with KI is very nonspecific, and whether or not the products are true peroxides has not been proved. In recent years hydroperoxides have been commonly named as the intermediates, and several theories have been developed to explain their formation. A number of investigators have been able to isolate hydroperoxide from methyl oleate after oxidation at or near room temperature.

The course of the oxidation is now believed to be a chain reaction and the current theory suggests that free radicals are the intermediates. It has been necessary to develop a theory which can explain how oleate with a double bond between carbon 9 and 10 can yield products in which the oxygen is attached to carbon 8, 9, 10, or 11. It is suggested that the initiating reaction, under the influence of light, is the formation of a double free radical as oxygen adds to the double bond, one on the methylenic carbon and one on the oxygen. In the equation, unpaired electrons indicate the site of the free radical.

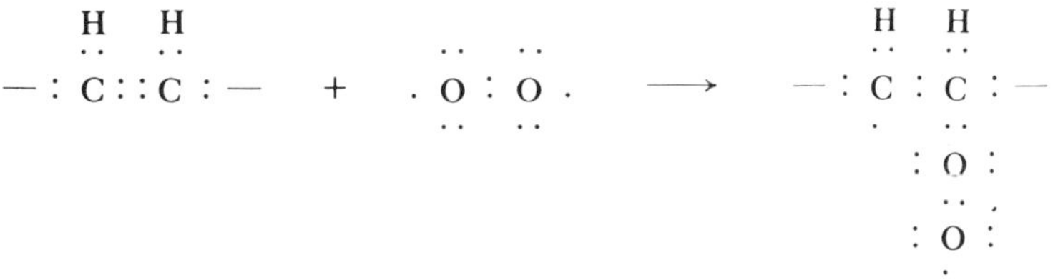

$$-CH=CH- + O_2 \rightarrow -CH-CH(OO^{\cdot})-$$

A free radical may react with a methylenic carbon adjacent to a double bond to form another free radical:

$$\dot{R} + -CH_2CH=CHCH_2- \longrightarrow$$

$$RH + -\dot{C}HCH=CHCH_2- \quad \text{or} \quad -CH_2CH=CH\dot{C}H-$$

$$\updownarrow \qquad\qquad\qquad \updownarrow$$

$$-CH=CH\dot{C}HCH_2- \qquad -CH_2\dot{C}HCH=CH-$$

Because each of the two free radicals formed is a resonance hybrid receiving contributions from two structures, there are four places at which an oxygen molecule can add to the free radical:

$$R^{\cdot} + O_2 \rightarrow R-OO^{\cdot}$$

The free radical peroxide may then react with another methylenic carbon to form a hydroperoxide and another free radical:

$$R—OO\cdot + R—H \rightarrow R—OO—H + R\cdot$$

And so a chain reaction has started. With oleate four products—8, 9, 10, or 11 hydroperoxide form. Linoleate and linolenate are more reactive than oleate and this theory explains that reactivity. A methylene group between two double bonds will be more reactive and the possible number of isomers formed will be quite large.

$$—CH_2—CH{=}CHCH_2\,CH{=}CH—CH_2—$$

Each carbon in the portion of the chain shown is capable of forming a free radical and adding oxygen. Aldehydes and acids will form by cleavage of the carbon chain at the point of attachment of the hydroperoxide:

$$\underset{\overset{|}{OOH}}{—CH}—CH— \rightarrow —CHO \text{ or } —COOH$$

The course of oxidation in a fat is marked by a slow period of oxygen uptake called the induction period, followed by a much more rapid rate of absorption of oxygen. Figure 2.3 shows this change in rate of the uptake of oxygen by peanut oil as measured by the peroxide value. The graph also demonstrates the induction period, the time during which the antioxidant is undergoing oxidation. The sample of peanut oil used in this study was antioxidant-free. Graded amounts of antioxidant were added to subsequent samples. The series of curves show that as the amount of antioxidant increased, the length of the induction period increased. However, notice that the point at which the odor of rancidity products could be detected (arrows) was not constant for the various samples.

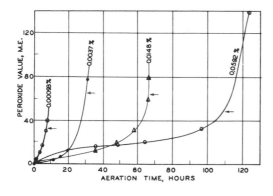

FIGURE 2.3. TYPICAL STABILITY CURVES OF FATS AERATED AT 99° C. Antioxidant-free fat containing different amounts of antioxidant concentrates (tocopherols) from peanut oil. Arrows indicate point at which oils smelled rancid. Reproduced from Bailey, A. E., "Industrial Oil and Fat Products," Interscience Publishers, Inc., New York, N. Y., 1945.

The presence in natural fats of minute amounts of compounds which protect them from oxidation has been recognized for a long time. But the identity of these compounds is still not entirely clear. The first compounds identified were the tocopherols, the vitamin E's discovered by Evans and his coworkers. Four different tocopherols, α-, β-, γ-, and δ- tocopherol occur. You will notice that α- tocopherol has three methyl groups in 5, 7, and 8 positions, while β- and γ- tocopherols have two (β- in the 5 and 8 positions, γ- in the 7 and 8), and δ- only one. The tocopherols are readily oxidized and consequently protect the fat from oxidation. The amounts of these compounds present in refined oils are very small. (See Table 2.3, p. 18). Other antioxidants are believed to occur in natural fats since some oils with relatively low concentrations of the tocopherols are relatively stable.

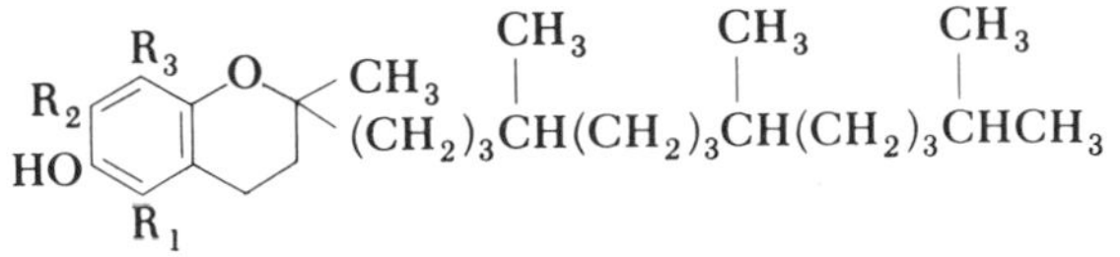

$$R_3 \quad O \qquad CH_3 \qquad CH_3 \qquad CH_3 \qquad CH_3$$
$$R_2 \qquad \qquad \qquad | \qquad \quad | \qquad \quad | $$
$$HO \qquad \qquad (CH_2)_3CH(CH_2)_3CH(CH_2)_3CHCH_3$$
$$R_1$$

Tocopherols

	R_1	R_2	R_3
α	—CH_3	—CH_3	—CH_3
β	—CH_3	—H	—CH_3
γ	—H	—CH_3	—CH_3
δ	—H	—H	—CH_3

Numerous compounds have been developed synthetically and may be added to fats to inhibit oxidation. Many of these compounds have been patented as antioxidants. The development of fats which do not require refrigeration during the few months that they are stored in the ordinary kitchen is chiefly the result of the use of these compounds. Some that have been tried include the o- and p- dihydroxy benzenes, such as hydroquinone and pyrogallol; the aromatic amines; glucamine; gums; and cereal flours. Sometimes antioxidants show a synergistic effect; two antioxidants may be much more effective than would be expected from the sum of their activities. Butylated hydroxy anisole (BHA) or butylated hydroxy toluene (BHT) are often combined with propyl gallate. An example is a solution for protection of nuts, and composed of 14 per cent butylated hydroxy anisole, 6 per cent propyl gallate, and 3 per cent citric acid in ethyl alcohol.[26] A small amount is used in nut candy.

The phosphatides, particularly the cephalins, seem to augment the antioxidant activity of other antioxidants such as the tocopherols. Since phosphoric acid and other acid products are capable of reinforcing the activity of some of the phenolic antioxidants, it may well be that the effect of the cephalins arises from the free hydrogen of the phosphate. Lecithins, however, do not have this effect.

The onset of rancidity in baked goods follows quite a different pattern from that of fat alone. The keeping quality of crackers and cookies is of great economic importance since these products are often stored for extended periods before they are consumed, and they are not protected from oxidation. Sugar in cookies and biscuits appears to have a marked inhibiting effect on the onset of rancidity. For example, McKinney and Bailey[20] found that biscuits made with hydrogenated vegetable oil became rancid in 80 days; whereas, when sugar was included in the recipe, they did not become rancid until 360 days. Oleo oil and hydrogenated lard gave similar results, although the difference with lard was not as marked. Rancidity occurred in the lard biscuits in 34 days, in those with added sugar in 50 days.

The effect of spices as antioxidants has been investigated, but the results are difficult to summarize briefly since considerable difference is obtained with differences in the food observed.[3] For the most part spices, and especially cloves, are more effective in a water-fat emulsion than in dry lard and more effective in dry lard than pie crust. In a study of ground pork, mayonnaise, and French dressing it was shown that allspice, rosemary, cloves, sage, oregano, and thyme improved the stability of the fats. Clove has the greatest effect in ground pork, while oregano is most active in mayonnaise and French dressing. In oil-water emulsions, clove has the greatest activity, while in lard, sage and rosemary are outstanding.

Prooxidants promote the onset of rancidity. It is doubtful whether prooxidants occur naturally in fats and oils in significant amounts. Metallic ions act as catalysts in rancidity reactions, and they are often introduced in small amounts during the refining of the oil. If a small amount of rust is formed in steel equipment, it readily dissolves; or if copper vessels are used, small amounts of copper oxide may be dissolved. Since these metals are active at concentrations measured by parts per million, the nature of the utensils, pipes, and valves with which the oil comes in contact during refining is of great importance in its resistance to rancidity.

Tests for Rancidity. Rancidity tests have developed over the years in order to establish sensitive control of the stability of fats or the foods in which they are used.[6] The more common tests are (1) Peroxide Value, (2) determination of carbonyl, (3) active oxygen determination, (4) thiobarbituric acid test (TBA), and (5) Schaal oven test, which is used primarily in the baking industry.

All of the tests for Peroxide Value measure the amount of iodine released when potassium iodide reacts with rancid fat. The Lea method uses 1 g of fat and 1 g of potassium iodide with an acetic-acid–chloroform (2:1) solvent. After heating, the iodine formed is determined by titration with standard thiosulfate. Other methods modify quantities of solvent, but the principle remains the same.

The measurement of carbonyl compounds has followed traditional methods for this group. The Kreis and Shibsted tests were early ones which have dubious reliability. Today 2,4 dinitro phenyl hydrazine is commonly used in the Lappin-Clark method.

Active oxygen is a method which measures the length of time required to produce 20 meq peroxide per 1000 g fat when air is bubbled through fats under standard conditions. A bubble train is set up so that clean, dry air bubbles through the oil at a constant rate such as 2.33 ml per sec. The oil is heated in a constant-temperature bath. Other samples are added at definite time intervals so that the time required for 20 meq can be calculated.

In the *TBA method for determining rancidity,* an oxidized or rancid fat will react with 2-thiobarbituric acid (TBA) to form a red color the intensity of which is proportional to the amount of rancidity. In recent years this has been developed into a method for measuring the extent of rancidity in a sample of fat. In the rancid fat the compound formed which reacts with the 2-thiobarbituric acid is malonaldehyde, $CH_2(CHO)_2$.

The *Oven Test* is widely used in the baking industry. It takes little equipment and is very easy to set up. Biscuits, cookies, or crackers are stored in beakers or jars with loose fitting tops at 63°C or 145°F. The number of days required to develop rancidity is measured by odor and taste. The temperature is slightly above what might be encountered during distribution of the food through regular commercial channels. Its greatest use is in comparing fats. It is difficult to equate shelf life of a product with the Oven Test where conditions are variable.

Reversion [5]

Many oils and fats undergo a change in flavor before the onset of rancidity which is known as *reversion*. Some investigators believe that it is characteristic of all fats, but that some fats show more pronounced changes. The name "reversion" came to be applied to the change because some marine oils which possess a fishy flavor before processing revert to this fishy odor on storage. The term is poor because the flavors which develop in most fats were never there before, but it has become firmly established and cannot be displaced. The flavors which develop are quite different from rancid flavors. Soybean oil, which reverts readily, is described as developing first a buttery or beany flavor, then grassy or hay-like, then painty, and finally fishy.

The conditions under which reversion occurs are those encountered in marketing and also the high temperatures experienced during baking and frying. The problem of reversion is therefore of considerable importance in edible fats and oils. The factors which are known to influence the onset and development of reversion are (1) temperature, (2) light, (3) oxygen, to a limited extent, and (4) trace metals.

As the temperature of storage is increased, the length of time in which reversion flavors can be detected decreases. Thus Bickford found that soybean oil stored in the dark at 5°C (41°F) did not revert for several months, but oil stored at room temperature showed flavor changes in 10 to 14 days. The effect of light of various wave lengths on hydrogenated vegetable shortening was studied by Gudheim. He found that it is light in the blue-violet and from 325 to 460 mμ which causes reversion changes. Light at 325 mμ is in the ultraviolet and below this wave length does not pass through glass, so it is not a problem in ordinary packaging. Oxygen is not necessary in large quantities for the development of reversion, but a small amount appears indispensible since flavor changes which occur in oils stored in vacuum or under inert gases are not characteristic of reversion.

The effect of traces of metallic salts on reversion has been demonstrated by adding small amounts of salts to refined oils and measuring their tendency to revert. For example, soybean oil containing 0.015 ppm of iron was treated with iron salts to bring the levels to 0.03, 0.3, and 1.0 ppm. After storage at 60°C (141°F) for four days, the sample containing 1.0 ppm had the poorest flavor, 0.3 ppm intermediate flavor, and that with 0.03 ppm about the same flavor as the control. It has been shown that copper, cobalt, chromium, and zinc cause an acceleration in reversion. Aluminum, tin, and nickel are not so active. During the processing of fats, compounds are often added which have the ability to form complexes with metal ions. These are called "metal scavengers" since they remove the ions from the field of action and inhibit their catalytic effect. An example is EDTA (Ethylenediaminetetraacetic acid).

$$\begin{array}{c} HOOCCH_2 \\ \\ HOOCCH_2 \end{array} NCH_2CH_2N \begin{array}{c} CH_2COOH \\ \\ CH_2COOH \end{array}$$

Rancidity and reversion are not the same thing. Indeed some oils and fats which are susceptible to rancidity, such as corn oil, are reversion-resistant. The flavor changes which occur during reversion vary with different fats, but in rancidity the final flavor is the same for all fats. The peroxide value is widely used as a measure of the development of rancidity, but it is impossible to show a correlation between this value and reversion.

The chemical reactions that occur in reversion are not completely known, but much research has been done on the problem. It is still not yet definite exactly which compound undergoes change, what the change is, or what products are formed. The flavorful products which impart the distinct flavors of reversion are steam-distillable, and it has been demonstrated that these products are present in reverted oils and fats in extremely small quantities. Numerous compounds have been suspected as reactants in reversion, but the data are impressive only for linolenic and isolinoleic acids.

In 1936 Durkee, in searching for an explanation of the reversion tendency of some oils and the reversion resistance of others, noticed that oils with reversion tendency were particularly rich in linolenic acid. Dutton altered corn oil, which is naturally reversion-resistant, to a product with a high linolenic content by interesterification with methyl linolenate in the presence of a catalyst. He also prepared a control by interesterification of the corn oil with methyl linoleate. After storage a taste panel compared these two products with corn oil and soybean oil. The corn oil high in linolenate was judged as soybean oil (reversion-susceptible) in five out of six trials. This indicated that reactions leading to the flavor development in a synthetically high-linolenate corn oil are similar to those in soybean oil.

Some interesting work on linseed oil, which is not a food fat, has implicated "isolinoleate" as the precursor of reversion. When linolenic acid or its esters are hydrogenated, the first product of the addition of one molecule of hydrogen is called "isolinoleic acid." Hydrogenation of methyl linolenate produces 8, 14; 9, 15; and 10, 14 isolinoleate.

$$C_{17}H_{29}COO + H_2 \rightarrow C_{17}H_{31}COO^-$$

Linolenic   Isolinoleic

Linseed oil is a reversion-susceptible oil and on hydrogenation it forms oils whose reversion tendency closely parallels the isolinoleate content. As hydrogen is added and isolinoleate begins to form, the reversion tendency rises. This continues until the point is reached at which hydrogen begins to add to the isolinoleate. As hydrogenation continues, the reversion tendency then falls. A concentrate of glyceryl tri-isolinoleate is very susceptible to reversion. In another study ethyl linolenate was mixed with sunflower seed oil which is reversion-resistant and hydrogenated to shortening consistency. The sample reverted in pastry.

It appears as if both linolenate and isolinoleate may be responsible for reversion. The flavor of all reverted oils is not the same and even the same oil has different flavors under different conditions.

Some attempt has been made to identify the products of reversion which apparently result from degradation of linolenate and/or isolinoleate by steam distilling reverted oil and concentrating the distillate. Compounds which have been isolated and identified include 2-heptenal, $CH_3(CH_2)_3CH=CHCHO$, di-n-propyl ketone, $(CH_3CH_2-CH_2)_2CO$, maleic aldehyde, $OCHCH=CHCHO$, $\Delta^{2,4}$-decadienal, $CH_3(CH_2)_4CH=CHCH=CHCHO$, and acetaldehyde, CH_3CHO. Exactly what contribution each compound makes to the subtle flavor blend in reversion has not been determined.

Much attention has been focused on practical methods of processing to prevent or minimize reversion. Avoidance of contamination with small traces of metals and the use of metal scavengers to minimize trace metal effects have been tried. Reversion is still an important problem in the refining and use of soybean oil, marine oils, rape seed oil, and a few others which have a marked tendency to revert.

THE TECHNOLOGY OF EDIBLE FATS AND OILS

Three principal methods are used for the extraction of edible fats and oils from the animal or vegetable tissues in which they occur. These are (1) rendering, which is chiefly applied to animal tissues; (2) pressing; and (3) solvent extraction. In general, there are more fats and oils in fatty animal tissue than vegetable tissue; in other words, less water, protein, and other nonfatty material are present. The problem in extracting the fat from any one tissue is always a little different than it is for another; and although a manufacturer of cotton seed oil may also make peanut oil in the same equipment, he is often limited to these two oils and does not attempt any other. Generalizations about methods of extraction of the fats can be made, but variations in procedure necessarily occur because of differences in the oil-bearing tissues. For details on methods of extraction of a particular oil or fat, see some of the books on fats.

Rendering

Rendering is a process by which fat is removed from a tissue by heat. It is also called "trying out." The tissue containing a high percentage of fat is carefully removed from the animal and chopped or minced. Heat allows the lipids to escape from the cells. If the heat is high, the cells are completely ruptured, a cooked flavor develops, and "cracklings" are left with the oil floating on top. Rendering can be carried out either in the presence of water—"wet rendering"—or in its absence—"dry rendering." In wet rendering the fat can be separated by gentle heat in an open kettle or in an autoclave in the presence of steam. In the first method, the well-chopped tissue is introduced into an open kettle along with a charge of water, stirred gently, and heated to about 50°C. The fat floats to the top and is carefully skimmed off. It has a bland flavor and requires little deodorization, but the process does not remove all of the fat from the tissue. It is much more common today to use steam in digesters or autoclaves at relatively high temperatures and pressures of 40 to 60 psi. In this way the tissue is quite well disintegrated and the separation of the fat is efficient.

Dry rendering is the process used in cooking bacon. The tissue is heated and the fat separates as the protein is denatured and the water is evaporated. Commercially the process is carried out under vacuum in steam-jacketed cookers.

Pressing

Pressing is the application of high pressures to the oil-bearing tissue to squeeze out the fat. In some cases, such as the pressing of olives, virgin oil is the first pressing of the fruit and is particularly bland in flavor. The fruit is then subjected to subsequent pressings to give other grades of oil.

Other oil products require that as much extraneous material as possible be removed from the seed or fruit before pressing. Thus cotton seeds are delintered and the kernel then separated from the hull, while in cereals the germ is separated before pressing is used. It is difficult to separate oils efficiently by pressing and the greater the bulk of extraneous material, the lower the efficiency. Often the oil-bearing tissue is rolled, crushed, or ground to flaky particles. Usually the flake is cooked to denature the protein and release the oil. It is then pressed in a filter with filter cloths to hold back the extraneous material or it is passed through an expeller. An expeller is constructed something like a meat grinder, with a worm screw which increases the pressure as the material is carried into it. At the end of the worm is an opening for carrying out the residue with holes in the bottom of the apparatus where the oil runs out. Sometimes the residue from the expeller is then placed in a filter press.

Solvent Extraction

This process has been used since Deiss extracted the press cake of olives

Courtesy of Procter and Gamble.

FIGURE 2.4. PRODUCTION FILTER PRESS USED IN FAT AND OIL PROCESSING. In addition to simple filtration of raw materials, these filters are used to separate bleaching earth and the carbon used for decolorizing.

with carbon disulfide in Marseilles in 1855. In the United States solvent extraction with petroleum ether has been used primarily for the production of soybean oil. (Garbage grease is also recovered by this method.) However, rapid improvements in continuous counter-current equipment has been such that in Europe solvent extraction is applied to many other oils. In solvent extraction the tissue is treated in a fashion similar to that used in pressing to form a thin coherent flake. The oil is extracted either in batch extractors in which the flake is gently agitated with successive portions of solvent, or by counter-current extraction where the flake and solvent move in continuous streams in opposite directions.

Common solvents are petroleum ether, benzene, chlorinated hydrocarbons, carbon disulfide, or Stoddard solvent. In the United States, petroleum ether boiling at 140°F to 185°F is used frequently, but hexane and other solvents are now coming into use. The solvent must then be removed from the oil. The method is quite efficient but is expensive because of the inevitable loss of the solvent through evaporation, and only in large plants is it practical. However, in the removal of oil from tissues which have a relatively low percentage of oil, it is the only practical method.

Sometimes a tissue is subjected to pressing and then the press cake with its low fat content is extracted with a solvent.

The residue from pressing or extracting is high in protein and is a valuable feed for cattle. If a solvent is used in extraction, this must be removed from the cake, of course, before it can be fed to cattle.

Refining

The crude oils extracted from tissues often contain material that must be removed before these oils can be put on the consumer market or sent to the hydrogenator. The crude oils may contain one or more of the following groups of substances: (1) cellular material or derivatives, both protein and carbohydrate; (2) free fatty acids and phosphatides; (3) pigments; (4) odorous compounds such as aldehydes, ketones, hydrocarbons, and essential oils; and (5) glycerides with high melting points.

The first step in refining most oils and fats is the removal of finely divided cell debris. This is usually accomplished by settling and then either filtering or centrifuging. When the particles are colloidal, adsorbing or filtering agents may be added before filtering.

Free fatty acids occur in some crude oils in goodly amounts. Thus palm oil usually has about 5 per cent free fatty acids. These can be fairly completely removed by steam refining and the remainder by alkali refining. Almost all edible fats and oils are subjected to alkali refining since it re-

(*Courtesy of Procter and Gamble.*)

FIGURE 2.5. SOLVENT EXTRACTION. Mill room flaking rolls.

moves not only free fatty acids but also phosphatides and some other resinous material.

Steam Refining. This consists of blowing steam through hot oil under vacuum The free fatty acids with molecular weights below myristic steam distill but the fat is relatively nonvolatile. The process is used for the deodorization of fats. Oils with a high free fatty acid content are usually first subjected to steam refining; this is followed by alkali refining. Those with low fatty acid content may simply undergo alkali refining.

Alkali Refining. In alkali refining hot oil is treated with a solution of an alkali, usually sodium hydroxide (caustic soda) but sometimes with sodium carbonate (soda ash) and occasionally with other alkaline sodium salts. The free fatty acids react and form sodium soaps which are dispersible in the water layer. If the process is carried out rapidly, the amount of saponification of the oil that occurs will be very small.

In the United States, the common method of alkali refining is a continuous one in which the solution of alkali and oil is mixed, passed through a heater, and then through a battery of primary centrifuges. These centrifuges discharge the soap and alkali solution on one side and the oil mixed with a small amount of soap and alkali solution on the other. The oil is then mixed with hot water and passed to a second battery of centrifuges that discharge washed oil relatively free of alkali or soap; the oil is then dried. Although alkali refining was once carried out by the batch method in which the alkali solution and oil are mixed in a conical kettle, heated, and then separated, this process has been largely superseded by the continuous method which is more rapid and efficient.

Bleaching. Pigments are removed by adsorption on fuller's earth, other clays, and sometimes through the addition of small amounts of charcoal. The process is called bleaching although it is a physical process. The oil is heated to 220°F to 240°F, agitated, and then filtered. It is returned to the kettle until the color of the filtered oil is sufficiently light. Some processors use vacuum since subjecting oils to high temperatures in the presence of air unfavorably affects their resistance to rancidity. Chemical agents, either oxidation or reduction agents, are not generally used for bleaching edible oils but only for industrial fats and waxes.

Steam Deodorization. The oils and fats demanded by the consumer today must be very bland in flavor. By steam deodorization fats and oils can be rendered so bland that it is difficult to detect their origin by taste or smell. Large kettles in which it is possible to maintain low pressures permit the use of only small quantities of steam for deodorization. The fat is introduced into the kettle and the pressure reduced to 1/4 inch or 6 mm of mercury. The fat is heated and steam is injected into the bottom of the vessel. Oils are rapidly deodorized at 425°F to 475°F. If lower temperatures

a. Soybeans.

b. Flaked beans ready for extraction.

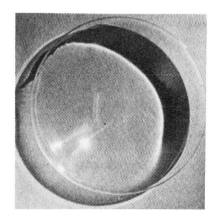

c. Soybean oil.

d. Residual meal after grinding.

FIGURE 2.6. STAGES IN SOLVENT EXTRACTION.   *Courtesy of Procter and Gamble.*

are used, longer periods are required. The oil is then rapidly cooled. It is most important to prevent the contact of hot oil with air, since oxygen is injurious both to its keeping properties and to its flavor.

When cottonseed oil was first manufactured, it was noticed that the oil drawn from the tops of tanks in winter did not become cloudy or tend to solidify when refrigerated. Thus the term "winterized" came to be applied to oils which have been processed so that they remain fluid at refrigerator temperatures. The process is applied to salad oils. The oils are slowly chilled without agitation so that large, filterable crystals are formed. The crystals are composed of glycerides that contain the higher molecular weight and more saturated fatty acids and that consequently possess higher melting points. The cold viscous oils are carefully filtered and the filter cake used for some products in which the temperature of solidification is not important. Cottonseed oil with an iodine number of 107 yields 65 per cent of an oil with an iodine number of 113 and a 35 per cent cake with an iodine number of 95. Lecithins can be added to oils to inhibit clouding and solidification. Thus a sample of cottonseed oil which clouds in 10 hours at 32°F will not cloud for 15 hours on the addition of 0.05 per cent soybean lecithin.

Hydrogenation

Hydrogenation of fats and oils is the process by which molecular hydrogen is added to double bonds in the unsaturated fatty acids of the glycerides. Economically, hydrogenation is a most important process since by its use the physical properties of a natural fat can be altered. Thus the physical properties of the products can be regulated so that many natural fats can be interchanged, so that liquid fats can be substituted for plastic fats, and so that an improvement occurs in the properties of natural plastic fats.

Hydrogenation occurs when hot oil saturated with hydrogen is brought in contact with an active catalyst, but the course of the reaction and its velocity is influenced by numerous factors. In most oils with a high content of C_{18} unsaturated acids the possible reactions are (1) linolenic to linoleic or isolinoleic, (2) linoleic to oleic, and (3) oleic to stearic. When the oil contains unsaturated acids with carbon chains of other lengths, hydrogen may be added to these chains. The products may be the natural linoleic and oleic acids or they may be the so-called iso acids in which the double bond is in a different position than that of the naturally occurring acid. Thus it has been found that hydrogenation of linoleic may form natural oleic with the double bond in the 9, 10 position; but it will also form some iso-oleic with the double bond in the 12, 13 position, which has a higher melting point.

$$CH_3(CH_2)_7\overset{10}{C}H=\overset{9}{C}H(CH_2)_7COO-$$
$$\text{Oleate}$$

$$CH_3(CH_2)_4\overset{13}{C}H=\overset{12}{C}HCH_2\overset{10}{C}H=\overset{9}{C}H(CH_2)_7COO-$$
$$\text{Linoleate}$$

$$CH_3(CH_2)_4\overset{13}{C}H=\overset{12}{C}H(CH_2)_{10}COO-$$
$$\text{An Iso-oleate}$$

Methyl linolenate on hydrogenation gives 8, 14; 9, 15, as well as 10, 14 isolinoleate. Whale and fish oils contain rather large amounts of unsaturated C_{20} and C_{22} acids; and consequently, more variation in the molecules hydrogenated can occur. There is also the problem of the selective hydrogenation of fatty acid residues which occur on the α or β position of the glycerol. Finally, during hydrogenation trans isomers are also often formed rather than natural cis isomers.

H   H
| |
C═══C
| |
$CH_3(CH_2)_7$ $(CH_2)_7COOH$

$CH_3(CH_2)_7$ H
      \ |
      C═══C
     / |
   H $(CH_2)_7COOH$

Cis   Trans

Mabrouk and Brown[18] examined the infra-red spectra of six commercial margarine samples and five shortenings and found a high (22.7 to 41.7) per cent of trans isomers in all except one sample. O'Connor[22] claims that hydrogenation of methyl oleate causes change to as much as 38 per cent of the trans isomer. So the possible products are numerous and the complete course of the reactions and the influence of all factors is not as yet known.

Nevertheless, the course which the reaction follows is of great importance as far as both the physical properties of the resultant fat and its keeping qualities are concerned. It will be remembered that both linolenic and isolinoleic acids have been implicated as active in promoting reversion so that the quantities of these acids remaining or formed influence the keeping quality.

When hot oil and catalyst are stirred together under an atmosphere of hydrogen, the properties of the final product are affected by temperature, rate of mixing, nature of the catalyst, concentration of the catalyst, and pressure of hydrogen.

A number of active metals are capable of catalyzing this reaction, but industrially nickel is used. Numerous methods are employed for producing the finely divided metal with many active centers on its surface. Even when the method of preparation is as carefully standardized as possible, to produce a catalyst with exactly the same activity in each batch is difficult. Since the catalyst is slowly poisoned during the hydrogenation, it needs to be reactivated after it has been used for some time. Sulfur compounds, such as hydrogen sulfide and sulfur dioxide which may contaminate the hydrogen, poison the catalyst rapidly even when their concentration is low. Carbon monoxide is a reversible poison that can be evacuated from the

Courtesy of Procter and Gamble.

FIGURE 2.7. EXTERIOR OF EQUIPMENT USED IN HYDROGENATION OF FATS.

catalyst in a vacuum. It is also a contaminant of hydrogen made by the steam-iron method.

Hydrogen of reasonable purity is desirable to protect the catalyst. In the United States the hydrogen used in this process is made by electrolysis in areas where electric power is relatively cheap. This gas is quite pure. In other areas the hydrogen is made by the steam-iron method. Iron ore is alternately oxidized by steam with the formation of hydrogen and then reduced. The reducing agent is usually "blue gas" or "coal gas," a mixture of hydrogen and carbon monoxide made from hot coke and steam.

$$C + H_2O \rightarrow CO + H_2$$
$$3Fe + 4H_2O \rightarrow Fe_3O_4 + 4H_2$$

Hydrogen synthesized by this process is purified to remove sulfur compounds formed from the coke, as well as carbon monoxide and carbon dioxide.

The hydrogenation is usually carried out by the batch method in large vessels with a capacity of 10,000 to 25,000 lbs of fat. The vessels are equipped with heating and cooling coils. Oil and catalyst are combined and heated under vacuum between 150°F and 400°F. Hydrogen is introduced under pressure in such a manner that the maximum amount is dissolved in the oil. Variations in the apparatus are not only designed to improve the agitation of the oil but also to facilitate the introduction of hydrogen. The temperature and pressure used vary with the oil being hydrogenated and the characteristics of the final product. Sometimes the hydrogenation is carried out in two stages at different temperatures. The reaction is halted when the fat has reached the desirable iodine number or consistency. (Iodine number and consistency parallel one another when the procedure is standardized; otherwise they do not.) The tank is evacuated and the oil cooled to about 150°F. The hot oil must not come in contact with air, of course. The oil is filtered to remove the catalyst and chilled rapidly so that small crystals are formed. Sometimes monoglycerides, antioxidants, and other compounds are added, depending on the product that is desired. Air is sometimes whipped into the product to impart a snow-white appearance.

Technology of Individual Fat Products

Several groups of fat products are as follows: (1) butter, (2) oleo, oil, and oleostearin, (3) lard, (4) salad, cooking, and frying oils, (5) shortenings, and (6) margarine.

Butter. In the United States most butter is now manufactured in creameries although some is still produced in farm kitchens. The tendency is to have creamery depots that collect cream or milk from the farms and then send it to the factory. Cream varies greatly in its quality and is usually bought on a graded scale that is based on acidity, flavor, odor, and the amount of foreign material present. High-quality butter cannot be made from putrid, dirty cream. A small amount of butter is churned from sweet cream, but by far the greater part is made from ripened cream. Salt is usually added to the extent of 2.5 to 3.0 per cent. The minimum butter fat content is established by law at 80 per cent; and the remainder is composed of buttermilk, water, and milk solids.

The common practice is to neutralize the cream with sodium bicarbonate, magnesium oxide, or calcium carbonate; pasteurize it to kill microorganisms; and inoculate it with a starter of selected bacteria. After ripening for 3 or 4 hr, during which time the bacteria grow and produce flavorful compounds and acids, the cream is churned. The butter separates and is washed with cold water. It is then kneaded to reduce the water content, salt is added and blended in, and the product is packaged.

The use of a starter for the ripening of the cream has made possible the production of a high-flavor, standard butter. It has been found that the flavor of butter is the result of a number of compounds, but the most important is diacetyl. This compound is synthesized by the action of *Streptococcus citrovorus* and *Streptococcus paracitrovorus* on citric acid present in the cream. It has been shown that acetyl methyl carbinol is an intermediate in the synthesis and that both compounds are present in the ripened cream and butter.

$$CH_3COCH(OH)CH_3 \rightarrow CH_3COCOCH_3$$

Acetyl Methyl Carbinol   Diacetyl

Lactic acid and small amounts of propionic and acetic acids are formed from fermentation of lactose present in the cream. *Streptococcus lactis* is added to the starter along with *Streptococcus citrovorus* and *Streptococcus paracitrovorus* so that these reactions can be assured.

The consistency of butter varies somewhat during the year with variation in feed. High fat diets of oil cakes have a particularly marked effect on the composition of the fat secreted in the milk. The age and breed of the cow are also of some importance in determining the composition of the glycerides present and consequently of the consistency of the butter.

Oleo Oil and Oleostearin. These substances are prepared by fractionation of the fresh, selected, internal fat tissues of beef. The tissues are carefully rendered at low temperatures and the fat produced held at about 90°F for several days until partial crystallization has occurred. A slow crystallization is desirable so that large, easily filtered crystals will be formed. The mixture of crystals and oil is then filtered in a filter press to yield oleo oil and cakes of oleostearin. Oleo oil is used in the manufacture of some margarines and also unmodified as a shortening by confectioners and bakers. Oleostearin is used in the manufacture of compound shortenings by blending with other fats and in the production of some specialty margarines.

Lard. This substance is the processed fat of hogs. The quality of the lard and its characteristics depend on the part of the carcass from which it is taken, the way in which it is processed, and the feed of the animal. Leaf lard is the highest quality lard; it is rendered from the leaf fat (the fatty tissues around the kidneys). Fatty tissue from the back of a hog gives second quality lard, while that from the intestines is lowest in quality. The lard may be either dry or wet rendered. A great deal of the lard produced in the United States is steam rendered.

Interesterification in Lard.[10,11] A recent development in the technology of lard may return this fat to its former position as a widely used and high-priced fat. This is directed interesterification of lard which improves its plastic range and diminishes its tendency to graininess. Natural lard tends to develop fairly large crystals of disaturated glycerides (fats with two saturated fatty acid residues) that give a grainy texture when the lard is chilled and make it difficult to cream in batters and doughs. Since the melting points of the disaturates are in the range of room temperature, the lard softens so much that it has little body and almost no plasticity at these temperatures. Lard has approximately 37 per cent saturated fatty acids, principally stearic and palmitic. Under suitable conditions the glycerides present undergo interesterification with the rearrangement of the fatty acids in the glycerides. In directed interesterification the reaction is carried out at a temperature where all the glycerides exist in the liquid phase except the trisaturated esters that are solids. As trisaturated glycerides are formed, they crystallize and are removed from the reaction mixture. The interesterification no longer is random, but is driven toward the formation of these solid trisaturates. The catalyst for the reaction in commercial interesterification is an alloy of potassium and sodium.

Lard is now interesterified commercially by a continuous process in which the catalyst is metered into the fat in very small particles. The temperature is carefully controlled and the fat is constantly agitated as crystallization occurs. The level of trisaturated glycerides formed can be controlled by the temperature of the fat and the length of time allowed for crystallization. As the percentage of trisaturates increases, the resistance of the lard to softening at room temperature is increased or, expressed in another way, the percentage of solids at a given temperature increases.

The catalyst is quenched by reaction with water and carbon dioxide. The amount of hydrolysis of fat is minimized by the introduction of carbon dioxide along with the water, which prevents the pH from rising too high. The soap formed is removed by centrifugation, washing, and further centrifugation. The lard is then dried in a continuous vacuum drier.

The lard is usually hydrogenated to increase its resistance to rancidity and increase its consistency. It is deodorized and supplemented with antioxidants and monoglycerides before packaging.

Although the process is applied commercially to lard, interesterification of all natural fats will occur at elevated temperatures. Often the rate of reaction is slow and a considerable period of time is necessary to reach equilibrium. The change in the composition of the glycerides that is brought about by interesterification changes the physical properties of the fat, of course. After interesterification cocoa butter has a much higher melting point and a very different consistency. Some fats are more radically altered than others. Many catalysts have been used but the alkali metals and their alcoholates are most active.

Salad, Cooking, and Frying Oils. These are prepared oils and, with the exception of virgin olive oil, they are alkali refined, bleached, winterized if necessary, and deodorized. Salad oil is an oil which will remain substantially liquid at a temperature of 40°F to 45°F and forms a mayonnaise emulsion stable at these temperatures. An oil labeled "cooking oil" will not have these properties. Most oil is sold in the United States as salad oil. The oils on the market in the United States are principally cottonseed oil, followed by corn, soybean, olive, peanut, and rather small quantities of sesame, sunflower seed, safflower and others. Cottonseed oil must be winterized before it can be sold as salad oil. Some corn oils and soybean oils deposit small amounts of crystals at refrigerator temperatures if they are not winterized. Peanut oil forms very small crystals that are difficult to filter, and it is therefore not easy to winterize.

In the United States, where factory methods are used for the production of fats and oils, the extent of extraction of the crude oils is high and the flavor is strong. Since the consumer market demands oils with bland flavor, oils must be deodorized before they are packaged. Olive oil is the exception, since it is customary to anticipate a flavor in olive oil. The olive oil is usually blended so that the variations in season and climate that affect the flavor can be minimized and a reproducible flavor obtained. The market for olive oil in the United States is, of course, relatively small.

Bleaching is likewise necessary because of the demands of the consumer. Most salad oils on the market are relatively light in color. A darker cottonseed oil is produced for a section of the market that is accustomed to the darker color of olive oil.

Oil that is packaged as cooking oil is sold for commercial deep fat frying. In hotels and restaurants almost all frying is of this kind. Also, in the preparation of doughnuts, deep fat frying is used. The oils sold for this purpose are alkali refined and deodorized but are sometimes not bleached since a light color is not important. Potato chips and products not to be consumed immediately are usually fried in solid fats rather than cooking oils.

Mayonnaise is an emulsion of oil in water that readily breaks if crystals begin to form in the oil. Therefore, the oil used in mayonnaise and salad dressing should be thoroughly winterized. The oil is alkali refined and deodorized, but the amount of bleaching may be small. Mayonnaise is expected to have a yellow color and the use of a dark oil in its preparation is therefore not objectionable.

Shortenings. Two types of shortening may be found on the market in the United States: compound shortenings, which are mixtures of high-melting fats such as animal fats and low-melting fats, hydrogenated fats, or oils; and those formed by the hydrogenation of an oil or a mixture of oils to the desired point. Compound shortenings now comprise only a small bulk of the shortenings sold in the United States. Compounding first began in the 1870's when beef fats were added to lard. Later cottonseed oil was added. Finally, although the products were still sold as "lard" or "refined lard," the amount of lard in them was sometimes small. After a Congressional investigation, the use of the term "lard compound" was made mandatory. Eventually the term "shortening" was introduced; since it has no implications as to the origin of the fat but only as to its use, this term has become firmly established.

In the United States, the style of cooking is based on the use of plastic fats. This arose because the early settlers were from northern Europe where butter and lard were the fats most commonly used. The cooking practices of southern Europeans, with their use of liquid fats such as olive oil, were introduced later when migration occurred in the nineteenth century. This did not alter the basic pattern of American cookery but simply expanded and enriched the cuisine, making it more cosmopolitan.

In 1900 all shortenings on the market were attempts to imitate lard as closely as possible. But with the introduction of hydrogenation about 1910, manufacturers of all-hydrogenated oils, usually cottonseed oil, began to recognize the advantage of developing their ware as a new product and not simply as a substitute for lard. The shortenings on the market today are far different from lard. They have much greater resistance to rancidity and are better creaming or emulsifying agents in batters. They are completely bland in flavor; and although this may not be an advantage to the tongue of every epicure, the American consumer has come to expect all fats except butter and oleomargarine to be flavorless. As a result, the use of lard has decreased and today lard is being processed in some plants to resemble the hydrogenated shortenings.

The number of oils and fats that can be used for the production of shortenings is quite large, since a suitable consistency can be obtained by blending various hard and soft fats or by hydrogenating soft fats or oils. The solid or hard fats used in compounded shortenings are oleostearin or edible tallow as well as lard blended with various oils, principally cottonseed oil. When soybean, whale, or fish oils are used, they are usually partially hydrogenated in order to reduce their marked tendency to flavor reversion. Hydrogenated all-vegetable shortenings are made principally from cottonseed, peanut, and soybean oil in the United States, although the amount of soybean oil is limited by its tendency to reversion. Cottonseed oil is the principal oil, but does not yield a hydrogenated product with as long a keeping period as peanut oil.

Usually the oils are carefully blended and the hydrogenation controlled so that the glycerides will form a product with the longest possible keeping period, plasticity over a long temperature range, and a consistency that makes for ready incorporation into doughs and batters. Since resistance to oxidation and the appearance of rancidity is related to the amount of linoleic acid present in the final product, the amount of this acid, instead of oleic acid, should be reduced as much as possible during hydrogenation. However, hydrogenation of linoleic acid yields not only oleic acid but also iso-oleic acid, which has a tendency to harden the fat and reduce the plastic range. A balance must be achieved, therefore, in which both linoleic and iso-oleic are kept at a minimum. Some manufacturers produce the shortening by blending two different hydrogenated fats; others control the properties of the product during hydrogenation and do not use blending.

Special hydrogenated shortenings are prepared for the bakers of sweet biscuits, cookies, and crackers. A long shelf life for the products made from these special shortenings is absolutely essential; the fat of all other ingredients is more likely to undergo flavor changes either from reversion or rancidity. Most of these products are prepared in factories and mixed in large batches where the temperature can be controlled. It is not essential for shortenings used under these conditions to be plastic over a wide temperature range. Therefore, biscuit- and cracker-type shortenings have a maximum resistance to oxidation and a short plastic range.

Since 1934 *superglycerinated* or *"high ratio"* (this term is copyrighted by Proctor and Gamble Co.) shortenings have been on the market. These are shortenings to which monoglycerides and diglycerides have been added to promote the emulsification of fat in water and to increase the amount of water and consequently the amount of sugar that can be incorporated in a batter and still produce a cake with sufficient strength not to fall during baking or cooling. Formulas for each of these two types are shown below.

$$
\begin{array}{cc}
\text{H} & \text{H} \\
| & | \\
C_{17}H_{35}COOCH & C_{17}H_{35}COOCH \\
| & | \\
HOCH & C_{15}H_{31}COOCH \\
| & | \\
HOCH & HOCH \\
| & | \\
\text{H} & \text{H} \\
\text{A Monoglyceride} & \text{A Diglyceride}
\end{array}
$$

A monoglyceride or diglyceride is both hydrophilic, because of the free hydroxyl groups, and lipophilic, because of the fatty acid residues. In other words, part of the molecule has an attraction for lipids but none for water; whereas part possesses an attraction for water. Consequently, these compounds are able to promote emulsification, to act as emulsifying agents in fat and water mixes. The practical result is that recipes can be used in which the amount of liquid is increased and the ratio of sugar to flour is high. This produces a sweeter cake. When monoglycerides and diglycerides are added to a shortening it is possible to prepare cakes with a sugar to flour ratio of as much as 1.4:1, while with ordinary shortenings a ratio of 1:1 cannot be exceeded. This is the origin of the term "high ratio shortening."

Monoglycerides and diglycerides are prepared by treating a fat with one fifth to one sixth its weight of glycerol and adding a small amount of sodium hydroxide as catalyst. The reaction mixture is heated with stirring under an atmosphere of an inert gas. When the reaction is complete, it is treated with an excess of phosphoric acid, dehydrated, and filtered to remove the fat-insoluble sodium phosphate. The product contains a mixture of monoglycerides, diglycerides, and triglycerides, a small amount of free glycerol, and some free fatty acids formed by the neutralization of the soap with phosphoric acid. This mixture is added to the shortening at the end of deodorization.

The addition of the monoglyceride and diglyceride mixture to shortenings increases their price by a small amount but also increases their usefulness in doughs and batters. They are therefore widely sold for this purpose. When the shortening is used for deep fat frying, the addition of these compounds is actually detrimental since they decompose readily and lower the smoke point. Antioxidants are commonly added to shortenings to increase their resistance to rancidity.

Margarine. Margarine is a manufactured fat that has shown a rapid gain in consumption during the past few years. It was first developed by the French chemist, Mege-Mouries, in 1870 when Napoleon III offered a prize for the development of a butter substitute. Mege-Mouries reasoned that body fat is converted to butter fat in a cow's udder. He therefore mixed the low-melting portion of beef fat (now called oleo oil) with milk and chopped tissue from udders. Of course, his reasoning was in error; but he did produce a product remarkably similar to butter and he won Napoleon's prize for his work. The oleo oil that he used has the property of melting at about body temperature so that when it is eaten it becomes liquid on the tongue as butter does. It also could be produced as a relatively bland fat at a time when technology had not advanced far enough for this to be true of other fats. The addition of milk and the udder tissue probably resulted in the introduction of microorganisms that soured the milk and produced flavors similar to those found in butter made from sour milk. The incorporation of the milk with the oleo oil also produced a product which, like butter, is an emulsion of water in fat. The name "oleomargarine" was based in the belief at that time that the fraction of beef fat used was composed of the glycerides of oleic and margaric acid. Margaric acid is the straight chain, C_{17} acid, which has not been found to occur naturally. Through the years many changes have occurred in the manufacture of margarine, but the basic process is still the same.

Today margarine is prepared from a variety of fats and oils. Oleo oil and lard—as well as the same vegetable oils that are suitable for high grade shortenings, i.e., cottonseed, peanut, sesame, palm oils, etc., and coconut oil—can be prepared in bland form. The amount of soybean oil must be limited because of its tendency to reversion, and fish oils are not suitable for the same reason. In Europe whale oils are used for margarine manufacture. The fats must be carefully extracted and refined so that the final product does not have a detectable flavor from the oils but only from the milk incorporated in it.

The consistency of margarine is of great importance in the success of the product. It must melt in the mouth as butter does, since a residue leaves a pasty sensation. Margarine is definitely a substitute for butter and, in the eyes of the consumer, must have properties similar to it in order to be acceptable. Margarine must be quite plastic at room temperature so that it spreads readily and be fairly hard at 40°F to 45°F refrigerator temperatures, as butter is. In the temperature range from 45°F to 60°F butter is too hard to spread easily, and much margarine is now produced which is superior to butter in its plasticity in this range. The consistency of margarine is the result of the fats used in its preparation, the extent of hydrogenation, and the course of the reactions during hydrogenation. Usually margarines are produced by carefully controlling the hydrogenation of the total body of fat rather than by blending.

A few specialty margarines are manufactured, and the consistency of these products does not require the rigid specifications demanded by table margarine. Margarine for puff-pastry, for example, has rather a high melting point and is tough and waxy at room temperature, so that the rolling in of the fat into the dough is facilitated.

In preparing margarine the natural fats and oils are carefully extracted, alkali refined, deodorized, and then hydrogenated to the desired consistency. The fat is then emulsified with ripened milk. In the United States skim milk is commonly used. It is pasteurized to destroy bacteria and then

FIGURE 2.8. MARGARINE INGREDIENTS ARE AUTOMATICALLY MIXED IN THESE TANKS.

inoculated with a strain of select bacteria that can produce compounds with desirable flavors in the milk and in the emulsion. These are the same strains of microorganisms that are used in the production of butter. The inoculated milk is held for 12 to 24 hr to permit the growth of the organisms. The ripened milk is run into the liquid fat and stirred vigorously. Emulsifying agents are often added at this point.

Emulsifying agents stabilize the margarine and prevent leakage, the separation of fluid during storage. They also prevent the rapid separation of fat and water when the margarine is melted, spattering, and the sticking of milk solids to the bottom of the pan. In butter natural emulsifying agents are present that hold the water in the emulsion and when the butter is heated, allow steam to escape by foaming rather than by spattering.

Lecithins, particularly those from soybeans, are widely used as emulsifying agents in margarines. A number of synthetic products are also used. The mono and diglycerides used in the formation of superglycerinated shortenings help stabilize the emulsion and prevent leaking, but do little to prevent spattering.

The sodium sulfoacetate derivative of mono and diglycerides are effective in minimizing spattering and are added to many margarines for this purpose. The fat-milk emulsion is cooled and the plastic, solid mass held for some time to allow bacterial action and the development of flavor. Salt is then added to the extent of 2.5 to 3 per cent of the total weight. Since the salt dissolves in the aqueous phase, the salt content of these tiny drops is much higher. It is so high that the activity and growth of the bacteria is stopped. The margarine is worked or kneaded during the operation of salting and the crystals are reduced so that no graininess occurs.

In the United States most margarine produced for the consumer market is fortified with vitamin A or provitamin A, the carotenes, to the extent of 1500 units per pound. A yellow dye is added to much of the margarine sold in this country since it has become legal to do this without the payment of a high excise tax. Sodium benzoate is occasionally added as a preservative.

The Shortening Value of Different Fats

Baked goods all use fat as one of their ingredients and its role in the cookery process is particularly important in those leavened by baking powders or those of the pound-cake type that are leavened by air. The amount of fat in most baked goods is considerable: in pie crust, 15–20 per cent; pound cake, 15–30 per cent; doughnuts, 15–20 per cent; cookies, 5–20 per cent; cake, 10–20 per cent; crackers, 5–12 per cent; but in bread only approximately 1 per cent. The fat is not only important for the flavor it imparts but also in the development of texture and the tenderness of the product.

In pound cake leavening occurs by incorporating air into the fat during the creaming process. When the batter is heated in the oven, the small air bubbles expand and fill with steam. Indeed, the data shown in Table 2.10 indicate that the greater percentage of leavening in pound cake comes from the steam that collects in the tiny air bubbles rather than from the air itself. The other components, principally the gluten in the flour, form the walls around each little bubble, and during baking they set to a fairly rigid structure. When the cake is removed from the oven and is cooled, the air in the small bubbles contracts and the steam condenses. But in a well-balanced recipe the walls are sufficiently strong to hold up and the cake does not fall. The role of the fat in the leavening process is in its ability to trap air during the mixing process. The aqueous phase composed of the milk and eggs and dissolved substances such as sugar and salt does not have this ability. So the fat is indirectly responsible for the texture of the cake. Its tenderness is likewise believed to be a function of the fat. The walls around each bubble are made of gluten and starch which can be very tough and hard. Fat is streaked through and serves as a lubricant so that when the cake is bitten, particles of gluten and starch slide on one another and the walls crumble.

In butter cakes fat has much the same role as in pound cakes although leavening does not depend on air and steam alone but is supplemented by the carbon dioxide produced from baking powder. If the batter is formed by creaming the fat first, the air bubbles may serve as the nucleus for carbon dioxide bubbles. Cakes with satisfactory grain can be produced, however, by other methods of mixing. Certainly the fat in butter cakes is important as far as tenderness is concerned and allows the gluten particles to slide on one another when the cake is bitten or cut.

In pastry a stiff dough is rolled out. The fat is flattened into sheets between the layers of flour and water and causes their separation. This produces flakiness. The fat is also squeezed in between gluten particles or strands so that a continuous tough matrix is not formed but rather one which is very brittle or short. This also occurs in high-fat cookies.

Fats differ considerably in their ability to render a cake tender or a pastry or cookie short. In commercial bakeries it is essential that products be uniform. A marked variation in the fat may have disastrous effects on the bakery goods.

The relative shortening value of a fat is determined by means of a shortometer. This apparatus measures the load necessary to break a wafer prepared under standard technique and with a standard recipe. If it takes a

TABLE 2.10. TYPICAL EFFECT OF INCORPORATING AIR IN A POUND CAKE FORMULA CONTAINING 21 PER CENT FAT AND NO CHEMICAL LEAVENING AGENT*

Experiment	A	B
Air Incorporation in Shortening	low	high
Calculated Volume of Dough Without Air, cc.	377	377
Actual Volume of Dough Including Air Incorporated, cc.	543	662
Volume of Air in Dough, cc.	166	285
Volume of Cake, cc.	1140	1520
Expansion of Dough During Baking, cc.	597	858
Calculated Expansion of Air Due to Heating in Oven (70°F to 212°F), cc.	42	71
Percentage of Total Expansion Due to Thermal Expansion of Air	7.0	8.3
Percentage of Total Expansion Due to Increased Vapor Pressure of Water	93.0	91.7

*From Bailey, A. E., "Industrial Oil and Fat Products," 278, Interscience Publishers, Inc., New York, N. Y., 1945.

very small weight to break the wafer, then the wafer is very short. The shortening value of a fat varies with a number of conditions and it is therefore absolutely essential in the comparison of two fats that a standard procedure be used. Factors other than the fat that influence the shortening power are (1) manipulation, (2) temperature, (3) ingredients other than flour and fat and their concentration, and (4) concentration of fat. When two fats are compared, all techniques are standardized as carefully as possible. In general, lard has a very high shortening value and butter a lower value, with hydrogenated shortenings intermediate. At least this is true in pastry made from water, flour, and fat. However a fat which ranks high as a shortening agent for pastry is sometimes lower on the list when used in sweet cookie dough.

Several theories to explain why one fat has a greater shortening power than another have been advanced, but none can explain all observations. The structure of doughs and batters is so complex and our knowledge is so incomplete that it is not possible to explain variation in shortening power.

IMPORTANCE IN DIET

Today the quantity of fat and other lipids in a typical American diet is quite high, often accounting for 40 per cent of the calories. This is caused by the relatively high caloric content of one gram of fat—approximately 9 calories (kilocalories) per gram. Although fat and other lipids are present in meat, the amount found in fruits and vegetables is usually rather low. These foods are often enriched with fat in the form of butter or margarine during their preparation. Desserts such as cakes, pastries, and puddings usually have considerable quantities of fat added in their preparation.

The chief contribution of fats and other lipids to the diet is their energy value. In America where at the present time want is seldom known, and where overeating is often a problem, the high contribution of fat may be undesirable. Some nutritionists and physicians now feel that the fat content of the American diet is too high and that for optimum health it must be lowered.

A high fat intake shows a positive correlation with blood cholesterol level in a number of studies. A high blood cholesterol level has been equated with incidence of atherosclerosis although this evidence is by no means conclusive. Unfortunately, it is difficult to measure atherosclerosis in a living man and to determine whether atheromas, the fatty placques in the walls of the blood vessels, are forming, but determination of blood cholesterol is easily run. Thus many experiments have been conducted on blood cholesterol level with the assumption that this measures atherosclerosis.

Much attention has been directed recently to the difference in effect of saturated and unsaturated fatty acids. Substituting oils with abundance of unsaturated fatty acids for hydrogenated fats and saturated fats such as butter and beef tallow often lowers blood cholesterol levels.

Fat and other lipids also have a role in the satiety value of a food. The feeling of satisfaction which we derive from a fat, the way it "sticks to our ribs," is the result of numerous physiological reactions, not all of which are known. The length of time food stays in the stomach is one reaction which gives a feeling of satisfaction and delays the onset of hunger. Fat retards the rate at which food leaves the stomach and in this respect increases the satiety value of the food. If too much food is eaten at one meal, this effect may prolong the feeling of being "stuffed" and may actually lead to discomfort. Nevertheless, when food is eaten in moderation, one of the valuable roles of fat and other lipids is in their satiety value.

Fats and other lipids also contribute essential fatty acids to the diet. While it has never been unequivocably demonstrated that these fatty acids are essential to man, the work with eczema in babies indicates that we may have a real need for small amounts of linoleic, linolenic, and arachidonic acids in the diet.

Fats are also solvents for the fat-soluble vitamins that are always naturally introduced into the diet in the fatty portion of the food. The fat-soluble vitamins are vitamins A, D, E, and K, and the provitamins A, the carotenes. All of these vitamins are important to optimum health.

The fats and other lipids are, therefore, important in the diet for a number of reasons. Whether or not the amount can exceed optimum is yet to be proved.

REFERENCES

1. BAILEY, A. E., "Industrial Oils and Fat Products," Interscience Publishers, Inc., New York, N. Y., 1945.
2. BAILEY, A. E., "The Chemistry and Technology of Food and Food Products," edited by Jacobs, M. B., 1, 581–582, Interscience Publishers, Inc., New York, N. Y., 1944.
3. CHIPAULT, J. R., MISUNO, G. R., and LUNDBERG, W. O., "The Antioxidant Properties of Spices in Foods," *Food Technol.*, 10, 209–211 (1956).
4. CHRISTENSEN, P. K. and SORENSEN, N. A., "Studies Related to Pristane, V: The Constitution of Zamene," *Acta Chem. Scand.* 5, 751–756 (1951) *C. A.*, 46, 3942h (1952).
5. "Consumption of Fats and Oils," *J. Am. Oil Chemists' Soc.* news sect. 8, Nov., 1957.
6. DAUBERT, B. F. and O'CONNELL, P. W., "Reversion Problems in Edible Fats," *Advances in Food Research*, 4, 185–207 (1953).

7. DETWILER, S. B. and MARKLEY, K. S., "Smoke, Flash, and Fire Points of Soybean and Other Vegetable Oils," *Oil and Soap*, **17**, 39 (1940).
8. DOLLEAR, F. G., "Vegetable Oils, Fats, and Waxes," *Ind. Eng. Chem.*, **50**, No. 6, 48A-51A (1958).
9. DUGAN, L., JR., "Stability and Rancidity," *J. Am. Oil Chemists' Soc.*, **32**, 605-609 (1955).
10. ECKLEY, E. W., "Vegetable Fats and Oils," Reinhold Publishing Corp., New York, N. Y., 1954.
11. HAWLEY, H. K. and HOLMAN, G. W., "Directed Interesterification as a New Processing Tool for Lard," *J. Am. Oil Chemists' Soc.*, **33**, 29-35 (1956).
12. HILDITCH, T. P., "The Chemical Constitution of Natural Fats," 3rd ed., Chapman & Hall, London, 1956.
13. JAMESON, G. S., "Vegetable Fats and Oils," 2nd ed., Reinhold Publishing Corp., New York, N. Y., 1943.
14. KOCK, R. B., "Fat Oxidation Mechanisms," *Baker's Dig.*, **30**, 48-53 (1956).
15. KUEMMEL, D. F., "Direct Determination of Saturated Fatty Acids in Fats," *J. Am. Oil Chemists' Soc.*, **35**, 41-45 (1958).
16. LANGE, N. A., "Handbook of Chemistry," 41st ed., Handbook Publishers, Inc., Sandusky, Ohio, 1959.
17. LANGE, W., "Cholesterol, Phytosterol and Tocopherol Content of Oils, *J. Am. Oil Chemists' Soc.*, **27**, 414-422 (1950).
18. MABROUK, A. F., and BROWN, J. B., "The Trans Fatty Acids of Margarine and Shortening," *J. Am. Oil Chemists' Soc.*, **33**, 98-102 (1956).
19. MCKERRIGAN, A., "Autooxidation and Flavor Deterioration of Fats," *J. Soc. Chem. Ind.*, **39**, 390-395 (1954).
20. MCKINNEY, R. H. and BAILEY, A. E., "Notes on the Keeping Quality of Fats in Baked Goods," *Oil and Soap*, **18**, 147-148 (1941).
21. MARKLEY, K. S., "Fatty Acids, Their Chemical and Physical Properties," Interscience Publishers, Inc., New York, N. Y., 1947.
22. O'CONNOR, R. T., "Fatty Acid Derivatives and Infrared Spectrophotometry," *J. Am. Oil Chemists' Soc.*, **33**, 1-15 (1956).
23. PISKUR, M. M., "22nd Annual Review of the Literature on Fats, Oils and Detergents," *J. Am. Oil Chemists' Soc.*, **33**, 203-218 (1956); "Report of the Literature Review Committee," *ibid.*, **35**, 208-225 (1958).
24. RALSTON, A. W., "Fatty Acids and Their Derivatives," John Wiley & Sons, Inc., New York, N. Y., 1948.
25. SINNHUBER, R. O., and YU, T. C., "TBA Method for Determining Rancidity," *Food Technol.*, **12**, 9-12 (1958).
26. STUCKEY, B. N., "Antioxidants in Candy and Candy Packaging Materials," *Mfg. Confectioner*, **34**, No. 6, 47 (1954); *C. A.* **50**, 8936g (1956).
27. YU, T. C. and SINNHUBER, R. O., "2-Thiobarbituric Acid Method for Measurement of Rancidity in Fishery Products," *Food Technol.*, **11**, 104-108 (1957).
28. WITTCOFF, H., "The Phosphatides," ACS Monograph 112, Reinhold Publishing Corp., New York, N. Y., 1951.

CHAPTER THREE

Carbohydrates

YOU HAVE HAD an introduction to the chemistry of some of the carbohydrates present in foods. You are familiar with the monosaccharides, the hexoses—D-glucose, D-galactose, D-mannose, and D-fructose—and perhaps with some of the pentoses—xylose and arabinose. Formulas are given for the common hexoses.

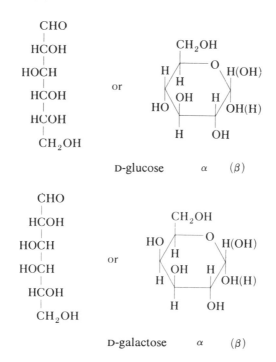

$$\begin{array}{c} CHO \\ | \\ HCOH \\ | \\ HOCH \\ | \\ HCOH \\ | \\ HCOH \\ | \\ CH_2OH \end{array}$$

or

D-glucose   α   (β)

$$\begin{array}{c} CHO \\ | \\ HCOH \\ | \\ HOCH \\ | \\ HOCH \\ | \\ HCOH \\ | \\ CH_2OH \end{array}$$

or

D-galactose   α   (β)

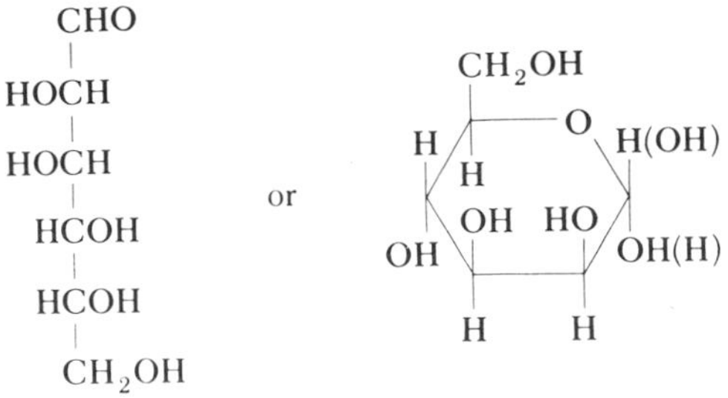

CHO
|
HOCH
|
HOCH
|
HCOH
|
HCOH
|
CH$_2$OH

or

D-mannose α (β)

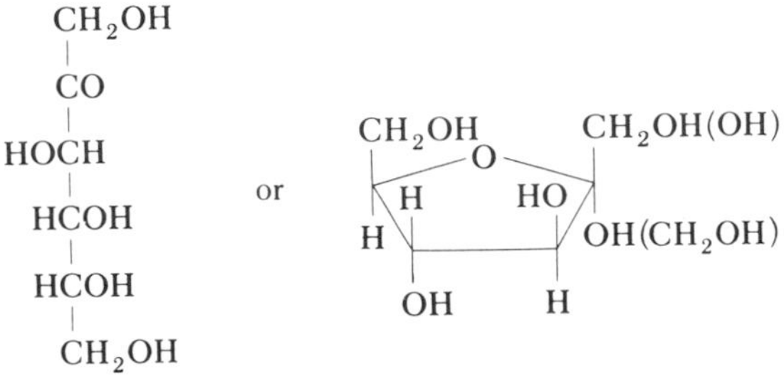

CH$_2$OH
|
CO
|
HOCH
|
HCOH
|
HCOH
|
CH$_2$OH

or

D-fructose α (β)

You have studied the disaccharides—sucrose, lactose, and maltose—and the polysaccharides—starches and glycogens. A review of the chemistry of these important compounds can be found in any organic chemistry or biochemistry book.

MONOSACCHARIDES

The pentoses are monosaccharides that contain five carbon atoms. Those that are important in foods, arabinose, xylose, and ribose, are aldoses

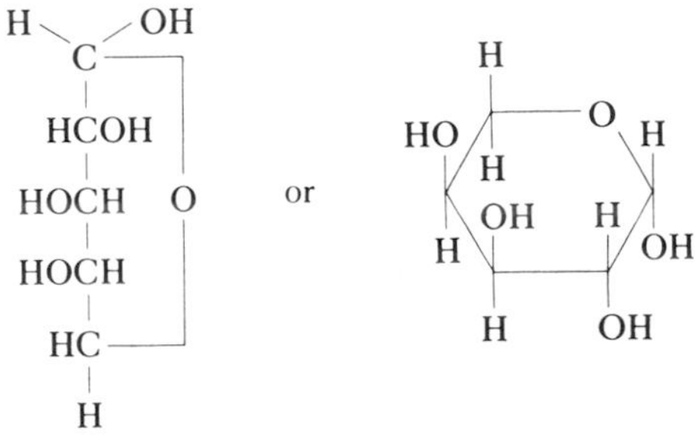

α-L-arabinose

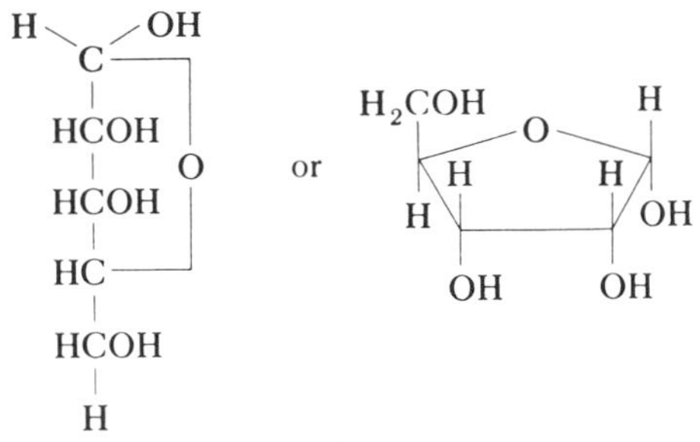

α-D-ribose

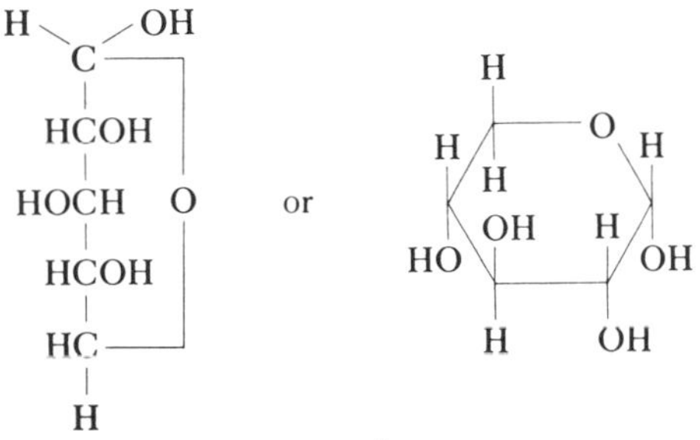

α-D-xylose

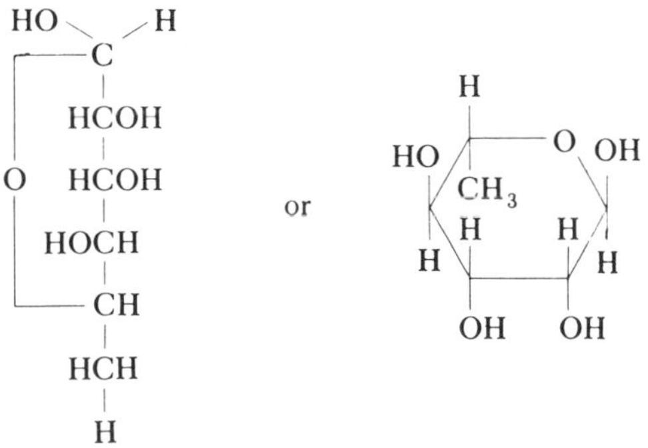

α-L-rhamnose

while rhamnose is a methyl aldose. Like the hexoses, these carbohydrates form ring compounds composed of five carbons and one oxygen atom (pyranose) or four carbons and one oxygen (furanose); and it is the ring form that occurs in polysaccharides.

The complexity of the mixture of monosaccharides in plant materials, the fruits and vegetables of our diet, is not always recognized. The presence of the most abundant carbohydrate is sometimes so thoroughly stressed that the presence of others is overlooked. Table 3.1 illustrates the large

TABLE 3.1. SUGAR CONTENT OF POTATOES GROWN IN 1954
AS DETERMINED BY PAPER CHROMATOGRAPHY
(mg/100 g RAW POTATOES), AV. 8 TESTS

Sugars	Katahdin	Red Kote	Red Pontiac	Russet Rural
Maltose	12.0	0.0	51.0	0.0
Sucrose	47.0	66.0	66.0	28.0
Fructose	42.5	70.0	35.5	40.5
Mannose	35.0	0.0	33.5	0.0
Glucose	47.5	45.5	75.0	26.7
Xylose	24.5	42.8	60.0	0.0
Total	208.5	224.3	321.0	95.2

From Habib, A. T. and Brown, H. D., *Food Technol.* **11**, 85–89 (1957).

number of monosaccharides present in a plant part, this time the potato. Data are given for the number of milligrams of each monosaccharide per 100 g of tissue for several varieties. The amounts of some are quite low.

DISACCHARIDES

The common disaccharides are already known to you. They are anhydrides of two monosaccharides (monoses) and include sucrose, lactose, maltose and cellobiose. Sucrose is widely distributed in the plant kingdom although sugar cane or sugar beets are the commercial source of most sugar. Maple sugar is also principally sucrose. Lactose occurs in the milk of all mammals while maltose and cellobiose occur in low concentrations in plants and processed foods.

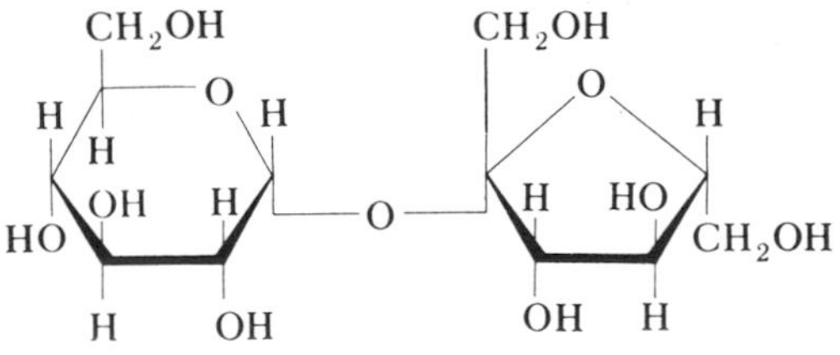

Sucrose

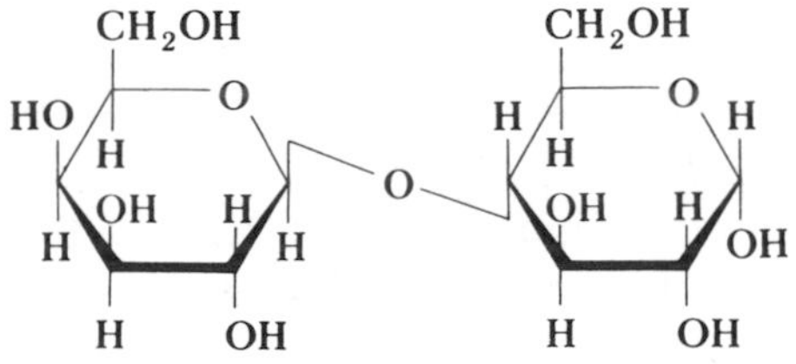

Lactose

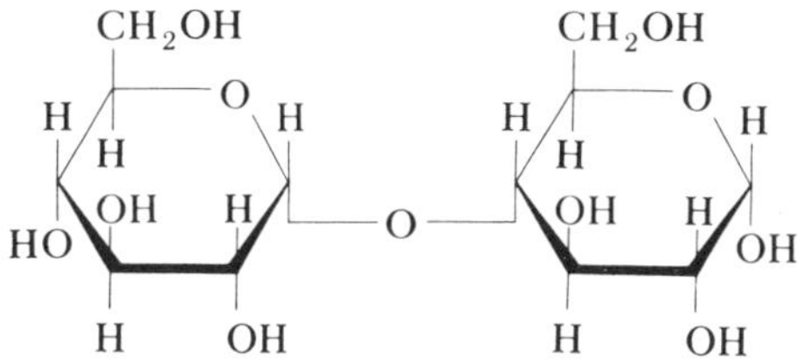

Maltose

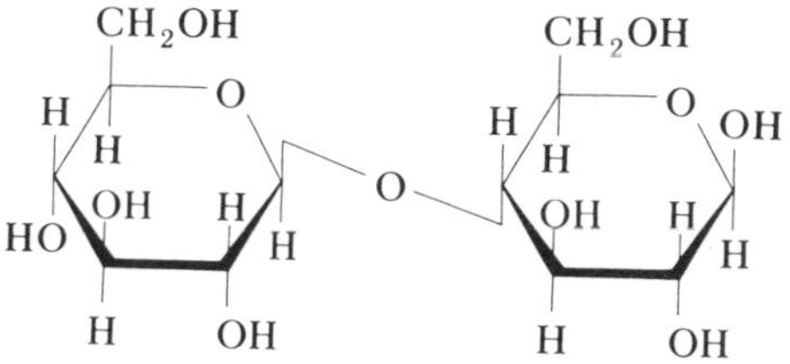

Cellobiose

Sucrose is particularly important in food processing and is available commercially in many crystal sizes from extremely fine to very coarse. It is also often used as a syrup sold as "liquid sugar." This prepared syrup is particularly useful for canners and other processors who use large quantities of solutions and is sold in containers of many sizes up to tank cars. Syrups are also available in which the sucrose has been partially inverted to glucose and fructose. These syrups can be prepared with a higher concentration of solids since fructose has a very high solubility and glucose does not readily crystallize. The graph (Figure 3.1) shows the improvement in solubility with increased inversion at various temperatures. Notice that after a certain critical concentration of invert sugar is reached, the solubility declines. The sweetness of these inverted sugars is comparable to sucrose. The graph (Figure 3.2) shows a comparison of hexoses with sucrose. Only fructose is above the equality line. Dulcin is a noncarbohydrate sweetner.

Refiners syrups are also available on the market. They contain not only sucrose but also some of the inorganic and organic compounds present in cane or beets or formed in processing. They vary in color from pale yellow to dark brown and possess considerable flavor other than the sweetness of sucrose.

Plants contain a number of other carbohydrates or carbohydrate derivatives that cannot be digested by man and which are, therefore, not of direct nutritional significance. However, many of these compounds are of great importance in the development of the plant and give those plant products

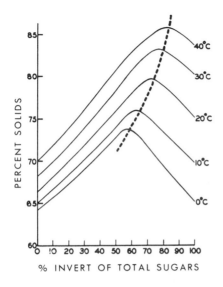

FIGURE 3.1. SOLUBILITY CURVES FOR SUCROSE-INVERT SUGAR MIXTURES AT VARIOUS TEMPERATURES. The dotted line shows peak solubility composition for various temperatures. Data of Junk, Nelson, and Sherrill. Reproduced from Davis, P. R., and Prince, R. N., "Liquid Sugar in the Food Industry," in "Use of Sugars and Other Carbohydrates in the Food Industry," *Advances in Chem. Ser.,* **12,** American Chemical Society, Washington, D. C. (1955).

used as foods much of their characteristic texture in the fresh and cooked states. These groups of compounds are the celluloses, hemicelluloses, and pectic substances. The lignins may be related to the carbohydrates, but these compounds will not be studied since they are deposited in appreciable amounts only in old, woody plant tissues that are not considered desirable for food.

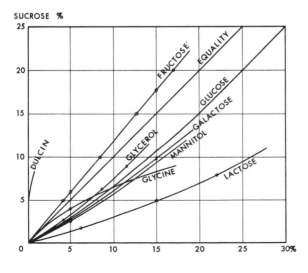

FIGURE 3.2. RELATIVE SWEETNESS OF VARIOUS COMPOUNDS COMPARED TO SUCROSE FROM DATA OF CAMERON AND DAHLBERG. Reproduced from Cotton, R. H., Rebers, P. A., Maudru, J. E. and Rorobaugh, G., "The Role of Sugar in the Food Industry," in "Use of Sugars and Other Carbohydrates in the Food Industry," *Advances in Chem. Ser.,* **12,** American Chemical Society, Washington, D. C. (1955).

OLIGOSACCHARIDES

The *oligosaccharides* are anhydrides of several monosaccharide residues. Just as the word "several" is not precise, neither is the prefix "oligo-." Usually any compound which contains ten or fewer monosaccharide units is classed as an oligosaccharide while those which contain more than ten are called polysaccharides. Actually, in nature the most abundant oligo-saccharide is sucrose, with two monosaccharide units and raffinose, with three. All compounds that are well known have less than six mono-saccharide units. An oligosaccharide may be composed of the same mono-saccharide units; however, often it is composed of different units.

The raffinose family is widely distributed and these compounds could readily be produced in large quantities commercially if there were a de-mand for them. They are found most frequently in seeds, roots, and under-ground stems. For example, the quantity of raffinose in the seeds of legumes is equal to or greater than the quantity of sucrose; and in cotton seed or soybean meal both raffinose and stachyose occur in abundance. French[7] shows the relationships between the structures of the individual oligosaccharides of the raffinose family by the following scheme:

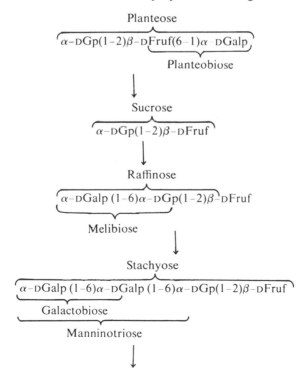

Planteose

α–DGp(1–2)β–DFruf(6–1)α–DGalp

Planteobiose

Sucrose

α–DGp(1–2)β–DFruf

Raffinose

α–DGalp (1–6)α–DGp(1–2)β–DFruf

Melibiose

Stachyose

α–DGalp (1–6)α–DGalp (1–6)α–DGp(1–2)β–DFruf

Galactobiose

Manninotriose

$$\underbrace{\alpha-\text{D}Galp(1-6)\alpha-\text{D}Galp(1-6)\alpha-\text{D}Galp(1-6)\alpha-\text{D}Gp(1-2)\beta-\text{D}Fruf}}$$

Verbascose

where:

G = glucose   Gal = galactose   f = furanose
Fru = fructose   p = pyranose

This scheme shows the ether link between the monosaccharide units by conventional means. Thus α or β refers to the configuration of the hydroxyl group formed in the ring structure of the monosaccharide; and the numbers refer to the carbons, numbering in glucose from the aldehyde car-

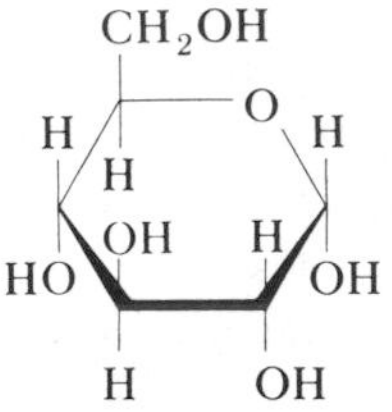

α-D-glucose β-D-glucose

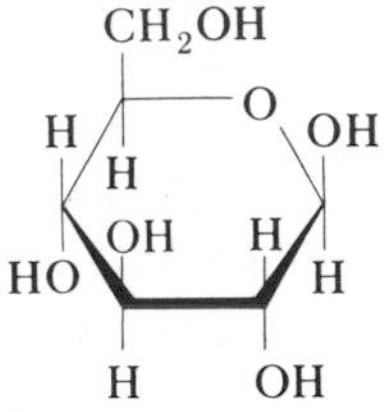

bon as 1 to the primary alcohol carbon as 6. The letters f or p refer to furanose or pyranose rings. Thus stachyose has a structure as shown:

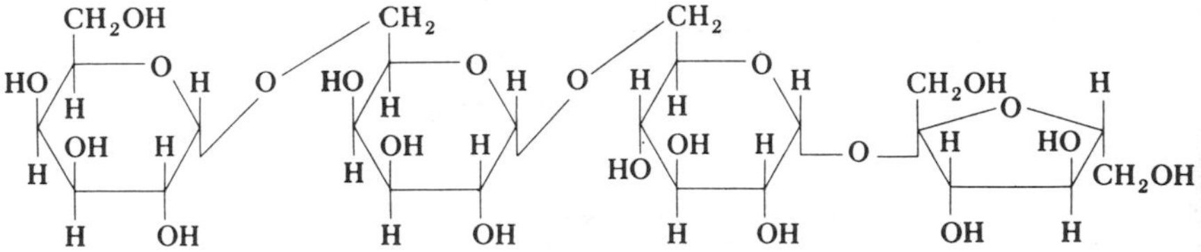

Stachyose

POLYSACCHARIDES

These compounds are anhydrides of one or more monosaccharides in which a large number of units are combined. These high molecular weight compounds are exceedingly difficult to study. Purification is an arduous and tedious task, and then, to demonstrate that a pure compound has been isolated is difficult. Even when the presence in a sample of only one type of molecule is reasonably certain, determining the molecule's structure presents problems. However, investigation of the structure of many classes of large molecules (they are called macro molecules) has progressed rapidly during the past two decades and advances in the study of one type of molecule often pushes forward the study of other molecules. Cellulose has been studied intensively; and although it is still impossible to describe with complete assurance a cellulose molecule, much is known about its structure. The hemicelluloses, gums, and pectic substances are not as well known, although many of the details of their structures have been ascertained. We can describe them with some reservations.

Many polysaccharides are composed of only one repeating monosaccharide, and even where a larger number exists they appear to be arranged in a systematic fashion. *Homopolysaccharides* possess only one repeating unit while *heteropolysaccharides* contain more than one. Most polysaccharides are composed of aldoses or their derivatives and the hydroxyl group formed in the ring structure on carbon-1 (the anomeric carbon) reacts to form the ether link that holds the monosaccharide residues together. This hydroxyl group may react with any other available hydroxyl on the adjacent monosaccharide, except the one on carbon-1, to form the polysaccharide chain. Usually the one with which it reacts is the same throughout the chain.

You are familiar with the structure of the two types of starches: amyloses with long chains of glucose residues and the amylopectins with branching chains that form a "bushy" molecule. You will remember that the glycogens are similar in structure to the amylopectins, although the number of glucose residues in the chain is smaller. (These can be reviewed in any organic chemistry textbook.)

Cellulose

Cellulose is the compound that gives strength to plant tissues. It is deposited in the cell walls of all seed-bearing plants in varying concentration, and forms long fibers that strengthen stems and ribs of leaves. On acid hydrolysis of cellulose, glucose is formed in yields of 95 to 96 per cent. This fact indicates that the structural unit in a cellulose molecule is glucose. When cellulose is treated with an acetylating agent, it forms cellobiose octaacetate, the only disaccharide produced. Cellobiose is a disaccharide of two glucose units which differs from maltose in the configuration of the 1-carbon. In maltose the alpha configuration occurs, while in cellobiose the opposite configuration, beta, exists. We can conclude from the formation of a cellobiose derivative that the beta configuration exists in cellulose. Other studies have shown that the link between the glucose residues is a 1-4 link, the same one that is shown in cellobiose. The exact number of glucose residues in the chain is still unknown, but it is certain that the number must be large. Attempts to determine the molecular weight of cellulose isolated from various plants, by viscosity measurements, by the ultracentrifuge, and by end group analysis have not given consistent results. Most determinations give ranges of values and indicate that the samples are composed of molecules of different molecular size.

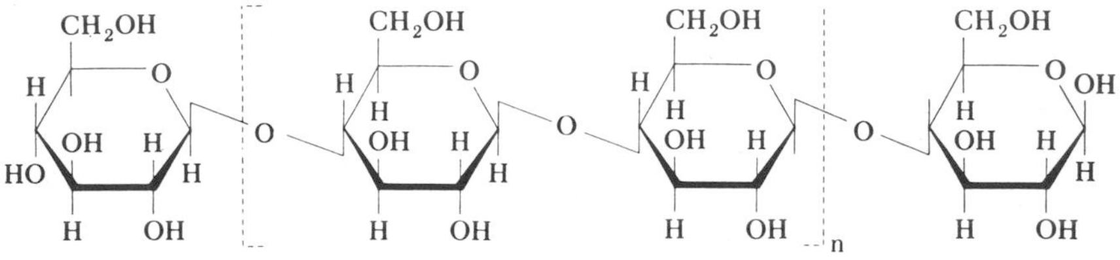

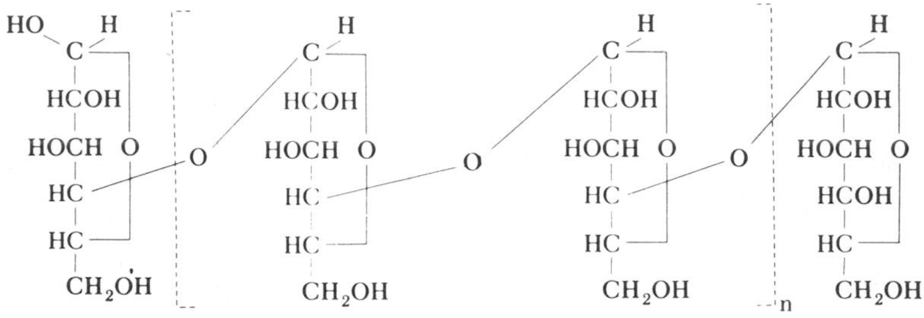

Schulz and Marx[24] have found that the degree of polymerization of cellulose from fibers in cotton, cotton linters, ramie, china grass, and flax is between 6500 to 9000 glucose units, while those from the first leaf past the cotyledon on plants, from root tips, or bacteria show considerable variation and are considerably lower in molecular weight. These celluloses are probably representative of those found in foods.

X-ray studies of cellulose fibers indicate that the fibers contain areas of crystallization and areas of disorganization called amorphous cellulose. These data indicate that in the crystalline areas the cellulose molecules are oriented parallel to the fiber axis. The dimensions of the cells have been calculated, and it is believed that one cellulose molecule lies in one direction, while the neighboring molecule lies in the opposite direction. The molecules are so close together that they probably are held by hydrogen bonds between oxygen atoms. In the amorphous portions of the fiber, a much looser arrangement exists; in this part adsorption of water and swelling readily occurs. Also this part of the fiber is believed to be flexible and capable of distortion without breaking. If this hypothesis is true, this property makes a cellulose fiber of peculiar value to the plant, strengthen-

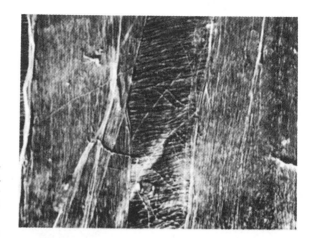

FIGURE 3.3. CELLU-
LOSE FIBER IN CELL WALL
OF *Valonia* LIE IN AL-
TERNATE LAYERS. Electron
micrograph × 15,000.

Courtesy of R. D. Preston.

FIGURE 3.4. CELLU-
LOSE FIBERS ON THE
INNER SIDE OF THE CELL
WALL OF *Valonia*. Granu-
lar material may be protoplasm. Electron micro-
graph × 22,500.

Courtesy of R. D. Preston.

ing a stem or leaf and yet not rendering it rigid and susceptible to breaking.

Starch

Starch is the reserve carbohydrate of plants and occurs as granules in
the cell in plastids, separated from the cytoplasm. Under the microscope
the plastids in some plant tissues can be seen filled with granules and
when the cells are ruptured, the granules stream out. The size and shape of
starch granules is characteristic of their origin, and anyone trained can
readily identify the origin of starch granules with a microscope.

Most starch granules contain both amylose and amylopectin molecules.
Waxy or glutinous starch from corn and other cereals contains little or no

amylose, while a sugary mutant corn and some of the legumes contain amylose in greater abundance than amylopectin. Amylopectin is usually the more plentiful type of starch. (See Table 3.2.)

Amylose molecules are straight chain polysaccharides in which α-D-glucose units are joined 1–4. Chain lengths vary from 250 to about 350 glucose units, and the long molecules appear to be coiled in an α helix. Amylopectin molecules are branched at carbon 6 on the glucose unit to form a 1–6 ether link. The length of the linear units in amylopectin is only 25 to 30 units but molecular weights show that 1000 or more glucose units are combined in a molecule.

The ability of a natural starch to form pastes and gels differs considerably with the source of the starch. Amyloses are believed to form gels more readily because the linear shape allows the formation of a three-dimensional network with ease. However, these molecules associate and crystallize readily and the starch dispersion undergoes retrogradation (see p. 107). Amylopectin molecules, with their bushy structure, do not crystallize readily. Waxy starches do not show retrogradation.

Enzymes and Starch. Three types of enzymes, widely distributed in nature, react with starches: α-amylases, β-amylases, and phosphorylases. The amylases are particularly interesting to the food chemist since they are often used to modify starches.

Alpha Amylase. Alpha amylases are also called "liquefying" or "dextrogenic" amylases from their action on starches. They occur widely distributed in the plant world often associated with β-amylases. Dormant, ungerminated seeds sometimes possess α-amylases in small quantities, but on germination all of these seeds rapidly develop these enzymes. Thus alpha amylases can be readily prepared from sprouting grains. Alpha

FIGURE 3.5. MICELLAR ORGANIZATION WITHIN SWOLLEN STARCH GRANULE, ACCORDING TO K. H. MEYER. Reproduced from Schoch, T. J. and Elder, A. L., "Starches in the Food Industry," in "Use of Sugars and Other Carbohydrates in the Food Industry," *Advances in Chem. Ser.*, **12,** American Chemical Society, Washington, D. C. (1955).

amylases also occur in saliva and pancreatic juice and are produced by a number of bacteria and fungi.

Alpha amylase catalyzes the hydrolysis of starches into low molecular weight dextrins with great rapidity. (See Figure 3.6.) In a short time a starch dispersion liquefies as the molecular weight of the colloid is decreased and soon the solution is filled with dextrins of approximately 6 glucose units along with a small amount of maltose. The bonds hydrolyzed are the 1-4 ether links. The 1-6 links in amylopectic molecules are bypassed. Any phosphate esters are likewise left intact.

Beta Amylase. This enzyme is also widely distributed. Beta amylase is called the "saccharifying amylase" because the chief product of its hydrolytic catalysis is maltose. (See Figure 3.7.) A relatively pure extract of β-amylase can be made from germinating soy beans. "Diastase," a commercial extract from barley malt which is widely used in industry both for food and many other products, is a mixture of alpha and beta amylases.

Beta amylase forms primarily maltose and the reaction will go almost to completion. The 1-6 link appears to be somewhat sensitive to hydrolysis in the presence of β-amylase as well as the 1-4 link between glucose units. The barrier link that prevents the formation of 100% maltose is probably the β-link which occurs infrequently in amylose and amylopectin on the 1-carbon of glucose.

Technology of Starch. Starch is produced in large quantities in the United States from a number of plant sources. Cornstarch is the most im-

TABLE 3.2. AMYLOSE CONTENT OF SOME STARCHES*

	Amylose Content, %
Sugary Mutant Corn	70
Steadfast Pea	67
Alderman Pea	65
Chick Pea	33
Buckwheat	28
Barley	27
Sorghum	27
Commercial Corn	26
Wheat	25
White Potato	23
Arrowroot	21
Sweet Potato	20
Tapioca	18
Waxy Barley	3
Waxy Corn	0-6

*From Whistler, R. D. and Smart, C. L., "Polysaccharide Chemistry," p. 242, Academic Press, Inc., New York, N. Y., 1953.

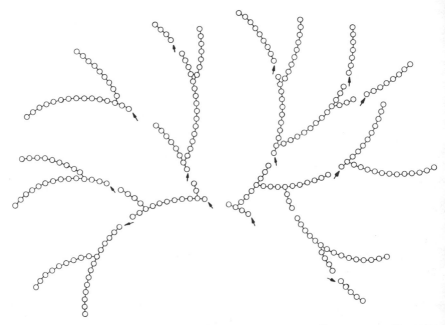

FIGURE 3.6. α-AMYLASE ACTIVITY ON AMYLOPECTIN WITH THE FORMATION OF DEXTRINS OF LOW MOLECULAR WEIGHT. *Advances in Enzymol.*, **12**, 390 (1951).

Courtesy of P. Bernfeld.

portant. The most common method is the one described below but many modifications exist.

The corn is steeped for 35 to 45 hr in water at 50°F with sufficient sulfur dioxide added (0.15 to 0.2 per cent) to prevent growth of microorganisms and perhaps start the disintegration of protein. During steeping the corn picks up water and is then readily ground to separate hulls and germ. The starch, protein, and water-soluble substances form a slurry that is separated from the germ. The slurry is then finely ground and separated from hulls and coarse particles. More sulfur dioxide is then added and the temperature increased to within the range of 85°F to 90°F to make the separation of the starch granules from the proteinaceous material easier. The granules are separated by various methods such as centrifugation or filtering and are washed and dried. The starch is sold in several forms: *Pearl* is rough-ground from the dryers while *powder* is finely ground and even sifted. *Lump* or *crystal* is the other form available on the market.

Many *modified starches* are on the market but those most commonly used in the food industry are the thin boiling types. They are the result of partial hydrolysis of the starch, usually by sulfuric acid or some other

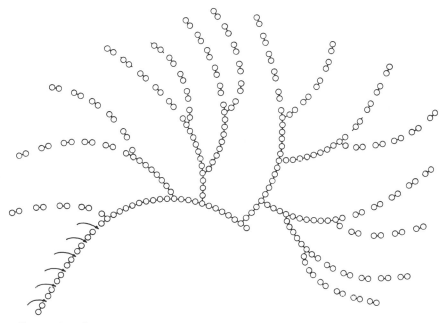

FIGURE 3.7. β-AMYLASE ACTIVITY ON AMYLOPECTIN WITH THE FORMATION OF MALTOSE AND A HIGH MOLECULAR LIMIT DEXTRIN. *Advances in Enzymol.*, **12**, 394 (1951).

Courtesy of P. Bernfeld.

acid. Before drying the starch slurry is treated with $0.1N$ H_2SO_4 at $50°C$ for 6 to 24 hr. Hydrolysis of some of the bonds in the starch results in a product that will disperse in boiling water to yield a dispersion with a viscosity not much greater than water. The more extensive the modification of the starch, the greater the effect on lowering viscosity.

Fructans

Fructans are polysaccharides composed of fructose residues and are deposited in the roots and tubers of a number of plants as well as in the leaves of some grasses. A number have been isolated and the structure of some is fairly well known. Some fructans are linear while others are branched, with either a 2–1 or 2–6 link between the fructose residues, and have a fairly low molecular weight. Since they are easily isolated and since fructose is the sweetest of the hexoses, fructose syrup has been produced commercially.

Mannans and Galactans

These as well as the heteropolysaccharides of galactose and mannose, the galactomannans, have been found in plant and animal tissues. These last are used as thickners and are produced commercially from seeds. They are briefly discussed under gums.

Chitin

Chitin is a homopolysaccharide that occurs as a structural compound in lobster, shrimp, and crab shells as well as in some insects and fungi. The shells contain calcium carbonate, proteins, and other compounds, but this polysaccharide is mainly responsible for the rigidity of the tissue. It is easily extracted. The repeating unit in chitin is 2-N-acetyl glucosamine, linked β-1-4.

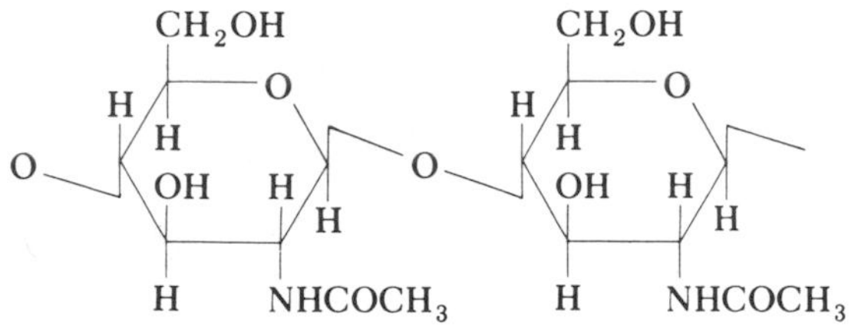

Portion of Chitin

Hyaluronic Acid and the Chondroitin Sulfates

The heteropolysaccharides of animal tissues have been studied for many years and several of them in the past decade have been well characterized. The problem of separation is a difficult one and there is still a great deal of confusion in this field in nomenclature and definition. Attempts in the past to determine structure with crude mixtures rather than with pure compounds have added to this confusion. Hyaluronic acid and the chondroitin sulfates are now well known and the structures at least partially de-

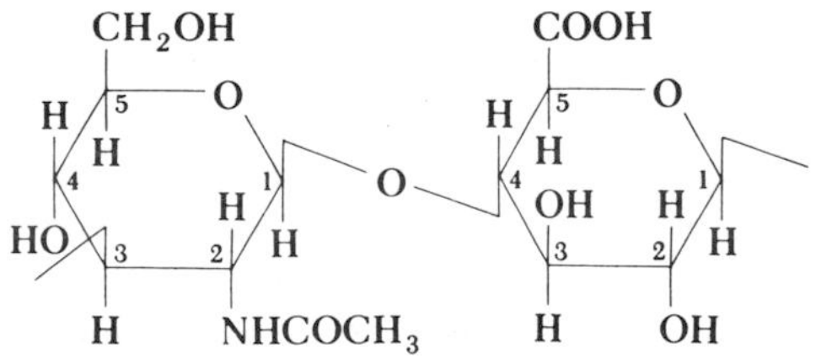

Portion of Hyaluronic Acid

termined. Both occur in the ground substance of connective tissue. Hyaluronic acid is more abundant in skin and soft tissues while the chondroitin sulfates are more abundant in cartilage. Hyaluronic acid can be isolated from umbilical cord and chondroitin sulfates from bovine nasal septum. Probably in connective tissue these carbohydrates occur linked to protein.

Hyaluronic acid appears to be a linear molecule or one in which there is little branching of alternate units of 2N-acetyl glucosamine and glucuronic acid. These units are joined β 1–3 and β 1–4. Some workers find that hyaluronic acid is difficult to methylate and believe this indicates a ramified structure for the molecule rather than a linear one. Viscosity studies indicate a molecular weight of 200,000 to 400,000.

Hyaluronic acid is found widely distributed. All connective tissue appears to contain it or derivatives and its jellylike quality when dispersed in water gives these tissues their particular softness and elasticity. Vitreous and aqueous humor, synovial fluid, and pleural fluid all contain it. In these fluids its rather high viscosity make it an excellent lubricant.

Three chondroitin sulfates have been isolated and are designated A, B, and C. Chondroitin sulfate A and C are polysaccharides composed of equal moles of 2N-acetyl galactosamine, D-glucuronic acid, and sulfuric acid. They are joined through beta 1–3 and 1–4 links.

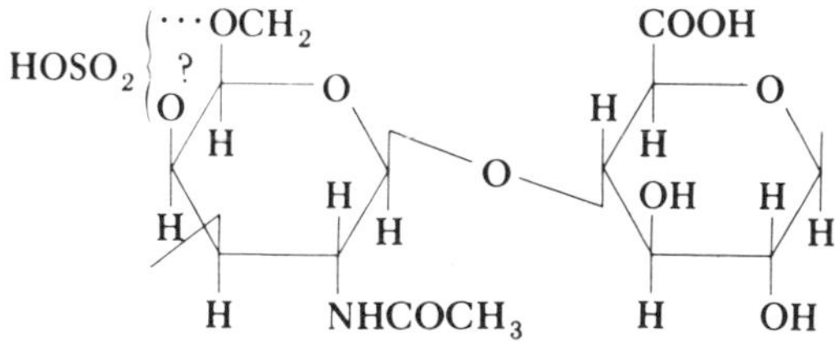

Portion of Chondroitin A and C Chain

Molecular weights as high as 2,000,000 have been found for chondroitin sulfate A, but chondroitin sulfate C is smaller in size. Chondroitin B is believed to differ from A and C by possessing L-iduronic acid in place of

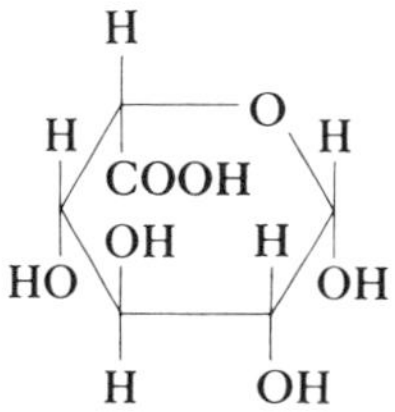

α-L-iduronic Acid

D-glucuronic acid. This differs from glucuronic acid in the configuration of carbon 5.

Chondroitin sulfates A, B, and C differ in their optical activity and their resistance to enzymatic digestion by either testicular hyaluronidase or the hyaluronidase of pneumococci. Chondroitin sulfuric acid has been isolated from cartilage with good yields: pig nasal septum contains 41 per cent while the cartilage from the epiphysis of an infant yielded 20 per cent.

TABLE 3.3. DISTRIBUTION OF HYALURONIC ACID AND
CHRONDROITIN SULFURIC ACID A, B, AND C*

Group	Tissue	Hyaluronic Acid	ChS A	ChS B	ChS C
I.	Vitreous Humor	+	–	–	–
	Synovial Humor	+	–	–	–
	Mesothelioma	+	–	–	–
II.	Hyaline Cartilage	–	+	–	+(?)
III.	Heart Valves	–	–	+	+
	Tendon (Pig and Calf)	±	–	+	+
	Aorta	–	–	+	+
IV.	Skin (Pig and Calf)	+	–	–	–
	Umbilical Cord	+	–	–	+

*From Meyer, K. and Rapport, M. M., *Science*, 113, 596–599 (1951).

Meyer and Rapport[17] give the distribution of hyaluronic acid and chondroitin sulfuric acid A, B, and C from ground substance polysaccharides in the tissues studied so far, as shown in Table 3.3. Chondroitin sulfuric acids are believed to be weakly linked to protein in connective tissue since they are readily extracted with solutions of calcium chloride.

Mucoitin sulfuric acid (hyalurono sulfuric acid) is a heteropolysaccharide which has been less extensively studied than chondroitin sulfuric acids or hyaluronic acid. It is closely related to them since on hydrolysis it forms glucuronic acid, 2-desoxy-2-amino D-glucose (glucosamine), acetic acid, and sulfuric acid. It has been isolated from cattle corneas and gastric mucosa. The compound from corneas is readily attacked by the enzyme hyaluronidase and is therefore believed to be the mono-sulfuric acid ester of hyaluronic acid. This is not true of the mucoitin sulfuric acid isolated from the gastric mucosa. It resists attack by hyaluronidase and must therefore possess a different structure.

Mucopolysaccharides and mucoproteins are complex compounds related to carbohydrates and are of considerable importance in mammals. The nomenclature is not yet precise, but the groups of compounds are usually classified either as mucopolysaccharides or mucoproteins on the basis of the amount of carbohydrate or protein present, although no exact dividing line can be given. These mucosubstances are slimy compounds that act as lubricants, protect epithelial surfaces, or form part of the cell wall and have many other functions.

The carbohydrate portion of these molecules is composed of hexosamine, monoses, and often sialic acid. The hexosamines are either glucosamine or galactosamine while the monose may be galactose, mannose or fucose or combinations of them. In bacterial cell walls the pentoses, rhamnose and arabinose, occur. The carbohydrate portion is often large but in some mucoproteins appears quite small as a prosthetic group. All these polysaccharides are bound tightly to protein. This is a different kind of bond than that which occurs between the chondroitin sulfates or hyaluronic acid and protein.

Sialic acid is the name now given to all acylated derivatives of neuraminic acid. The one most commonly encountered appears to be N-acetyl neuraminic acid. Neuraminic acid is now considered to be the condensate of mannosamine and pyruvic acid.

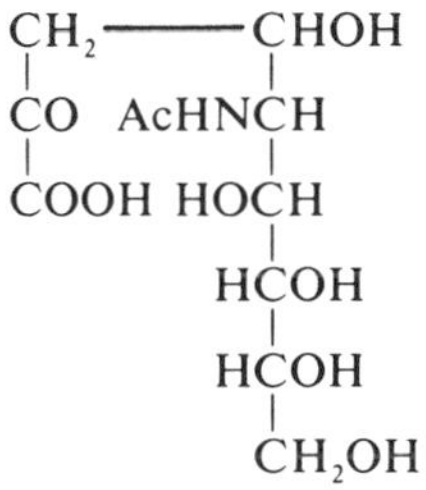

$$
\begin{array}{l}
CH_2 \rule{1cm}{0.4pt} CHOH \\
\;|\qquad\qquad\; | \\
CO \quad AcHNCH \\
\;|\qquad\qquad\; | \\
COOH \; HOCH \\
\qquad\qquad\; | \\
\qquad\quad HCOH \\
\qquad\qquad\; | \\
\qquad\quad HCOH \\
\qquad\qquad\; | \\
\qquad\quad CH_2OH
\end{array}
$$

Sialic Acid (N-acetyl Neuraminic Acid)

Many of these polysaccharides are present in small amounts in meat and meat products.

Hemicellulose

Plant cell walls are readily visible under a microscope because of thickness and the frequent occurrence of numerous layers. The most abundant compound present is α cellulose. Long fibers are embedded in a matrix of cementing compounds and in older tissues which have become woody are encrusted with lignins. In analysis of plant tissue and attempts to separate the compounds that make up the cell wall, lignins are first removed. This leaves "holocellulose," which appears to be an intimate mixture of cellulose and other compounds which are soluble in aqueous alkali. These alkali-soluble substances are called hemicelluloses and are not well defined. No satisfactory classification of the hemicelluloses has been possible yet although some generalizations are given.

It is generally agreed that the following differences exist between cellulose and hemicellulose molecules:

(1) Celluloses have a higher degree of polymerization (more monosaccharide units) per molecule than hemicelluloses.

(2) Celluloses are less soluble in alkali and less readily hydrolyzed by dilute acids than hemicelluloses.

(3) Celluloses are fibrous, while hemicelluloses are nonfibrous.

(4) Celluloses yield D-glucose on hydrolysis, while hemicelluloses yield predominantly D-xylose and other monosaccharides.

(5) Celluloses have a higher ignition temperature than hemicelluloses. These criteria are only general, qualitative means of differentiation.

Hemicelluloses are widely distributed in nature, but research has centered for the most part on fractions isolated from wood. Most hemicellulose is composed primarily of the pentose, xylose, often combined with a methyl uronic acid, although hemicellulose fractions from conifers contain mannose, one of the hexoses, as the principal monosaccharide residue. Table 3.4 shows that many other monosaccharides, both hexoses and pentoses, have been isolated from hemicellulose fractions. Plant hemicelluloses other than wood which have been studied have a similar composition with xylose the principal monosaccharide residue. This is shown in the data in Table 3.4. The uronic acids found have not been well characterized, although some have been identified as glucuronic or galacturonic acids, or their methyl ethers. Hirst[11] summarizes our knowledge of the structure of a few fairly well defined hemicelluloses in schemes which represent polysaccharides composed of a number of monosaccharide units. See Table 3.5. In this table where the number is known, it is given.

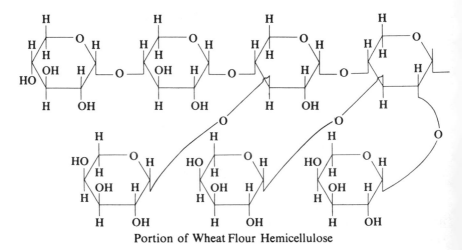

Portion of Wheat Flour Hemicellulose

TABLE 3.4 COMPOSITION OF HEMICELLULOSES*

	Xylose	Methyl-uronic	Hexoses			Pentoses		
			Glucose	Galactose	Mannose	Arabinose	Rhamnose	Glucurone
Oakwood								
Hemi A	82%	10%	8%					
Hemi B	60%	17%	23%					
Black locust			in sap-					
Lemonwood	78-93%	7.8-19.5%	wood but					
White birch			not heart-					
Cottonwood			wood					
Mesquite	11 moles	1 mole						
Wood	6	1						
Aspenwood	chief		sm.	sm.	sm.	sm.	sm.	sm.
Black spruce	✓		✓	✓	✓	✓	✓	✓
White Pine	44-50	10.5-15	sm.		36-46%			
Sapwood less sol.	7.7%	3.3%			chief			
Slash Pine								
Fract. A					11.5			
Fract. B					27.2			
Fract. C					45.2			
Black spruce								
A					9.3			
B					26.4			
C					44.9			
Alfalfa Hay	chief	some				sm.		
Sugar Cane Fiber	22 moles	0.9 moles (no me.)				1 mole		
Cornstalk	19 moles	2 (no me.)				7		
Oat Hulls	✓	1 (no me.)		2 mol		✓		
Esparto grass	18-20					1		
Sugar maple		uronic	methoxy					
Fract. IA	46.1	17.1	2.7					
IB	48.7	15.8	2.3					
II	54.7	22.8	2.6					
III	79.2	12.2	2.1					
IV	80.9	9.3	2.3					

*From Wise, L. E. and John, E. C., "Wood Chemistry," 2nd ed., Chap. 10, "The Hemicelluloses," Reinhold Publishing Corp., New York, N. Y. (1952).

TABLE 3.5. STRUCTURE OF HEMICELLULOSES*

Source	Structure	No. of Monosaccharide Units
Esparto grass	X1 ...4X1 ...4X1 ...4X1 ...4X 3 X1 ..4X1	$\xrightarrow{\frac{x\ y}{z}}$ $x + y + z = 80$
Corn cob xylan	similar	
Wheat flour	X1 ...4X1 ...4X1 ...4X1 ...4X 3 3 2 1A 1A1A	
Esparto grass	X1 ...4X1 ...4X1 ...4X 3 3 1A 1X4 ..1X	
Wheat straw	X1 ...4X1 ...4X1 ...4X1 ...4X1 ...4X 3 3 1A 1GA	$50x$
Oat straw	X1 ...4X1 ...4X1 ...4X1 ...4X 3 2 1A 1GA	
Pear fruit	X1 ...4X1 ...4X1 ...4X1 ...4X1 ...4X 3 3 X1 ...4X1 1GA	$\xrightarrow[z/GA]{x\ y}$ $x + y + z = 120$
Beechwood	X1 ...4X1 ...4X1 ...4X1 ...4X 2 1MGA	
Algal xylan Phododymeia palmata (Red seaweed)	X1 ...4X1 ...3X1 ...X	Ratio 1:12:3
Barley gum	...4Gβ1-3Gβ1 ...	
Larchwood	...Galβ1-6Galβ1 ... 3 1βGal6	
Guar	4Mβ1.-4Mβ1 ... 6 ...1αGal	

where:

X =	Xylopyranose
A =	Arabofuranose
G =	D-Glucopyranose
Gal =	D-Galactopyranose
M =	Mannopyranose
GA =	D-Glucuronic Acid
MGA =	4Methyl Glucuronic Acid

*From Hirst, E. L., *J. Chem. Soc.*, 2974–2984 (1955).

Hirst's scheme attempts to show not only the branching of the molecules but the carbons at which bonds are established. Thus wheat flour hemicellulose has a "back bone" of a chain of xylose residues, joined 1-4 with branching by arabinose moieties on either carbon 3 or occasionally carbon 2. The formulas show this. (See the formula presented above on p. 84.) The rather low molecular weights are demonstrated by rather low number of monose residues in esparto grass, wheat straw, pear fruit hemicellulose (80, 50, and 120).

Pectic Substances

These consist of a group of carbohydrates which are of considerable interest in foods of plant origin. They comprise not only those substances (pectins) in fruits and vegetables which are capable of forming gels with sugar and acids but a number of other compounds as well. The exact location of these compounds in a tissue is still a matter of dispute since the methods of chemical morphology are not yet specific enough. Most workers are agreed that pectic substances occur in the middle lamella between cells and perhaps in the cell wall, but whether they are ever contained within the cell is questionable. The structure of the pectic substances has been the object of a great deal of research and considerable knowledge has accumulated. The chief products of the hydrolysis of pectic substances are galacturonic acid, a derivative of galactose in which the 6-carbon is oxidized to a carboxyl group, and methyl alcohol.

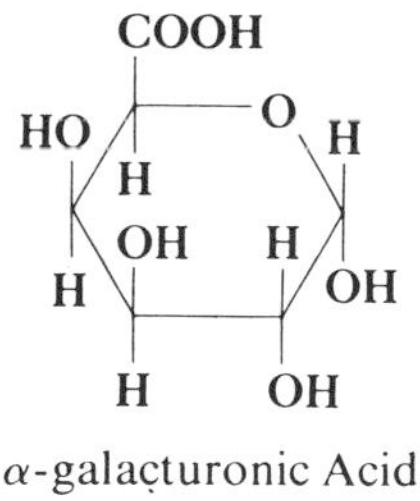

α-galacturonic Acid 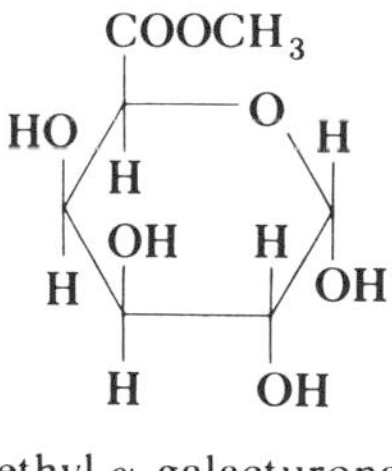   Methyl α-galacturonate

This indicates that pectic substances are polysaccharides of galacturonic acid or of its methyl ester. Whether other groups are present in the molecule is difficult to determine because pectic substances are hard to purify. Since some apparently pure samples of pectic substances have been prepared which contain only galacturonic acid or its methyl ester, when other moieties are found among the hydrolysis products, it is difficult to decide whether they are contaminants or intrinsic parts of the original molecules.

The pectic substances have been studied for many years and the nomenclature in the early days became very confused. In 1926 a committee of the

American Chemical Society attempted to standardize this nomenclature. Its report was revised and adopted (see Kertesz[14]) as official in 1943. These definitions are given here:

Pectic substances "Pectic substances" is a group designation for those complex colloidal carbohydrate derivatives which occur in, or are prepared from, plants and contain a large proportion of anhydrogalacturonic acid units which are thought to exist in a chainlike combination. The carboxyl groups of polygalacturonic acids may be partly esterified by methyl groups and partly or completely neutralized by one or more bases.

Protopectin. The term protopectin is applied to the water-insoluble parent pectic substance which occurs in plants and which, upon restricted hydrolysis, yields pectinic acids.

Pectinic acids. The term pectinic acids is used for colloidal polygalacturonic acids containing more than a negligible proportion of methyl ester groups. Pectinic acids, under suitable conditions, are capable of forming gels (jellies) with sugar and acid or, if suitably low in methoxyl content, with certain metallic ions. The salts of pectinic acids are either normal or acid pectinates.

Pectin. The general term pectin (or pectins) designates those water-soluble pectinic acids of varying methyl ester content and degree of neutralization which are capable of forming gels with sugar and acid under suitable conditions.

Pectic acid. The term pectic acid is applied to pectic substances mostly composed of colloidal polygalacturonic acids and essentially free from methyl ester groups. The salts of pectic acid are either normal or acid pectates.

The molecular weights of various pectic substances extracted from plant material have been determined by osmotic methods, the ultracentrifuge, end group analysis, and by viscosity, but no agreement is apparent as yet. This is not surprising since separation of individual molecular species in the pure state is very difficult, and the pectic substances present in plants are probably complicated mixtures. End group analysis indicates a range of molecular weights from 2500 to 7500 while measurements of pectic substance derivatives (acetyl and nitropectins) by osmotic pressures give molecular weights from 30,000 to 100,000. By use of the ultracentrifuge crude extracts give a range from 16,000 to 50,000, although some scientists have found for highly purified products a variation from 33,000 to 117,000. Viscosity measurements give results from 27,000 to 115,000 with one sample as high as 280,000. Probably pectic substances occur in plant tissues as a mixture of molecules with a variety of molecular weights. At the moment it is impossible to estimate the number of galacturonic acid residues present in any one type of pectic substance.

There have been numerous structures proposed for the pectic substances, but today the most popular hypothesis is that galacturonic acid residues are joined in linear units by α 1–4 links like those which occur in amyloses. Pectic acids are the polygalacturonic acids which are essentially free of methyl ester groups, while pectinic acids contain some methyl ester groups.

Kertesz[14] shows the difference between them in the following scheme:

```
G—COOH              GCO—O—CH₃
|                   |
G—COOH              G—COOH
|                   |
G—COOH              G—CO—O—CH₃
|                   |
G—COOH              G—COOH
|                   |
G—COOH              G—COOH
|                   |
    . . .           G—COOH
                    |
    . . .           G—CO—O—CH₃
                    |
    . . .               . . .

    . . .               . . .
|                   |
G—COOH              G—COOH
|                   |
G—COOH              G—CO—O—CH₃
|                   |
    . . .           G—CO—O—CH₃
                    |
    . . .               . . .

    . . .               . . .
   etc.                 etc.
Polygalacturonic      Pectinic
Acid or Pectic          Acid
    Acid              (Pectin)
```

The 1–4 link is an α ether link between the first carbon on one galacturonic residue and the fourth carbon on another.

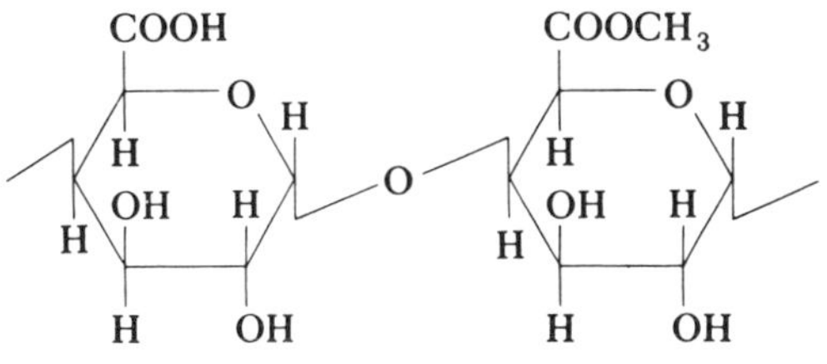

Pectic Substance Chain

The angle between the 1– and 4– carbons has been shown by X-ray study to be 90° to the plane of the galacturonic acid ring. This differs from that for other polysaccharides and gives the linear unit a screwlike configuration. The number of galacturonic acid residues in each linear unit is not known. Since pectic acids can be prepared from pectinic acids by hydrolysis, it is assumed that these two groups of compounds have essentially the same structure.

The free carboxyl groups in both pectic and pectinic acids will react with metallic ions to form salts. If the acid is completely neutralized, that is, if all carboxy groups react, the product is a normal pectate or pectinate. If only part of them react, the product is an acid pectate or pectinate. Many of the pectic substances exist in plants as acid calcium or magnesium salts.

Both pectic acids and pectinic acids are predominantly linear. However, many workers in the field believe that the screw-shaped linear units of galacturonic acid residues are associated as macromolecules, either through some sort of branching similar to that which occurs in starches and glycogens or by lamination or bundle formation. Three possible types of links which hold the linear units have been suggested: (1) anhydride formation between the carboxyl groups, (2) ester formation between carboxyl and alcoholic hydroxyl, and (3) hydrogen bonding.

(1) Some evidence points to the participation of the carboxyl group in links between linear units of galacturonic acid residues. For one thing, the equivalent weight is not high enough to indicate one carboxyl for each glucuronic acid residue. Also pectic substances are especially vulnerable to alkaline degradation, a characteristic of acid anhydrides. The anhydride link is the following:

$$GC\underset{|}{\overset{O}{\diagup}}\!-\!-OH + G\!-\!C\underset{|}{\overset{O}{\diagup}}\!-\!-OH \rightarrow GC\underset{|}{\overset{O}{\diagup}}\!-\!O\!-\!C\underset{|}{\overset{O}{\diagup}}\!-\!G + H_2O$$

(2) The link may be ester in nature although this group is more difficult to hydrolyze.

$$GC\underset{|}{\overset{O}{\diagup}}OH + GOH \rightarrow G\!-\!C\underset{|}{\overset{O}{\diagup}}\!-\!O\!-\!G$$

Either link (1) or (2) would form branched or laminated molecules with considerable stability since these are primary valence bonds.

(3) The hydrogen bond which weakly holds many macromolecules together and is readily disrupted by warming may hold linear units of galacturonic acid residues in laminated molecules. Hydrogen bonding can occur between hydroxyl groups or between hydroxyl and carboxyl. This type of macromolecule will dissociate readily into linear molecules.

Protopectin. Protopectin has been assumed for many years to exist as an insoluble molecule in plant tissues, but its exact nature is far from solved. There is always a great problem in separating a large insoluble molecule from the complex mixtures which occur in plants, without producing a change in composition. Attempts to study it by staining techniques has likewise not progressed far enough to give any information that is not tempered with doubt.

Many investigators in the past assumed that protopectin is a large insoluble molecule formed by the reaction of pectic substances with cellulose. Cellulose occurs in the cell walls of plants and protopectin is believed to be in these walls as well as in the middle lamella between the cell walls. It is certainly possible that such a combination exists, and such a huge molecule would be insoluble.

Other workers have suggested that protopectin is insoluble because it exists as a calcium-magnesium salt. Protopectin is rendered soluble by hydrolysis with acid and they have regarded this action as the substitution of hydrogen for the calcium and magnesium ions. In support of this hypothesis there are also the observations that isolated pectic and pectinic acids of rather high molecular weight form insoluble salts with calcium and magnesium ions. In other words, these workers consider protopectin as calcium-magnesium pectinate. Finally there are some researchers who consider protopectin a macromolecule built up of numerous pectinic acids and they have speculated on the type of linkages which hold the pectinic acids together. It may be that ether or ester links form unusually large branched or laminated molecules similar to those described for pectic and pectinic acids. It is also possible that the pectinic acids are held together by calcium ion bridges. Since the calcium ion is divalent, it is possible that it reacts with a carboxyl group on each of two pectinic acids.

$$
\begin{array}{ll}
| & | \\
GCOO—CH_3 & H_3C—OOCG \\
| & | \\
GCOOH & HOOCG \\
| & | \\
GCOO\text{————————} Ca^{++} \text{————————} OOCG \\
| & | \\
\cdot\ \cdot\ \cdot & \cdot\ \cdot\ \cdot \\
etc. & etc.
\end{array}
$$

The reactions of pectic substances are very evident in cooking fruits and particularly in making jelly. When fruits rich in pectic substances are boiled to hydrolyze the protopectin they form a juice which will gel when sufficient sugar and acid is added. Overripe fruit in which the pectinic acids have been extensively hydrolyzed to pectic acids, make very weak jellies or none at all. Boiling the juice for long periods often produces syrupy gels for the same reason. The physical-chemical problems encountered in preparing pectin gels are complicated; and although the companies which make commercial pectin either from apple pomace or from citrus fruits have standardized their products and have established procedures which insure successful jellies, many of the principles behind these procedures are still obscure.

Pectin Gels. The pectin which is present in most fruits has been used for many years to produce sugar-acid gels. Some fruits such as apple and quince are widely known to have excellent jellying power. The pectic substances which produce firm jellies are relatively high molecular weight molecules with a relatively high per cent of methyl ester groups and consequently a low per cent of free carboxyl. When a hot aqueous mixture of sugar, acid, and pectins is cooled, it sets into a gel. The sequence of events and the effects of ingredients is complex. Several theories have been proposed to explain the gel formation. The factors which influence gelling will be discussed, but their significance will be omitted.

A firm gel depends on (1) per cent pectin, (2) molecular weight of the pectins, (3) per cent methylation, (4) per cent sugar, and (5) pH. A desirable jelly must be firm enough to stand without appreciable deformation and yet tender enough to spread readily on bread. As the percentage of pectins increases in mixtures, the firmness of the jellies produced on cooling increases. A satisfactory jelly is produced with approximately 1 per cent pectin. The amount varies with the quality of the pectin preparation, the average molecular weights of the molecules present, and the degree of methylation.

The molecular weight of pectic substances is important in determining jellying power. Molecules must be colloidally dispersed before a gel is possible. Thus the protopectins cannot act in gelation until they have been changed to pectinic acids. As a gel forms, the molecules develop a three-dimensional network which traps solution in the interstices. If the molecules are too short, the network is not continuous in many spots and the gel is runny or soft.

The methyl ester content of the pectinic acid is another factor which influences the jellying power of the product. Excellent jellies can be prepared from pectins with a wide range of methyl content, but maximum jellying appears at about 8 per cent. This represents esterification of half of the carboxyl groups.

The effect of pH on jellying has been recognized for many years. In the home preparation of jellies it is well known that dead-ripe fruit with its lower acid content does not yield as good a jelly as partially ripe fruit. Most pectic products do not form a jelly until the pH is lowered to 3.5, and the firmness of the jelly increases as the pH is lowered. With very low pH's, the amount of pectin can be decreased and a satisfactory gel still formed. An optimum pH is usually found; and if the pH is lowered below this point, the firmness of the jelly diminishes and excessive syneresis develops.

Sugar (mainly sucrose) is necessary for the formation of the pectin gels and must be present in a minimum concentration. Most jellies are made with approximately 65 per cent sugar. If the amount is increased much above this point, crystallization tends to occur on the surface of the jelly and occasionally even within the jelly. We find that the same methods which prevent or diminish crystal growth of sucrose in sugar cookery are effective in fruit jellies. Cooking the sugar with acid fosters hydrolysis of sucrose to form glucose and fructose (invert sugar), or addition of corn syrup or other glucose or fructose products in small amounts decreases the tendency to crystallization.

Jelly Grade. The jelly grade of pectic products determines their commercial value but it is difficult to measure. A number of methods have been suggested to evaluate the ability of a pectin product to form a jelly. Before a manufacturer buys a product, he wants to know how much jelly can be made from 1 lb of the material. Jelly grade is defined in terms of the amount of sugar which 1 lb of the pectin will "carry" or gel. If a pound of pectin will carry 100 lb of sugar and form a satisfactory jelly, it is classed as 100 grade pectin. One of the most difficult points is the definition and measurement of "a satisfactory jelly."

Setting Time. The time which elapses between the addition of all components and formation of a gel, called the "setting time," often has a marked effect on the quality of the final jelly. If a gel sets too rapidly before pouring is complete, the jelly never achieves the firmness possible with slow setting. When setting is very rapid, the mixture is said to "curdle," as small lumps of gel are formed in the pot. A curdled mix is difficult to pour and does not fill evenly. With jams and preserves, however, slow setting allows the chunks of fruit to settle rather than remain evenly distributed through the jam. Commercial pectins are available either as "rapid set" or as "slow set."

A hot mixture does not set until it begins to cool. The rate of cooling and all of the physical factors which effect rate of cooling, such as the pot size and shape, have an effect on setting time. But when this is controlled, different pectin products still show a difference in the setting time. A "rapid set" begins at about 88° C (190° F), while a "slow set" forms a jelly below 54° C (130° F). A number of patents have been issued for methods of treatment of pectin products to prolong setting time. They are methods which by acid or enzyme treatment alter the pectinic acids by partial hydrolysis so that both molecular weights and total methyl are slightly changed.

The pH of the mixture has an influence on setting time. Setting is more rapid at lower pH's. Salts tend to delay setting; and some, such as disodium phosphate, are regularly added to "slow set" pectins. Buffer salts decrease the ionization of the organic acids present in fruit juices and by this action raise the pH and prolong setting time.

Adjustment of the pH of a fruit juice or mixture is allowed by the USDA. If the fruit has too high a pH, it is permissible to add acid. Citric and lactic acids are the most common ones used since they are believed to impart good flavor; but phosphoric, tartaric, and malic are also used. These acids are either added at the end of the cooking time so that too much hydrolysis of the sugar is prevented, or in jellies they are sometimes placed in the containers and mixed through the hot product immediately after pouring. To raise the pH of a fruit which is unusually acid, buffer salts are used. The USDA allows as much as 3 oz of either sodium citrate or sodium potassium tartrate per 100 lb of sugar.

Gel Formation with Low Ester Pectins. Low ester pectins which have a methyl content below 7 per cent are able to form gels in the presence of small amounts of divalent ions, even though the per cent of solids is very low. Commercial products are now on the market and are used for other types of gels than fruit jellies or jams. They are produced by de-esterification with either acid, enzymes, or alkali. The products formed from these three methods are evidently not identical since they differ sharply in their physical characteristics and particularly in their reaction to divalent ions. They are often called "low methoxyl pectins."

Low ester pectins do not require the presence of sugar for the formation of a gel as do the high ester pectins. The significance of the requirement for divalent ions probably lies in the ability of the ions to react with the carboxyl group on two molecules of pectic acid and form a bridge between them.

The low ester pectins are used in the preparation of low solid salads and desserts. Tomato aspic, for example, can be formed with low ester pectins but will not gel with the high ester pectins. They can also be added to fruit before freezing to reduce the amount of "run-off" when the fruit is thawed.

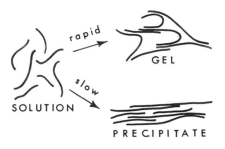

FIGURE 3.8. MECHANISM OF REGROGRADATION OF LINEAR FRACTION OF STARCH. Reproduced from Schoch, T. J. and Elder, A. L., "Starches in the Food Industry," in "Use of Sugars and Other Carbohydrates in the Food Industry," *Advances in Chem. Ser.*, **12**, American Chemical Society, Washington, D. C. (1955).

Coating for many foods is possible with a bath of low ester pectins to which a small amount of calcium chloride is added. The coating can be sprayed with calcium chloride to "tan" the surface. This is useful with fruit, meats, and other foods. It has been used to prevent freezer burn in fish and stickiness in candy.

Pectins in Alcoholic Fermentation. Alcoholic fermentation of fruits and grains with yeast, usually *Saccharomyces cerevisiae*, yields ethanol, C_2H_5OH, and very small amounts of other organic compounds. Occasionally methanol will contaminate the fermentation beer; and unless it is carefully removed by distillation, it will be in the final product. This methanol arises by demethylation of pectins by pectin esterase.[5] Yeast does not form an enzyme capable of hydrolyzing pectins, and consequently the reaction does not commonly occur in fruit fermentations. But pectin esterase is abundant in the peel of citrus fruits and also in fungi. The activity of the enzyme rises as the pH increases from 1 to 6 and the production of methanol goes up. If citrus fruit is limed before fermentation, or if fruit or grain with relatively high pH becomes contaminated with molds, the amount of methanol formed may be fairly high.

Gums and Mucilages

A number of plants form gums on their bark or fruit when they are injured or when they are stung by insects. An exudate forms which contains a high per cent of polysaccharide. This can be readily gathered and dried and will absorb water and form a sticky mass. Many natural gums are sold commercially and used as adhesives or thickeners. Gum arabic from the acacia tree, gum ghatti, gum karaya, and gum tragacanth are commonly used.

Seaweeds and a number of seeds contain polysaccharides which are similar in composition to those produced as exudates and which are used in similar fashion. Alginic acid and its salts are now widely used in the United States as thickeners. Agar is produced from seaweed (*Gelidium algas*) and carageenan from Irish moss (*Chondrus crispus*). Locust bean gum and quince seed gum are examples of those from seeds.

For the most part, fruit gums, on hydrolysis, form pentoses, hexoses, and uronic acids. Table 3.6 shows the distribution of monosaccharides and uronic acid in gums.

Gum Arabic. This gum is produced on the bark of Acacia trees when they grow in hot, dry, and high locations. It is extensively used industrially as a thickener and emulsifying agent and has, therefore, been the subject of a number of studies. Gum arabic occurs as a calcium salt with an equivalent weight between 1000 and 1200. The branched molecule appears to have a backbone of galactose residues held 1–3 with other monosaccharides or uronic acids in the side chain.

$$
\begin{array}{l}
\text{Araf3} \text{---} \text{1Galp} \\
\quad\quad\quad\quad\quad 6 \\
\text{Galp1} \text{- - -} \text{3Galp1} \text{- - -} \text{3Galp1} \text{---} \text{3Galp1} \text{------} \\
\quad\quad 6 \quad\quad\quad\quad\quad\quad\quad\quad\quad 6 \\
\quad\quad | \quad\quad\quad\quad\quad\quad\quad\quad\quad | \\
\text{Rhap1} \text{- - -} \text{3Galp} \quad\quad \text{Rhapl} \text{---} \text{3Galp} \\
\quad\quad 6 \quad\quad\quad\quad\quad\quad\quad\quad\quad 6 \\
\quad\quad | \quad\quad\quad\quad\quad\quad\quad\quad\quad | \\
\quad\quad \text{GA} \quad\quad\quad\quad\quad\quad\quad\quad \text{GA} \\
\quad\quad 4 \quad\quad\quad\quad\quad\quad\quad\quad\quad 4 \\
\quad\quad | \quad\quad\quad\quad\quad\quad\quad\quad\quad | \\
\quad\quad \text{Araf} \quad\quad\quad\quad\quad\quad\quad\quad \text{Araf}
\end{array}
$$

where:

Galp = Galactopyranose
Araf = Arabofuranose
Rhap = Rhamnopyranose
GA = Glucuronic acid
(Rhamnose is 6-deoxy mannose)

Gum arabic is widely used, particularly in confections, where strong stable gels are needed.

Seaweed Polysaccharides. The seaweed polysaccharides are produced in large amounts from various seaweeds throughout the world. Some seaweeds have a relatively low cellulose content and a high per cent of other polysaccharides. Often the surface of the plant is covered with these compounds which give it a slippery feel. They are readily extracted by boiling water or alkali and appear on the market in either a very crude state or partially purified. In some parts of the world they are used for food, but in the United States they appear in foods and other products as thickeners and adhesives.

Although many of these polysaccharides have been used for centuries by man, most of them are not as yet well explored chemically and only tentative or partial formulas can be assigned.

Agar. Agar, which is produced by boiling red seaweed in water, is a linear galactan sulfate. The galactose units are held together with 1–3 links with about every tenth link 1–4. The sulfate occurs in the ratio of 1:53 galactose, as the calcium salt of the galactose ester. Both D- and L-galactose occur in the molecule. Agar has an enormous ability to absorb water and form gels at low concentrations. The Japanese have used it as a food for centuries although it is questionable if it has any food value for man. In the United States the food industry uses large quantities as a stabilizer and

TABLE 3.6. APPROXIMATE COMPOSITION OF SOME PLANT GUMS*
(Moles of Constituent per Repeating Unit)

Gum	D-Glucuronic Acid or Methoxyl Derivative	D-Galactose	D-Mannose	L-Arabinose	D-Xylose	L-Rhamnose
Arabic	1	3	nil	2	nil	1
Damson	1	2	1	3	ca. 3%	nil
Cherry	1	2	1	6	ca. 3%	nil
Egg Plum	2	6	nil	7	1	nil
Purple Plum	1	3	nil	3(?)	1(?)	nil
Almond Tree	1	3	nil	4	2	nil
Lemon	+	+	nil	+	nil	nil
Orange	+	+	nil	+	nil	nil
Grapefruit	+	+	nil	+	nil	nil
Cholla	+	+	nil	+	nil	+
Mesquite	D-Galactu-ronic Acid	+	nil	+	nil	–

*From Pigman, W. W. and Goepp, R. M., "Chemistry of the Carbohydrates," Academic Press, Inc., New York, N. Y., 1948.

gelling agent. It is used in icings on baked goods, in cream cheese, and in fruit and vegetable jellies. Agar is often used in place of gelatin in jellied candies and marshmallows. It makes a fine jellied glaze for meat or fish.

Alginic Acid. Alginic acid or sodium alginate, is produced from the giant kelp of California, *Macrocystis pyrifera,* by extraction with sodium carbonate. It is a polysaccharide made up of units of mannuronic acid, joined β 1–4. It is used in ice creams, sherbets, syrups, processed cheese, salad dressings, and many other food products.

Carageenan. Carageenan is isolated from Irish moss by extraction with hot water. It is a mixture of polysaccharides made up of galactose mono- or disulfates. The composition of a few of these polysaccharides has been studied and they appear to be almost linear with one unit branching. The crude carageenan is also widely used in food products. Most chocolate milk in the United States is stabilized with carageenan, and it is often used in milk desserts and other products.

Locust Bean Gum. This is added to ice cream powders, pudding powders, preserves and jams, salad dressings, and pie fillings, to increase the viscosity or even gel the product. It is effective in low concentrations.

IDENTIFICATION

Color Reaction

Carbohydrates can be identified in mixtures of biological materials by a number of color reactions. One of the most widely used general tests is the

Molisch reaction which is carried out by adding a drop of an alcoholic solution of α-naphthol to a solution of the unknown. Concentrated sulfuric acid is then slowly poured down the side of the tube so that a layer is formed below the unknown solution. In the presence of saccharides a purple color develops at the interface. Carbohydrate acids or amines do not give the test, but mono- and oligosaccharides as well as many polysaccharides are positive. The sulfuric acid reacts with carbohydrates to form furfural or one of its derivatives. These compounds produce colored products with phenols and phenol derivatives. Alpha-naphthol used in the Molisch test can be considered a phenol derivative.

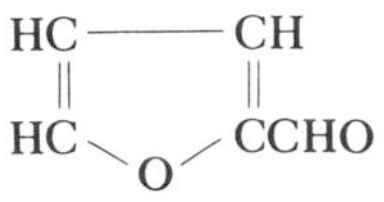

Furfural 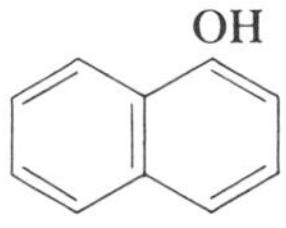 α-naphthol

Some of these color tests can more or less differentiate groups of carbohydrates from one another. Ketoses, pentoses, and uronic acids can be differentiated from aldohexoses since they react more readily and often form different colors with phenols and their derivatives. Seliwanoff's reagent is prepared by dissolving resorcinol in alcohol and adding hydrochloric acid to not more than 12 per cent. Fructose, a ketose, will give a red precipitate in 20 to 30 sec, while aldohexoses react much more slowly. The formation of a red color with glucose after boiling for some time is believed to be caused by a transformation of glucose to fructose. The color is the result of the formation of a furfural derivative and its subsequent reaction with resorcinol. Sucrose will give a positive test since it hydrolyzes in the acid solution. Variations of the method include substitution of the 12 per cent hydrochloric acid by alcoholic sulfuric acid or acetic acid. Phloroglucinol forms a cherry red color rapidly with pentoses, and slowly forms a dark red color which changes to brown with hexoses. Orcinol forms greenish blue colors with pentoses and is somewhat more valuable than other reagents for differentiating pentoses and hexoses. All of these tests must be used with caution and the time of heating and concentrations

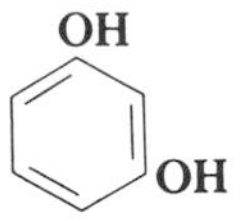

Resorcinol 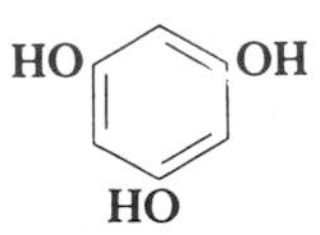   Phloroglucinol 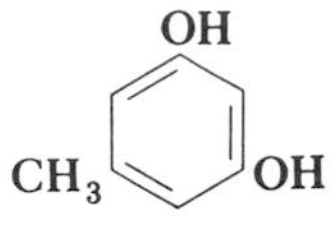   Orcinol of reagents carefully controlled. In Tauber's test benzidine in glacial acetic acid produces a cherry red color with pentoses and glucuronic acid, while

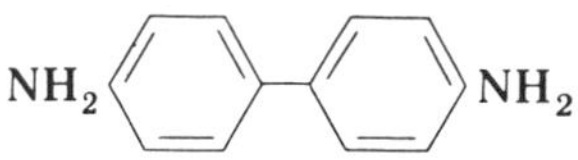

Benzidine 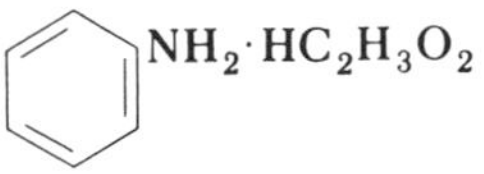   Aniline Acetate 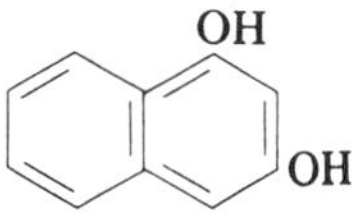   Naphthoresorcinol

hexoses give yellow to brown. Aniline acetate on filter paper turns bright red when held in the fumes from a mixture of pentoses and hydrochloric acid but not when hexose is present. The volatile compound is furfural. Uronic acids give a purple color when they are boiled with hydrochloric acid and then treated with naphthoresorcinol. These colored compounds are very soluble in ether and can be readily extracted from aqueous solutions. Anthrone is used in quantitative tests to give colors whose depth is proportional to the amount of carbohydrate present.

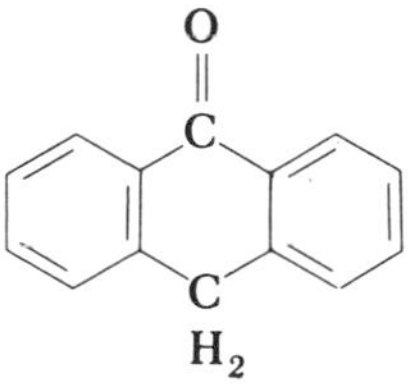

Anthrone

Sucrose can be identified by a color reaction when the solution is free of raffinose, gentiobiose, and stachyose. It reacts with an alkaline solution of diazouracil to form a green color.

The ease with which mono- and disaccharides which contain free or potential aldehyde and ketone groups are oxidized gives methods for differentiating them from sucrose, which has no potential aldehyde or ketone groups, and from polysaccharides which have few reducing groups in a large molecule. Many reagents have been devised using copper, silver, mercury, and other metal ions which are reduced to the free metal or oxide by the aldehyde or ketone groups of sugars. Benedict's reagent, which is composed of copper sulfate, sodium citrate, and sodium carbonate, is widely used for the detection of glucose in biological fluids. The deep blue color of the copper ion complex changes to the yellow or to the brick red color of cuprous oxide, Cu_2O, precipitate when glucose is present. Other substances which are readily oxidized can interfere; but when Benedict's solution is used cautiously, it is a very valuable reagent.

Some colored organic compounds which on reduction change colors are also used in detecting reducing sugars. Yellow picric acid changes to deep

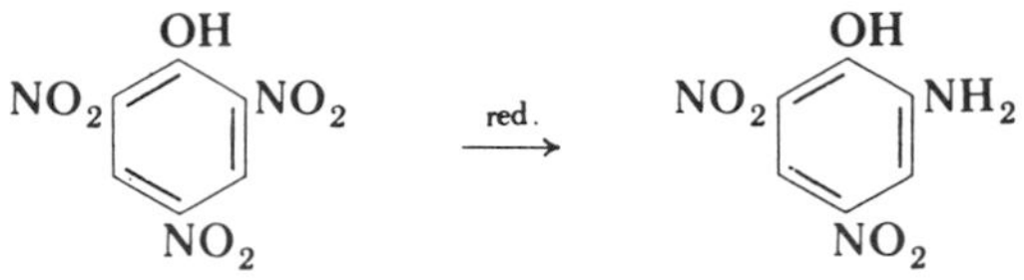

Picric Acid   Picramic Acid

red picramic acid. Orthodinitrobenzene, methylene blue, and safranin are other reagents.

Dilute solutions of iodine are used to detect some polysaccharides; and although it is difficult to differentiate them by this reaction, the method does have some usefulness. Amyloses give a blue color while amylopectins give violet. Glycogens give red brown and dextrins form colors from violet through reddish to colorless depending on their molecular weight.

The ability of iodine in neutral, slightly acid, or slightly alkaline solutions to oxidize aldoses can be used to determine them quantitatively. If the conditions are carefully controlled, aldoses can be quantitatively determined in the presence of ketoses which do not react.

Reaction Products of Mono- and Disaccharides with Hydrazines

The products of the reaction of mono- and some disaccharides with hydrazines are useful in differentiating many of them. Phenylhydrazine is the most commonly used reagent although dinitrophenylhydrazine and others sometimes form products which are easily isolated and purified. One molecule of phenylhydrazine reacts with a molecule of the carbohydrate to form a phenylhydrazone and when the product is highly insoluble as in the case of mannose, it precipitates. If the hydrazone is soluble, further reaction occurs and the osazone is formed.

Fermentation with Yeast

Some groups of carbohydrates can be differentiated from others by the ability of yeast (*Saccharomyces cerevisiae*) to ferment them with the formation of carbon dioxide and ethyl alcohol. Ordinary bakers' yeast will ferment D-glucose, D-fructose, D-mannose, D-maltose, and sucrose, but forms little or no carbon dioxide if supplied with D-galactose, lactose, or the pentoses. If the yeast is grown on a medium rich in D-galactose, the yeast then develops the ability to ferment the galactose. However, with ordinary commercial yeast, it is very easy to carry out a fermentation test and differentiate these sugars. Other strains of yeasts and other micro-

```
    HCO                                    HC=NNH─⬡
    |                                      |
    HCOH                                   HCOH
    |                                      |                        ⬡─NHNH₂
    HOCH         ⬡─NHNH₂                   HOCH
    |        +              ⟶              |                   ──────────────⟶
    HCOH                                   HCOH
    |                                      |
    HCOH                                   HCOH
    |                                      |
    HCOH                                   HCOH
    |                                      |
    H                                      H

  D-glucose      Phenyl              D-glucose
(Extended Form)  hydrazine           Phenyl  hydrazone
```

```
    HC=NNH─⬡                                   HC=NNH─⬡
    |                                          |
    CO                                         HC=NNH─⬡
    |                                          |
    HOCH                                       HOCH
    |                                          |
    HCOH                                       HCOH
    |                                          |
    HCOH             ⬡─NHNH₂                   HCOH
    |            ────────────⟶                 |
    HCOH                                       HCOH
    |                                          |
    H                                          H

    +                                        Glucosazone
    NH₃

    +

    ⬡─NH₂
```

organisms have been used to differentiate sugars and even to measure them quantitatively in carbohydrate mixtures. For example, D-galactose can be determined by the method of Wise and Appling in carbohydrate mixtures through the use of *Saccharomyces carlsbergensis*, a yeast which ferments it, and *Saccharomyces bayanus*, a yeast which does not. A table on the fermentation characteristics of a number of microorganisms is given by Pigman and Geopp.[22]

Separation by Chromatography

In many foods complex mixtures of carbohydrates exist and attempts to demonstrate the presence of one carbohydrate when others are also present is often difficult. Separation of the carbohydrates by chromatography allows the application of reactions for their identification without interference. When small samples are available, paper chromatography is usually used; but if the samples are large and if it is desirable to isolate the pure or relatively pure carbohydrates, column chromatography is used.

The ion exchange column can be used with ionized derivatives of carbohydrates. Thus the products of reaction of carbohydrates with borates are anions and are taken up by anion exchange resins. (See Boeseken.[4])

Optical Activity

When solutions of pure mono- or disaccharides are available, the determination of the optical rotation serves to identify or to determine quantitatively the amount of a sugar present. This method is applied to raw and purified cane and beet sugars but is not very useful in natural foods where complex mixtures occur. A polarimeter is used in measuring optical activity of a solution containing an accurately weighed sample. The specific rotation is calculated from the formula:

$$(\alpha)_D^{20} = \frac{100\,\alpha}{l \times c}$$

where $(\alpha)_D^{20}$ is the specific rotation at $20°\,C$, and, using the D line of the sodium spectrum as a source of light, α is the observed rotation, l is the length of the polarimeter tube in decimeters, and c is the weight of the sugar in grams per 100 ml of solution. When the sugar is known, since its specific rotation is also known, the equation can be rearranged to solve for the concentration of the sugar:

$$c = \frac{100\,\alpha}{l \times (\alpha)_D^{20}}$$

The method can sometimes be applied to mixtures of carbohydrates where the rotation can be made to vary under uniform conditions. For example the Clerget method for sucrose in the presence of glucose and fructose depends on the fact that hydrolysis of sucrose results in a change in rotation. The rotation of the sample is measured before and after hydrolysis and since the specific rotations of sucrose and of the equimolar mixture of glucose and fructose formed on hydrolysis is known, the percentage of sucrose can be calculated.

$$S = \frac{100(P - P_1)}{133 - 0.5(t - 20)}$$

S is the percentage of sucrose while P is the original rotation of the mixture, P_1 is the rotation after hydrolysis and t is the temperature in degrees centigrade. Because fructose is unstable in acid solution, enzymic hydrolysis is sometimes used.

Other Quantitative Determinations

The most widely used chemical methods for the quantitative determination of sugars depend on their ability to reduce alkaline solutions of metals such as silver, mercury, bismuth, and copper. Of all the methods which have been devised, those depending on the reduction of cupric ions to cuprous oxide are the most widely used. A monosaccharide such as glucose reacts with mild oxidizing agents to form a large number of products. The aldehyde is oxidized to a carboxyl group, but the alcoholic hydroxyls are also capable of oxidation. If conditions are right, the chain may be completely fragmented and a number of moles of cupric ion reduced. Since this is true, it might appear that such a reaction would be an unlikely prospect for the development of a quantitative method. Nevertheless, when conditions of reactions and reagents are standardized, surprisingly accurate results can be obtained. The amount of cuprous oxide formed is proportional to a given amount of glucose under standardized conditions. The cuprous oxide is determined by numerous methods. In some procedures it is weighed directly, in others it is ignited and changed to cupric oxide and then weighed. It can be dissolved and determined volumetrically by adding a ferric salt and titrating with permanganate, by adding a definite amount of iodine and titrating the excess iodine with thiosulfate, as well as with thiocyanate and silver salts. In some procedures the cuprous oxide is dissolved in the reaction mixture and determined directly. For example, in some colorimetric methods the cuprous oxide reduces molybdic acid to blue molybdic oxides. The depth of color is proportional to the amount of cuprous oxide formed and hence to the amount of carbohydrate present in the unknown. There is also a method which depends upon the determination of the excess amount of cupric ion remaining in the reaction mixture.

In some methods the sugar solution is titrated directly into the boiling copper reagent and the end point—when all of the cupric ion has reacted—is determined by the disappearance of the blue color, by spot tests, or with methylene blue as an internal indicator.

Other methods depend on the oxidation of the sugar molecules with ferricyanide in the presence of a rather high pH. Here again the extent of the oxidation depends on the nature of the reagent, the alkalinity, the temperature, and the length of time used. But again, if the conditions are standardized, the reduction of the ferricyanide is proportional to the amount of carbohydrate present. Here, too, reducing substances other than the reducing sugars interfere with the determinations and give high results. During the reaction, ferricyanide is reduced to ferrocyanide. The amount of ferricyanide remaining can be measured by reducing it with an iodide and measuring the amount of iodine released with standard thiocyanate. In some procedures the sugar solution is directly titrated with the ferricyanide by use of an oxidation-reduction indicator such as methylene blue. Here, again, conditions must be carefully standardized in order to calculate the concentration of the reducing sugar from the amount of ferricyanide reduced.

A number of color reactions of monosaccharides have been used for the quantitative estimation of these sugars. A sensitive micromethod which has been widely used is the color reaction with anthrone. The carbohydrate is treated with sulfuric acid to form a furfural derivative which gives a color with anthrone. The depth of the color is proportional to the amount of carbohydrate present under standard conditions. Either a colorimeter or a spectrophotometer is used to measure the color.

Carbazole and some of the phenols such as orcinol (3,5-dihydroxy-toluene) which are used in the qualitative detection of carbohydrate have been used in the development of quantitative methods.

Iodine in dilute alkaline solutions will oxidize aldoses without markedly affecting either ketoses or nonreducing sugars and this reaction can be used in determining quantities. Many other compounds will reduce iodine and the method must be used with care and under strictly standardized conditions. But when compounds are separated in chromatograms, the method can be used for the quantitative determination of aldoses. Iodine in alkaline solution forms hypoiodate which reacts with the aldose. Excess iodine is then back titrated by acidifying and titrating with standard thiosulfate.

CHANGES OF CARBOHYDRATES ON COOKING

Carbohydrates in food show some general changes on cooking or processing of the food. In general, these may be summarized as:

(1) Solubility

In some cookery processes soluble carbohydrates are dissolved. This physical change is usually of importance in mixtures where sucrose is one of the components.

(2) Hydrolysis

Some polysaccharides are hydrolyzed during cooking or processing. This action is particularly noticeable in foods which contain enough acid so that they are sour to taste and consequently are called acid foods. Pectic substances undergo this hydrolysis with the result that the fruit or vegetable becomes mushy and the juice thickens as the pectic and pectinic acids disperse in it.

Hydrolysis of starches also occurs to a limited extent during cooking. Occasionally the extent of hydrolysis may be so great that the thickening power of the starch is decreased. Thus when lemon or cherry pie filling is thickened, the mixture must not be cooked too long after the starch is added or the viscosity decreases again. Dextrinization of starch occurs on the crust of bread during baking but this is not simple hydrolysis. The starch molecule is degraded and the reactions which occur are much more complicated than hydrolysis.

(3) Gelatinization of Starch

When starch is mixed with cold water, no apparent change occurs; but when the water is heated, the viscosity of the mixture increases and if the concentration of the starch is sufficiently great, a gel is formed. Often if the starch suspension does not gel at the elevated temperature, it will when the mixture is cooled. This process is called the gelatinization of starch. It depends on a number of factors. The temperature at which gelatinization starts and the exact changes during the course of gelatinization are characteristic of, first, the variety of starch. Thus wheat and cornstarch show different behavior patterns, and even floury cornstarch from the crown region of most common corn acts differently from horny cornstarch from the region of the endosperm. (See p. 76). Second, the pH at which gelatinization is measured is most important. This fact was not recognized during the early work on starches and consequently the determinations of gelatinization made at that time are sometimes worthless. Third, the temperature at which observations are made and the length of heating are important. And fourth, gelatinization is influenced by the size of the granule. The temperature of gelatinization decreases as granules decrease in size.

The microscopic appearance of the starch granules changes markedly on heating, in three stages. The first stage, in cold water, is marked by the imbibition of approximately 25 to 30 per cent of water. This is apparently a reversible effect because the starch may be dried again with no observable change in structure. The viscosity of the starch-water mixture does not change during this phase. The second stage occurs at approximately 65° C for most starches, when the granules begin to swell rapidly and take up a large amount of water. Thus Meyer and Bernfeld[19] report that cornstarch takes up 300 per cent water at 60° C, 1000 per cent at 70° C, and at the point of maximum swelling 2500 per cent, based on the original weight of the starch. The granules change in appearance during this second phase, and some of the more soluble starch molecules are leached out of the granule. The liquid surrounding the granules will show a color with iodine even though the granules are still intact. This stage is not reversible. The third stage is marked by more swelling. The granule becomes enormous, often a void is formed, much more starch is leached out, and finally the granule ruptures, spilling more starch out into the surrounding fluid. The viscosity of the fluid increases markedly, and the starch granules stick together so that they can no longer be picked apart.

The swelling of starch, particularly amylose, which results in an increase in viscosity of a starch-water mixture and the formation, under proper conditions, of a gel is now believed to occur through the binding of water. In a starch granule, amylose and amylopectin molecules are loosely bound together by hydrogen bonds of the hydroxyls. A hydrogen on a hydroxyl of one molecule is attracted by the negative charge of the oxygen of a hydroxyl on another molecule, and this attraction forms a weak link between the molecules, as shown in Diagram A.

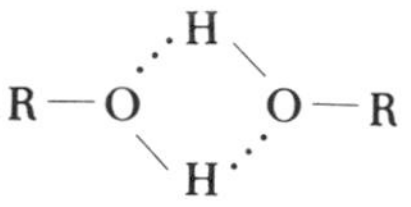

Diagram A

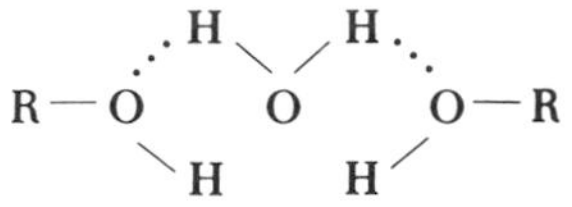

Diagram B

These aggregates of molecules, held together weakly, form micelles. As the temperature of a water-starch mixture rises, hydrogen bonding decreases for both the starch-starch bonds and water-water bonds and the size of the particles diminishes. The tiny water molecules begin to freely penetrate between starch molecules. Conversely, as the temperature decreases, water molecules are bound between the starch molecules (Diagram B) and there is an increase in size or swelling. Where two starch molecules were originally bound together, there are now the two starch molecules with water molecules in between.

The sticking together of granules is believed to be the result of molecules from adjacent granules becoming attracted and enmeshed in one another.

Gel formation occurs through the formation of a three-dimensional network of starch molecules, particularly the long straight chain amylose molecules. (See p. 126.) These molecules become interlaced through attractive forces between the molecules and particularly through hydrogen bonding on water molecules. Highly branched glycogen does not form gels or crystals, and gel formation in starch is consequently believed to be primarily the function of amylose rather than amylopectin.

Starch gels are readily cut by shearing forces and reduced to liquid. They are consequently called *thixotropic* gels. This phenomenon is often important in food preparation since stirring disrupts the gel. On standing the gel forms again.

On aging, most starch gels show marked syneresis or weeping in which the water gradually passes out of the interstices of the gel.

Another process also occurs both in gels and in viscous but still liquid dispersions. Part of the starch aggregates and forms microcrystals that precipitate. This is called the *retrogradation of starch*. Thus most starch dispersions which have stood for several days or weeks contain a deposit at the bottom of the vessel. The process also occurs in gels and is believed to be one of the important factors in the staling of bread. Figure 3.8 (see page 94) attempts to show the change in organization of molecules which occurs as starch precipitates.

CRUDE FIBER

Mangold[16] defines crude fiber as the sum of all those organic components of the plant cell membrane and supporting structures which in chemical analysis of plant foodstuffs remain after removal of crude protein, crude fat, and nitrogen-free extractives. Thus the crude fiber should be composed of the cellulose, hemicellulose and some of the materials that encrust the cell walls such as lignins and pectic substances. Actually the quantity of crude fiber found is a function of the method used in its determination. Crude fat is removed by ether extraction while crude protein is dissolved by hydrolyzing with dilute acid. During this hydrolysis, other compounds will likewise be hydrolyzed. However, when a standard method is used, reproducible results can be obtained and a given food will fall within a range of values. Crude fiber values for a given food show the same variation with climate, soil conditions, and degree of maturity as do other values.

The method which has been adopted by the Association of Official Agricultural Chemists employs boiling the sample in 1.25 per cent sulfuric acid, followed by boiling in 1.25 per cent sodium hydroxide. The residue is composed of the crude fiber and some salts. The crude fiber is determined by drying to constant weight, weighing, and then ashing, and weighing again. The difference between the two weights is the crude fiber.

Interest in crude fiber determinations centers either in the detection of adulteration or from the nutritional standpoint in the bulkiness of the diet. Some foods are occasionally adulterated with inert material by unscrupulous sellers. For example, spices may be adulterated and consequently extended with finely ground sawdust or with the waste from the spices. Crude fiber determinations will detect this increase. Mustard has a range of crude fiber by the official method of 4.2 to 6.5 per cent, while cloves ranges from 7 to 9 per cent. If the crude fiber determination is much out of line, the product has probably been adulterated.

The compounds in crude fiber make up most of the bulk or residue of the diet, and are not hydrolyzed by the digestive fluids of human beings. However, bacteria present in the paunch of ruminants and even those present in the colon of human beings may bring about considerable hydrolysis. The fibers that escape hydrolysis are excreted in the feces. They serve a very real function here, however, because most of them are capable of absorbing water, and they render the feces soft enough to pass out of the body readily and bulky enough to induce defecation.

BROWNING REACTIONS

The browning reactions are complex reactions which occur when many foods are processed. In some the brown flavor is highly desirable and is intimately associated in our minds with a delicious, high-grade product. In coffee, maple syrup, the brown crust of bread and all baked goods, potato chips, roasted nuts, and many other processed foods controlled browning is necessary. Yet in other foods, browning during processing is undesirable and forms off-flavors and dulled or even objectionable colors. In drying fruits or vegetables and in canning or concentrating orange juice, it is highly desirable to avoid browning. The presence of carbohydrates in foods is intimately connected with the browning which occurs. Other compounds are sometimes important, but they are ones which have some of the reactive groups of the reducing sugars and which are similar to them in their chemical properties. The pigments which are formed are high molecular weight polymers whose constitution is difficult to determine. The browning reactions appear to be complicated not only as to the final product but also as to the course of the numerous reactions. It has been exceedingly difficult to assess the chemistry of this change in the complex mixtures encountered in almost every food. During the past fifty years, and particularly during the last twenty, study of the browning reaction has been carried forward by the use of model systems. In this type of study one, two, or sometimes three compounds are allowed to react and the intermediates, products, and course of the reaction followed. Even this method of excluding and simplifying has not yielded all of the answers by any means, since the possible reactions are numerous. To a student not completely familiar with the field, the many investigations appear at first to be wholly unrelated and the state of the problem all confusion. However, in 1953 Hodges[12] attempted to correlate and integrate knowledge which had accumulated up to that time about browning reactions. An attempt will be made to briefly review his ideas.

In the past, three general types of browning reactions have been recognized to occur in foods during processing: (1) The reaction of aldehyde and ketones, among them the reducing sugars, with amino compounds such as amino acids, peptides, and proteins. This is independent of the presence of oxygen. (2) Caramelization, the change which occurs in polyhydroxycarbonyl compounds such as reducing sugars and sugar acids when they are heated to high temperatures and which is also independent of

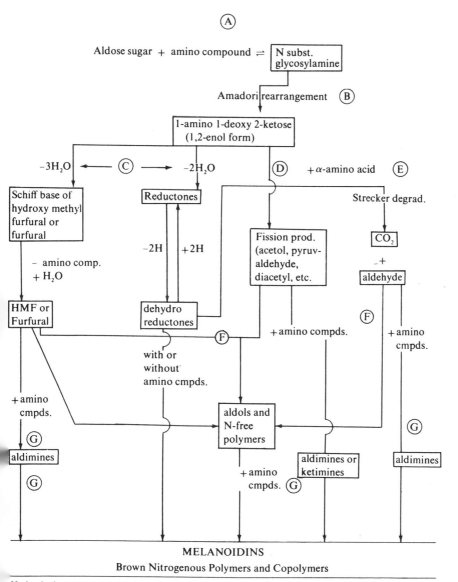

Hodges' scheme.

oxygen. (3) The oxidative change of polyphenols to di- or polycarbonyl compounds and possibly the oxidation of ascorbic acid. This may be partially or wholly enzymatic. No matter what the type of reactants the brown pigments, called melanins or melanoidins, are unsaturated polymers. You will notice that in each of the broad types of reaction except the third, a carbonyl or polycarbonyl compound is important, and in the third the first step in the browning is the formation of carbonyl compounds. Since the reactants of the third type furnish carbonyl compounds, it is correct to consider these compounds indispensible to any type of browning reaction. When a food is extracted to remove carbonyl compounds, browning is retarded or eliminated. It is therefore believed that although the course of these three types of reactions are incompletely understood, they are the reactions of importance in the browning of foods.

Hodges has given a scheme for browning reactions and has included all three general types in one scheme. See the diagram on page 109. Browning may occur by compounds entering the scheme at any point. Although the first reaction, is a reaction of an aldose sugar with an amino compound, there is some evidence that ketoses can cause browning and that they may react in a similar fashion. This first reaction of an aldose with an amine is shown in the following Amadori reaction:

$$
\begin{array}{ccc}
\text{HCO} & & \overset{\text{RNH}}{\underset{|}{\text{HC}}}\!\!\!-\!\!\!\rule{1em}{0.4pt} \\
\underset{|}{(\text{CHOH})_n} + \text{RNH}_2 \xrightarrow{\quad\textcircled{A}\quad} & & \underset{|}{(\text{CHOH})_n} \xrightarrow{\ +\text{H}^+\ } \\
\underset{|}{\text{CHOH}} & & \underset{|}{\text{HC}\!-\!\text{O}}\!\!-\!\!\rule{1em}{0.4pt} \\
\text{CH}_2\text{OH} & & \text{CH}_2\text{OH}
\end{array}
$$

Aldose   Amine

$$
\left[\begin{array}{c}
\overset{\text{RNH}}{\underset{\parallel}{}} \\
\text{HC} \\
| \\
(\text{CHOH})_{n+1} \\
| \\
\text{CH}_2\text{OH}
\end{array}\right]^{+}
\xrightarrow{\ -\text{H}^+\ }
\begin{array}{c}
\text{RNH} \\
| \\
\text{HC} \\
\parallel \\
\text{COH} \\
| \\
(\text{CHOH})_n \\
| \\
\text{CH}_2\text{OH}
\end{array}
\underset{\textcircled{B}}{\rightleftarrows}
\begin{array}{c}
\text{RNH} \\
| \\
\text{CH}_2 \\
| \\
\text{CO} \\
| \\
(\text{CHOH})_n \\
| \\
\text{CH}_2\text{OH}
\end{array}
$$

    Cation   1-2 Enol   Carbonyl
                                      Compound

The product of the Amadori rearrangement can then undergo a number of fates depending on the conditions of the reaction. It can, in neutral or acid media, lose water and form a ring compound of the Schiff's base of hydroxy methyl furfural, or furfural, and then eliminate the amine to form the free hydroxy methyl furfural, or furfural. Ⓒ In the dry state it can form reductones. These are compounds with a high reduction potential that have not been identified as yet. Or finally, the product can undergo fission to form small molecules such as acetol, CH_3COCH_2OH; pyruvaldehyde, CH_3COCHO; diacetyl, $CH_3COCOCH_3$; and others. Ⓓ In the

CH—CH
HOCH₂C O CCHO

Hydroxy Methyl Furfural

CH—CH
CH O CCHO

Furfural

scheme shown (page 109), Hodges has indicated that all of these carbonyl compounds react with amines to form aldimines or ketimines, Ⓖ or polymerize to aldols and similar large molecules which subsequently react with amines. Ⓕ The final brown pigments are products that contain nitrogen. This scheme covers all of the reactions of the first broad general type of browning reactions—the reaction of aldehydes and ketones with amino compounds.

The second broad general type of browning reaction is caramelization. This type has not been as extensively studied. It is, of course, a reaction which does not require an amine. Sugars will show caramelization when heated to relatively high temperatures. This type of reaction is markedly affected by high pH, as is the first general type. While the browning of these carbohydrates is not as rapid as in the presence of amino compounds, it is accelerated by the presence of carboxylic acids, the salts of these acids, phosphates, and metallic ions. These accelerators are commonly present in foods. The nitrogen-free intermediates formed in carbonyl-amino browning reactions are also produced in nonamino browning. The formation of 1,2 enolization, furfural and hydroxymethyl furfural by dehydration, and sugar fission products has been demonstrated in a few model systems. It has also been shown that these intermediates will form colored polymers in the absence of amino compounds. So in general this type of browning reaction fits into Hodge's scheme, joining it after the Amadori rearrangement and forming different (nitrogen-free) polymers.

The third type of browning reaction is also believed by Hodges to have, at the least, a possibility of fitting into the scheme. This is the browning which occurs in the presence of oxygen when fruits or vegetable tissues are cut or bruised. It is completely discussed in Chapter 7 on the fruits and vegetables. Enzymes bring about the transformation of tyrosine, catechols, or other polyphenols to quinones. Since catechols are ortho di-phenols, they are therefore conjugated enediols and are reductones. Hodges considers that they enter the scheme at this point. Ⓒ Not a great amount is known of the reactions of quinones as they brown. Browning may be enzy-

Catechol → Quinone (−2H, oxidation)

matic, but some steps may be nonenzymatic and may well pass through reactions common to the other intermediates in types (1) and (2).

Browning is a problem of general interest in food technology. The large number of studies now being conducted both on model systems and on food products makes it possible that these reactions will soon be more completely understood.

We see then that both carbohydrates and amino compounds are important in nonenzymatic browning. Habib and Brown[8] have tested the validity of this hypothesis in potato chips which are notable for their variation in browning. Browning during frying is desirable in potato chips not only for the effect on appearance but also for the flavor which the browning reaction imparts. Habib and Brown[8] separated the amino acids and sugars present in four varieties of potatoes by paper chromatography. The potatoes were studied at three different times and their browning reactions compared to the content of carbohydrates and amines: 20 days after harvest, after 4 weeks of cold storage at 40° F, and after a further reconditioning for 4 weeks at 75° F. The reconditioned potatoes had the lowest level of amino acids, with the drop in quantity particularly in the basic amino acids. These potatoes also produced the lightest-colored chips, supporting the hypothesis that browning depends on reactions of carbohydrates and amino acids.

REFERENCES

1. BERNFELD, P., "Enzymes of Starch Degradation and Synthesis," *Advances in Enzymol.*, **12**, 379–428, (1951).
2. BETTELHEIM-JEVONS, F. R., "Protein-Carbohydrate Complexes," *Advances in Protein Chem.,"* **13**, 35–105 (1958).
3. BEVENNE, A. and WILLIAMS, K. T., "Hemicellulose Components of Rice," *J. Agr. Food Chem.*, **4**, 1014–1018 (1956).
4. BOESEKEN, J., "The Use of Boric Acid for the Determination of Configuration of Carbohydrates,"*Advances in Carbohydrate Chem.*, **4**, 189 (1949).
5. BRAVERMAN, J. B. S. and LIPSHITZ, A., "Pectin Hydrolysis in Certain Fruit During Alcoholic Fermentation," *Food Technol.*, **11**, 356–358 (1957).
6. DEVEL, H., and STUTZ, E., "Pectic Substances and Pectic Enzymes," *Advances in Enzymol.*, **20**, 341–382 (1958).

7. FRENCH, D., "The Raffinose Family of Oligosaccharides," *Advances in Carbohydrate Chem.*, **9**, 149–184 (1954).

8. HABIB, A. T. and BROWN, H. D., "Role of Reducing Sugars and Amino Acids in the Browning of Potato Chips," *Food Technol.*, **11**, 85–89 (1957).

9. HAYNES, W., "Cellulose: The Chemical That Grows," Doubleday and Co., Garden City, New York, N. Y., 1953.

10. HEUSER, E., "The Chemistry of Cellulose," John Wiley & Sons, New York, N. Y., 1944.

11. HIRST, E. L., "Some Problems in The Chemistry of the Hemicelluloses," *J. Chem. Soc.*, **1955**, 2974–2984.

12. HODGES, J., "The Chemistry of Browning Reactions in Model Systems," *J. Agr. Food Chem.*, **1**, 928–943 (1953).

13. KERR, RALPH, ed., "Chemistry and Industry of Starch," 2nd ed., Academic Press, Inc., New York, N. Y., 1950.

14. KERTESZ, Z. I., "The Pectic Substances," Interscience Publishers, Inc., New York, N. Y., 1951.

15. LALIKAINEN, T., JOSLYN, M. A. and CHICHESTER, C. O., "Browning Reaction in Ascorbic -citric Acid Glycine Mixture," *J. Agr. Food Chem.*, **6**, 135–139 (1958).

16. MANGOLD, E., "Digestion and Utilization of Crude Fibre," *Nutr. Abstr. & Rev.*, **3**, 647 (1934).

17. MEYER, K. and RAPPORT, M. M., "The Mucopolysaccharides of the Ground Substance of Connective Tissue," *Science*, **113**, 596–599 (1951).

18. MEYER, K. H., "Present Status of Starch Chemistry," *Advances in Enzymol.*, **12**, 341–378 (1951).

19. MEYER, K. H. and BERNFELD, P., "Reserches Sur l'amidon VII Sur la Structure Fine Du Grain D'amidon et Sur les Phénomènes du Conflement," *Helv. chim. Acta*, **23**, 890–897 (1940).

20. MORDIN, P., and KIM, Y. S., "Browning Reaction and Amadori Rearrangement," *J. Agr. Food Chem.*, **6**, 765–767 (1958).

21. MORI, J., "Seaweed Polysaccharides," *Advances in Carbohydrate Chem.*, **8**, 315–350 (1954).

22. PIGMAN, W. W., ed., "The Carbohydrates: Chemistry, Biochemistry, Physiology," Academic Press, Inc., New York, N. Y., 1957.

23. PIGMAN, W. W. and GOEPP, R. M., JR., "The Chemistry of the Carbohydrates," Academic Press, Inc., New York, N. Y., 1948.

24. SCHULZ, G. V. and MARX, M., "Molecular Weights and Molecular Weight Distribution of Native Celluloses," *Makromol. Chem.*, **14**, 52–95 (1954); *C. A.*, **49**, 9269d (1955).

25. SMITH, F. and MONTGOMERY, R., "Chemistry of Plant Gums and Mucilages," ACS Monograph No. 141, Reinhold Publishing Corp., New York, N. Y., 1959.

26. VANHOOK, A., "Sugar," Ronald Press Co., New York, N. Y., 1949.

27. WHISTLER, R. and BEMILLER, J., "Industrial Gums, Polysaccharides and Their Derivatives," Academic Press, Inc., New York, N. Y., 1959.

28. WHISTLER, R. L. and SMART, C. L., "Polysaccharide Chemistry," Academic Press, Inc., New York, N. Y., 1953.

29. WISE, L. E. and JAHN, E. C., "Wood Chemistry," 2nd ed., Chap. 10, "The Hemicelluloses," Reinhold Publishing Corp., New York, N. Y., 1952.

CHAPTER FOUR

Proteins in Foods

WHILE PLANTS ARE ABLE TO UTILIZE inorganic sources of nitrogen such as ammonia, nitrates, and nitrites, man and other higher animals are for the most part dependent on a source of amino acids to build their body proteins. Although some bacteria can utilize atmospheric nitrogen, these organisms are low in the scale of life and far removed from man. Higher animals are directly or indirectly dependent on plant protein. Often this plant protein is consumed by one animal, digested and synthesized into its proteins, and reaches man after passage through one or more animals. Thus, large fish rely on plant foods by way of a long line of smaller and smaller marine animals down to the tiny ones actually eating the plant food.

Every living cell contains protein. Indeed, the name "protein" comes from the Greek root which means "to be first." In man synthesis of the many proteins that make up body tissue requires the presence of amino acids. Approximately 20 different amino acids are in the body proteins, although certain proteins such as thyroglobulin contain one or more unusual amino acids as well. All 20 of the amino acids are not required in the diet since, as has been demonstrated, man is capable of synthesizing all of the amino acids except 8 (leucine, isoleucine, lysine, valine, threonine, tryptophan, phenylalanine, and methionine).

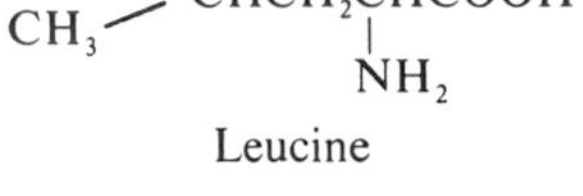

$$\begin{array}{c} CH_3 \\ CH_3 \end{array}\!\!\!> CHCH_2\underset{\underset{NH_2}{|}}{C}HCOOH$$

Leucine

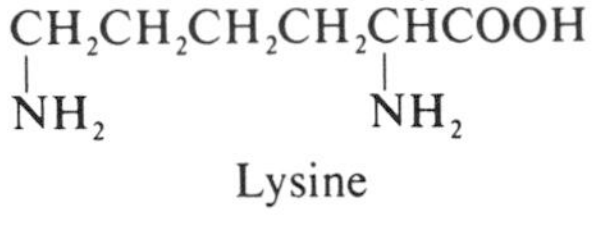

$$\underset{\underset{NH_2}{|}}{C}H_2CH_2CH_2CH_2\underset{\underset{NH_2}{|}}{C}HCOOH$$

Lysine

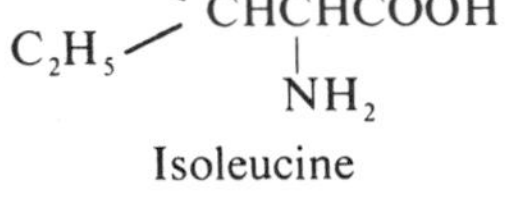

$$\begin{array}{c} CH_3 \\ C_2H_5 \end{array}\!\!\!> CH\underset{\underset{NH_2}{|}}{C}HCOOH$$

Isoleucine

$$\begin{array}{c} CH_3 \\ CH_3 \end{array}\!\!\!> CH\underset{\underset{NH_2}{|}}{C}HCOOH$$

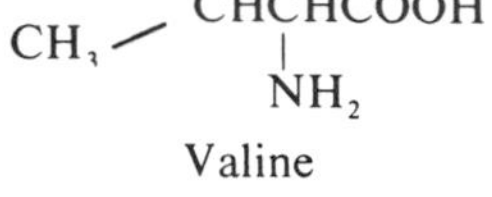

Valine

$$CH_3CHCHCOOH$$
$$OHNH_2$$

Threonine

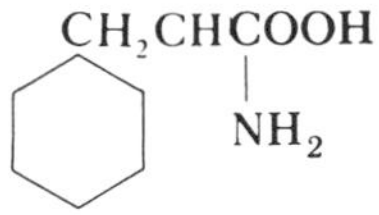

$$CH_2CHCOOH$$
$$NH_2$$

Phenylalanine

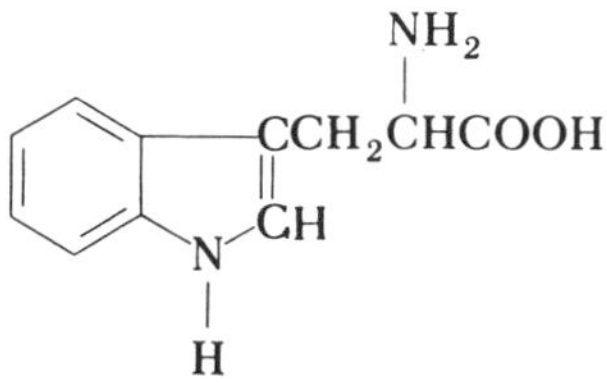

$$NH_2$$
$$CCH_2CHCOOH$$
$$N \quad CH$$
$$H$$

Tryptophan

$$CH_2SCH_3$$
$$CH_2$$
$$CHNH_2$$
$$COOH$$

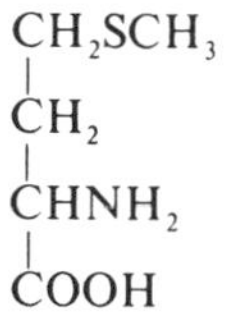

Methionine

However, some source of nitrogen in the amino form must be available. The amino group from other amino acids enters the nitrogen pool and can be used for the synthesis of any amino acid whose carbon skeleton can be made. Man's food must supply sufficient amino acids for the purpose of building proteins. It must allow for occasional synthesis of nonessential amino acids, as well as supplying adequate amounts of the essential amino acids that he cannot synthesize.

PROTEINS IN MAN'S DIET

Although man's need for protein and its amino acids is rather stringent, his method of fulfilling his need varies considerably in different parts of the world. Primarily the difference has depended on the availability of foods. In the tropics many peoples have developed dietary patterns based primarily on plant foods with the cereals most abundant. In most of these diets animal protein in the form of meat, fish, eggs, and milk is added when available. Although it is difficult to build an optimum diet on plant products alone, it is possible if the diet is varied.

Many of the proteins present in plant tissues are deficient in one or more of the essential amino acids. For example, zein, one of the proteins of corn, is lacking in lysine and tryptophan; while gliadin, one of the proteins of wheat, is low in lysine. However, both wheat and corn contain other proteins that possess these amino acids. A diet restricted entirely to wheat or to corn is low in lysine and, under the stringent demands for rapid synthesis, is inadequate. If the wheat or corn is supplemented with proteins that are relatively rich in lysine, then the amino acids supplied are adequate.

Although the food mixtures eaten by primitive man and handed down as diet patterns were probably adequate nutritionally, modern man does not appear to have an instinct for fulfilling his nutritive needs. Where he is restricted to a vegetarian diet, malnutrition is often very common. Kwashiorkor, the deficiency disease of many children in Africa and the Latin American countries, is a protein deficiency complicated with other deficiencies. In the part of Africa where this disease is common, yams are the staple food of the people, while in Latin America it is corn.

In the Arctic regions the diet of the Eskimos is principally animal. Not only are fish and meat the principal food, but almost the complete animal is eaten, often uncooked. This diet is unusually high in protein and the assortment of amino acids in these proteins meets the requirements for the synthesis of protein in man.

In the temperate zones most diet patterns include both animal and vegetable products. The common pattern when the economic level allows is adequate in the amount of protein as well as in the distribution of the amino acids. Inadequacy is almost always the result of economic factors although it is occasionally caused by faulty eating habits.

CHEMICAL AND PHYSICAL PROPERTIES

Proteins are high molecular weight compounds that yield amino acids as their principal hydrolysis product. The structure of proteins and the classical classification of proteins on the basis of solubility and structure is adequately described in textbooks in organic or biochemistry and can be reviewed there. A few of the properties of proteins which are not always studied in organic chemistry and a brief discussion of the methods of determining molecular weight and testing homogeneity will be presented.

Proteins are large molecules with some reactive groups, such as carboxyl and amino, on their surfaces. Because of the size of the molecules, they do not form true solutions in water. When solubility of a protein is mentioned, dispersibility is intended. A dispersion of a protein in water consequently has the properties of a colloidal dispersion rather than a true solution. The reactive groups and the size cause a marked sensitivity not only to the pH of the solution but also to the presence of many different electrolytes. Protein molecules not only associate with small molecules but also with one another, sometimes to form tightly bound and sometimes loosely bound products. The size of the molecule is such that adsorption can occur readily and because of this, purification of proteins is extremely difficult. Protein molecules are quite sensitive to many reagents and conditions and readily undergo small transformations in structure. The difficulty of isolating a protein from its natural source, which occurs in a cell or biological fluid, without any alteration in the molecule is great.

The field of protein chemistry is consequently a difficult one. Chemists are learning how to handle better, determine, and evaluate large macro molecules. Protein chemistry is profiting from work with other large molecules and is also contributing to it. Despite the difficulty of working with proteins, many investigators are actively engaged in the field and the number of publications which appear each year, is very large. Organic chemists are studying structure and reactions, physical chemists are attempting to make measurements and define the molecule, biochemists are describing the role of proteins in all aspects of life.

The shape of protein molecules varies greatly. Some exist as fibers, others as spheres while intermediate are many shaped like spindles, cigars, etc. The manner in which a polypeptide chain can achieve a spherical shape or an intermediate one, is the object of considerable speculation at present. Obviously the chain must be folded or coiled in some fashion and there must be some bonds which hold the chain in a relatively permanent shape. A considerable amount of study is at present directed to the solving of this problem.

Molecular Weights and Homogeneity

Molecular weights of proteins are so large that the classical methods used for determining the molecular weights of compounds that will form true solutions or vaporize are useless. New methods have been devised and are yielding significant information. The most widely used are those of electrophoresis and sedimentation, with some work on osmotic pressure and a little on light scattering. These methods also give some information on the homogeneity of a protein preparation.

The production of a crystalline protein is no more proof of the purity of the preparation than is the preparation of any crystalline compound. With low molecular weight organic compounds, determining the melting point of the compound is usually possible, and if the melting point is sharp or is not depressed by a mixed melting point, we are reasonably satisfied of its purity. Proteins do not melt. If they are heated, they undergo decomposition. Some other methods must therefore be applied in order to demonstrate the purity of a preparation.

Electrophoresis. It will be remembered that because of the presence of available free amino and carboxyl groups, proteins are amphoteric. In the presence of alkali, they act as acids and achieve a negative charge through the ionization and neutralization of free carboxyl groups. But in the presence of acids, the amino groups react with hydrogen ions and assume positive charges. On the pH scale most proteins are negatively charged at high pH and positively charged at low pH. It will also be remembered that at some intermediate pH, called the *isoelectric point*, the total charge on the protein molecule is close to zero, with the available amino and carboxyl groups neutralizing one another and forming zwitterions. In an electric field the protein molecules in a solution at high pH will move to the positive pole (anode) since they are negatively charged, but at a low pH they will migrate to the negative pole (cathode).

$$\text{Low pH} \qquad \text{Prot} \begin{cases} NH_3^+ \\ COOH \end{cases} + H^+Cl \rightarrow \text{Prot} \begin{cases} NH_3^+Cl^- \\ COOH \end{cases}$$

$$\text{High pH} \qquad \text{Prot} \begin{cases} NH_3^+ \\ COOH \end{cases} + Na^+OH^- \rightarrow \text{Prot} \begin{cases} NH_2 \\ COO^-Na^+ \end{cases} + H_2O$$

$$\text{Isoelectric point} \qquad \text{Prot} \begin{cases} NH_3^+ \\ COO^- \end{cases} \qquad \text{"zwitterion"}$$

They will not move if the solution is buffered to the isoelectric point of the proteins. Migration in an electric field at a definite pH is called electrophoresis and is a valuable tool in determining molecular weights of proteins as well as in separating mixtures or proving the uniformity of a crystallized or purified protein. The rate of migration in a constant electric field depends on the charge on the molecule, the size of the molecule and, to a certain extent, on the shape of the molecule.

Tiselius has devised an apparatus in which measurements can be made at constant temperature and with a minimum of boundary diffusion. A buffered protein solution is covered by a layer of buffer solution; the whole cell is immersed in a constant temperature bath, and the electrodes inserted. Protein molecules that are alike will move at the same rate and will form a sharp boundary in the cell. A colored protein such as hemoglobin can be seen moving through the cell. However, most proteins are not colored and other methods must be used to follow their path. Usually the refractive index is used and in a Tiselius apparatus provision is made to record the changes in refractive index photographically. In a mixture of proteins such as that present in plasma, a series of peaks will be present on the record indicating the presence of a series of proteins moving at different rates. When the method is used to demonstrate that only one molecular species is present in a fraction, it is necessary to determine th

Proteins in Foods

FIGURE 4.1. TISELIUS ELECTRO-
PHORESIS APPARATUS. The elec-
trode vessels are on each side with
the migration chamber in the middle, constructed so that fractions
can be cut out and migration
visualized. *Courtesy of The Upjohn
Company.*

FIGURE 4.2. OPTICAL SYSTEM
FOR THE TISELIUS ELECTROPHORE-
SIS APPARATUS. A light source
passes through the sample and the
diffraction of light gives a measure of the protein passing. A cam-
era is mounted at the left hand end
of the tube and a power source is
below it. *Courtesy of The Upjohn
Company.*

FIGURE 4.3. AN ELECTRO-
PHORETIC PATTERN OF NORMAL
SERUM SHOWING THE PROTEIN
FRACTIONS PRESENT. The ordinate
is time and the abscissa from the
light diffraction gives the amount
of protein. This is an oversimplified description. *Courtesy of The
Upjohn Company.*

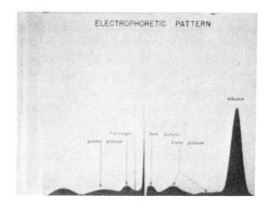

electrophoretic mobility at several pH's. Occasionally two proteins will migrate at the same rate at a given pH when the *balance* of forces caused by the charge on one species of molecule, the weight of this molecule, and its shape are exactly matched by that of the other species. This balance would be impossible at other pH's and consequently, if the sample still shows a sharp boundary at a second pH, it is considered pure.

Paper electrophoresis is widely used to demonstrate the presence of numerous proteins in small samples of biological materials and also to give evidence of the homogeneity or heterogeneity of a given protein. A strip of paper is wet with buffer solution and arranged so that one end is in a vessel of the buffer solution with the cathode and the other end is in a vessel of the buffer solution with the anode. The protein is placed near one end of the strip. Under the influence of electric current, the protein moves on the paper; and if it is homogeneous, a thin zone will be maintained as each molecule moves at the same rate. However, a mixture of proteins separates into zones. When the paper is suitably developed with reagents that give a color with proteins, the number of zones shows the number of proteins. Sometimes a sample must be run at more than one pH on separate pieces of paper to demonstrate all of the different proteins present.

Sedimentation. The ultracentrifuge has proved to be an important instrument in determining the molecular weight of large molecules such as proteins. It is an apparatus invented in 1925 by the Swedish chemist, The Svedberg, for applying a very strong centrifugal force to a dispersion. In recent modifications of Svedberg's ultracentrifuge, speeds have been attained in which the rotor revolves at the rate of 60,000 times per minute with centrifugal forces developed equal to 500,000 times gravity (500,000 g). The rotor is operated in a high vacuum to minimize air resistance and cut down the amount of heat developed by air friction.

The rate at which a particle will travel in the field depends on the shape, size, and density of the particle; on the density and viscosity of the dispersion medium; and on the centrifugal force. If the particles are identical, they will fall through the solution at the same rate and produce a sharp boundary between the colloidal dispersion and the dispersion medium. A mixture of molecules will form several boundaries depending on the number of molecular species present. The method is consequently useful in proving the homogeneity or heterogeneity of a protein fraction.

A single boundary is evidence of the presence of molecules all of the same size. However, it is not certain that the protein is pure if it gives a single boundary unless other methods support this. Occasionally proteins are found to give a single boundary in the ultracentrifuge but to act as mixtures under other conditions. The boundary is followed in the same manner that is used for electrophoresis, an optical system which records the change in refractive index photographically. The apparatus which has been devised to record this change is by no means simple.

Molecular weights can be calculated from measurements of the variable in sedimentation. While estimates of the molecular weights by this means compare favorably with those of other methods, they are only approximate. When good agreement by several methods is obtained, confidence in the results is enhanced.

Osmotic Pressure. Colloidal dispersions show small osmotic pressures; this fact has been used to determine the molecular weights of a number of proteins. If a protein dispersion is separated from water by a semipermeable membrane making osmosis possible, the water tends to migrate in greater amounts into the protein dispersion. However, osmotic pressure is one of the properties of solutions that depends on the number of molecules present. Thus since the number of protein particles is low, the osmotic pressure is low. When the weight of the protein dispersed is known, it is then possible to calculate the molecular weight.

Osmotic pressure measurements of protein dispersions are technically rather difficult. They are usually made at the isoelectric point of the protein. The dispersions often attain equilibrium very slowly so that days must be allowed for a single experiment. In this length of time it is difficult to prevent denaturation of the protein and the growth of microorganisms. Measurements by this method compare very favorably with other molecular weight determinations and are of about the same accuracy.

Light Scattering. When a beam of light is passed through a colloidal dispersion, part of the light is transmitted and part is scattered. This is the well-known Tyndall effect which is seen with every airport beacon on foggy nights. The amount of scattering depends on the number of colloidal particles and also on their size. Hence a measurement of the amount of light scattering gives a method for calculating the molecular weights of protein particles. It has been used effectively in many protein fractions and gives good agreement with other methods of determining molecular weights of large molecules. Its accuracy is of approximately the same order as other methods.

Analysis of Unusual Component. Occasionally a protein contains an unusual component which can readily be detected and estimated, and hence can serve to give a minimal molecular weight. For example, the iron of hemoglobin has been used for such a purpose. When other methods showed values for the molecular weight approximately four times that based on iron determinations, it was apparent that each hemoglobin molecule contains four iron atoms. Since iron can be determined with great accuracy, the molecular weights approximated by other methods could hence be refined. However there are few proteins that contain an unusual element such as iron.

Chemical Properties of Proteins

Amphoterism of Proteins. The ability of proteins to react either as acids or as bases has been mentioned. The free carboxyl groups of proteins are very weakly ionized but, nevertheless, are available for reaction with bases. The free amino groups, on the other hand, are hydrogen acceptors and are available for reaction with acids. The number of these groups available for reaction on each protein molecule is small since it is only the side chain carboxyls of glutamic acid and aspartic acid and the side chain amino group of lysine which can react. The carboxyl and amino groups of an amino acid such as glycine are tied up in the peptide bonds. The imino groups of histidine, tryptophan, proline, and hydroxyproline are not as basic as amino groups, but can accept hydrogen ions. The ability of the protein to act as base or acid will depend not only on the number of these groups, but probably on their placement, whether they are close to the surface of the molecule, or covered.

Binding of Ions. Proteins can bind both cations and anions through reaction with either the carboxyl group or the free amino groups. At pH values above the isoelectric point a protein exists as negative ion and binds or reacts with cations. At pH's below the isoelectric point it will exist as a positive ion and will react with or bind anions. The precipitation of protein by heavy metal ions depends on the formation of these protein salts. A number of metallic ions have the ability to form coordination compounds (complex ions) and these will readily form through reaction with amino N, the substituted amine of the peptide link, and imidazole N. Copper, nickel, and iron are examples of metals which form these complexes.

In most foods mixtures of proteins occur, and at a given pH some proteins may be below their isoelectric point and others above theirs. Under these conditions cations will be bound by some of the proteins and anions by others.

Hydration of Proteins. Proteins can form hydrates with water and this type of reaction is often important in food chemistry. A protein molecule contains a number of groups in which a nitrogen or an oxygen atom contains a pair of unshared electrons and is therefore capable of forming a hydrogen bond. The nitrogen in the peptide link as well as the nitrogen in the free amino group are in this condition, and by their relative negativity can attract the hydrogen of a molecule of water.

$$
\begin{matrix}
\text{H} & & & \\
| & & & \delta^{+} \\
-\text{N}^{\delta^{-}} & \cdots & \cdots & \text{H} \\
| & & & \diagdown \\
\text{H} & & & \quad \text{O} \\
& & \text{H}^{\diagup} &
\end{matrix}
\qquad\qquad
\begin{matrix}
\text{H} & & & \\
| & & & \delta^{+} \\
-\text{N}^{\delta^{-}} & \cdots & \cdots & \text{H} \\
| & & & \diagdown \\
& & & \quad \text{O} \\
& & \text{H}^{\diagup} &
\end{matrix}
$$

The double bonded oxygen of the carboxyl (—COOH) group or of the carbonyl (CO) of the peptide link is more strongly negative and has a greater attraction for the hydrogen than the nitrogen. The ionized groups formed at pH levels below or above the isoelectric point have a greater affinity for water and consequently hydrate formation is greater at pH's other than the isoelectric point. The water molecule which has been bound can attract another molecule of water since it possesses an oxygen with an unshared pair of electrons. Aggregates of water can therefore build up around each polar group on the protein molecule. In protein dispersions in which other compounds that form hydrates are present—a condition frequently occurring in food chemistry—competition for water may occur between the molecules. Electrolytes, sugars, alcohols, and many other substances have this tendency to combine with water and form hydrates and may compete with protein for the water. The extent of hydration of a protein dispersion therefore depends not only on the concentration of the protein dispersed but also on the pH, the presence of other substances which combine with water, and the temperature.

Precipitation with Antibodies. Although it is now possible to determine the approximate molecular weights of proteins, there is no chemical method for differentiating most protein molecules. If a protein contains an unusual element such as the iron present in hemoglobin, it is possible to use that to differentiate the protein from those which do not possess iron. Many proteins are very similar in molecular weight and amino acid composition but are nevertheless different structurally. So far, the only method for precise identification of proteins is a biological reaction. If a foreign protein (antigen) is injected into the blood stream of an animal, a substance (antibody) which precipitates that protein is developed. For example, if a rabbit is given several injections of egg albumin, an antibody will develop in the blood of the rabbit and later, even a minute amount of egg albumin will result in the formation of a precipitate. These antigen-antibody reactions are highly specific. The blood of the rabbit which is highly sensitive to egg albumin will not give a precipitate with any other albumin. This must mean that egg albumin differs structurally from all other albumins even though their molecular weights and chemical and physical properties are similar. Thus, although no chemical methods have been developed as yet to demonstrate subtle differences between protein molecules, biological tests are able to establish them.

Native and Denatured Proteins

The proteins that occur in the tissues, whether within the cells or in the fluids, of living plants and animals are called *native* proteins. These large molecules are quite fragile and many reagents and conditions cause slight or extensive changes in the structure of the protein. The properties of the protein change and this, of course, indicates an alteration in the structure of the molecule. The resultant changed protein is called *denatured protein,* and it usually shows a very different solubility from the native protein since the change in the structure allows the molecules to aggregate and precipitate. Frequently the denaturation is irreversible, but occasionally under very mild conditions reversible denaturation occurs.

Denaturing agents include acids, alkalis, alkaloids, heavy metal salts, and other compounds such as urea and ethanol. Most of these are not important in food chemistry although the possibility of denaturation of protein must never be overlooked. Many ions such as iodide, bromide, and chloride are denaturation agents and the synthetic detergents are active at low concentration.

Although denaturation in foods is usually the result of increase in temperature, sometimes denaturation can be produced by mechanical means. Extensive whipping of an egg white foam will produce some denaturation and the foam will begin to break as the protein starts to precipitate. Ultraviolet light, high pressures, and ultrasonic vibrations are all denaturation agents. Spreading of protein films at interfaces usually results in denaturation.

In studies of protein structure where the native protein is the center of interest, it is imperative that denaturation be avoided. This is often exceedingly difficult during the tedious separation of protein from other substances in the tissue.

Reversible denaturation of protein occurs under very mild conditions or very short exposure and may occur in living systems. Reversibly denatured protein is able to return to the native state and show the original physical properties and biological activity. If the protein is rapidly removed from the reagent or condition that causes denaturation and placed in a solution at the isoelectric point where its charge is at a minimum, rapid precipitation occurs. But if it is placed in a solution with a pH different from the isoelectric point, it will develop a charge, molecules will repel one another, and it will rapidly return to the native state.

The nature of the transformation that occurs when a protein is denatured is now believed to be an unfolding of the molecule. It may well be that in reversible denaturation only a limited unfolding occurs while in irreversible denaturation it is more extensive. It scarcely seems probable that a long polypeptide chain which has become completely disentangled would readily be able to reassume the spatial relations of the native protein. Many studies have attempted to find differences other than solubility between native and denatured protein. The sulfhydryl (—SH) and disulfide (—SS—) groups are readily detected in proteins and have been extensively studied. Often the denatured protein shows a different content of sulfhydryl groups and some years ago it was thought that the process of denaturation was fundamentally one in which sulfhydryl groups are formed from disulfide. Today the results are more commonly interpreted to mean that more sulfhydryl groups are exposed as the molecule unfolds. There is also some evidence that more of the phenolic groups present in tyrosine, and more of the indole groups in tryptophan are available for detection in denatured protein. These observations are also interpreted as exposure on unfolding of the molecule.

In many respects native and denatured protein are very similar. But in some of the biological properties which are more discriminating of structure than the methods of chemistry now available, marked changes are observed. In general denatured proteins are more readily attacked by proteolytic enzymes. When the protein studied is an enzyme or an antibody, denaturation is accompanied by a loss of activity. This surely indicates a change in the molecule.

Since details of protein structure are still not completely known, it is not surprising that denaturation cannot be fully described. If native protein is a coiled or folded molecule in which certain parts of the polypeptide chain fit together with definite other parts, it is not unlikely that denaturation represents some disorganization of this pattern.[9]

Gel Formation

Gel formation is a very important process in food chemistry. Not only do the properties of living cells, both animal and vegetable, depend on the gel structure, but in food preparation the stiffening which occurs during meat and flour cookery, the rigidity of pectin and starch gels, the high viscosity of many plant juices, the changes that occur in egg cookery and many other processing operations are a function of the gel.

Although colloid chemists have studied gel formation for many years, and although the literature is full of papers describing gels, the number of compounds investigated is relatively small compared with the number about which we should like information. As in many other topics in food chemistry, we must thank the colloid chemist for the work and insight so far gained and then attempt to apply the limited results to a large number of situations, where not only has the gelling compound escaped study but the number of other compounds present produces a very complicated situation. Therefore, it is highly worthwhile to review briefly the present state of knowledge concerning gel formation and the applications to food chemistry now known.

A gel is a remarkable phenomenon, displaying the property of rigidity, sometimes at quite low concentrations of solute and yet often showing the properties of the solvent practically unchanged. For example, most gels in water show vapor pressure and electrical conductivity very close to water. Gel formation occurs in gelatin dispersions in water with as low a concentration as 1 per cent gelatin and in plasma with a fibrinogen concentration of 0.04 per cent. The phenomena displayed by gels are complex and are not always the same with different gels.

When colloidal dispersions of some relatively large molecules are cooled, the viscosity increases to a point at which some rigidity is attained. This point is called the *gel point.* Usually on further cooling or sometimes on standing, the rigidity of the gel increases. Many gels on continued standing lose solvent and the gel shrinks in a process called *syneresis* or *weeping.* Gels differ considerably in their rigidity. Some are deformed under pressure and others even flow.

Theories of Gel Formation. Many attempts to explain the properties of a gel have been made, and through the years a number of theories have been proposed and argued back and forth. Three major theories are now supported by colloid chemists. With some gels one theory is more likely than another. These theories are: (1) adsorption of solvent, (2) three-dimensional network formation, (3) particle orientation.

(1) Adsorption of Solvent. This theory postulates that adsorption of solvent molecules by the solute particles results, on cooling, in the formation of larger and larger particles with increasing layers of solute. The enlarged particles eventually touch or overlap enclosing more solvent, so that the entire system is immobilized and rigidity occurs. Support of this theory depends on demonstrating that adsorption of solvent molecules is very extensive and that the adsorption increases with decreased temperature.

(2) Three-dimensional Network. This theory postulates that the compound capable of gelation is either fibrous in structure or can react with itself to form a fiber. On cooling the fibers form a three-dimensional network by reacting either at widely separated intervals on the chain or at relatively small distances. The bonds established which tie the fibers into the three-dimensional network can be either primary bonds between func-

tional groups, secondary bonds, such as hydrogen bonds, or nonlocalized secondary attractive forces such as might occur between alkyl groups. (a) The first type of bond would yield a network that would possess considerable permanence. This type could only occur in gels that are not readily dissociated by cutting or beating, for example. They would be capable of swelling and shrinking, within limits, because it would be possible for the fibers to be pushed farther apart without disrupting the network, by straightening the chain of atoms. There would, of course, be a limit to the amount of solvent that could be taken up in swelling, or the amount that could be lost. (b) A secondary type bond is one of considerably lower strength than a primary valence bond, and these gels would consequently be broken by relatively small forces. A gel that can be disrupted by beating might possibly have a network structure where secondary valence bonds hold the fibers together. (c) The third type of bond, a nonspecific attraction between portions of the molecules or along the entire molecule, can explain gelation in a few situations. The properties of this type of gel would depend on a nice balance of forces—the solute molecules would attract one another at some spots but would be separated by solvent molecules that are attracted at others. The result would be gelation. If the solvent molecules were too strongly attracted to the solute, then the solute molecules would have no opportunity to contact one another and a network would be impossible. Under these conditions a gel would not form. On the other hand, if the attraction of the solute molecules for one another were too strong, dispersion in the solvent would not occur. A precipitate or an insoluble compound will be formed rather than a gel. This theory can explain gel formation in systems that are very demanding as far as temperature, concentration, pH, and salt concentration are concerned.

(3) Particle Orientation. This theory postulates that in some systems there is a tendency for the solute and solvent particles to orient themselves in definite spacial configurations through the influence of long range forces such as occur in crystals. Certain protein crystals have the ability to take on or lose water without distortion of the crystal. They may form structures of this type. Tobacco mosaic virus has been studied and found to form gels under a large range of concentrations. X-ray diffraction studies indicate that the gels possess a two-dimensional lattice. This is interpreted to mean that the particles are oriented in one direction.

The gelation of a few proteins has received considerable study in the past, but many have as yet not been investigated. Gelatin and fibrin clots have probably received the largest share of attention, while myosin and actomyosin from muscle, and denatured egg albumin have also been studied. Ferry[7] finds that the three-dimensional network theory of gel formation most closely fits the data that have accumulated for proteins.

Gelatin. Gelatin is a partially degraded protein. It is prepared from the collagen of skin, ligaments or bone by alkali or dilute acid hydrolysis. When bone is used, the bone salts are usually dissolved with acid and the protein then hydrolyzed with $Ca(OH)_2$, lime. The native proteins that are hydrolyzed are collagens (see p. 176), and it is usually assumed that the collagen molecules are broken into shorter fragments that are nevertheless still fibrous in structure. A study of gelatin from bone[19] indicates that the collagen is first hydrolyzed into molecules of approximately 110,000 molecular weight and then more slowly hydrolyzed at random into smaller fragments. The gelatin is therefore composed of an assortment of molecular species and cannot be considered a single compound. This assortment of molecules apparently occurs in all gelatin samples although the assortment will not be identical in each sample.

Gelatin gels have been studied extensively for many years and much data have been accumulated concerning the properties of these gels under many conditions, some applicable to food chemistry and some not. The factors which influence the rigidity of a gelatin gel include concentration of the gelatin, temperature, molecular weight, added reagents, and pH in some ranges—although the influence of these factors is limited. The rigidity is roughly proportional to the square of the concentration. The proportionality is not strictly valid for some samples and under all conditions. Temperature has a marked effect and the rigidity increases rapidly with decreasing temperature. This relationship is linear. Molecular weight has a direct effect on rigidity so that samples that have high average molecular weights form more rigid gels under controlled conditions than those with lower average molecular weights. The effect of pH is only slight in the range close to the isoelectric point of the sample. Thus a series of gels at 3 per cent concentration showed little change in rigidity over a pH range of 4.6 to 8.2, while at a concentration of 1.5 per cent, constancy of rigidity occurred at pH 4.3 to 6.7. The influence of salts on rigidity has been studied but in ranges of concentration of both gelatin and salt which are of little significance in food work. For example, a series of isoelectric gelatin gels with a concentration of 14 per cent showed a decrease in rigidity with increasing sodium fluoride concentrations up to 0.5 M and then increasing rigidity.

Denatured Protein Gels. The gels of denatured protein that have been studied show very marked effect by many conditions. They form only under very specific conditions. There have been only a limited number of definitive studies of denatured protein gels, and those on denatured egg albumin are perhaps most numerous. In all of these gels the dependence on pH and salt concentration as well as on the presence of nonpolar solvents indicates that the gel is produced by a type 3 (see p. 127) three-dimensional network. The nice balance of forces that is necessary to achieve gelation indicates that the molecules must show a relatively nonspecific type of attraction between the polypeptide chains. There must be a balance between forces of attraction in part of the chain and forces of repulsion in others. The solvent must neither be too strongly nor too weakly attracted to the chain.

The conditions under which egg albumin has been studied have not been directly applicable to food preparation. One comprehensive study used n-propyl alcohol and water as the solvent, another used relatively high concentrations of acetic acid in water. But all studies indicate the sensitivity of albumin gels to various conditions of pH and salt concentration. There appears to be two steps in the formation of the gel (1) denaturation and (2) gelation. In the denaturation reaction a fibrous molecule is formed. This is followed by the setting up of a network through the attraction of portions of molecules to one another.

DETERMINATION OF PROTEIN IN FOODS

Accurate determinations of the protein content of the complex mixtures which are encountered in foods is a difficult job. Kirk[12] in his discussion of the "unsatisfactory state of analytical development in this field" makes the following statement: "The reasons are apparent. Proteins form a very diverse group of similar compounds of extraordinary complexity, with widely different compositions and properties, yet difficult to separate completely, to purify and to dry. Their amphoteric nature, high adsorptive capacity, hydration properties, and sensitivity to electrolytes cause them to vary widely in behavior depending on the composition, pH, and temperature of the solvent medium."

In the determination of proteins in food, interest is commonly on the total protein content of the food, rather than on an accurate determination of the presence of a specific protein. Occasionally in some research problems it is of interest to determine specific proteins. An example is the collagen content of various types of meat, or the total amount of gluten capable of development in various grains. Often rough determination of total protein content is sufficient for the study. This is well and good, if the rough protein determination is not then used as if it were an accurate determination.

Since proteins are so readily precipitated from solution, it might be concluded that it would be relatively simple to precipitate the protein from a food, filter it off, dry, and weigh it. But the problem is far from simple. It has not been possible so far to find a satisfactory reagent that will precipitate all proteins free of contamination with other substances. Lipides, salts, and many organic molecules are carried down in protein precipitates. Some can be removed by washing, but the process is difficult to achieve quantitatively and is laborious and time consuming. Drying the protein to constant weight is at present impossible. Proteins have a high degree of hydration and the sample slowly loses water over a long period of time. Although the hydration is a reversible reaction, complete removal of water is very difficult, if not impossible, without protein decomposition. The freeze-dry (lyophile) method has so far not been widely applied to proteins but may be a possible method. So far, direct methods of estimating protein show considerable variation and present difficulties in the reproduction of consistent data.

A few methods have been investigated for precipitating protein and then measuring the amount of precipitating agent that has combined with the protein under the conditions of the experiment. None of these methods is free from a considerable margin of error; and none has been widely adopted.

One method of determining protein is analysis for carbon. After all organic compounds except protein have been removed as well as possible, analysis for carbon can be attempted. With a few biological systems it has been successful. Analysis of sulfur has not been used extensively. The most widely used method of determining protein is the analysis for nitrogen.

When a food is analyzed, provision must be made to remove nitrogen-containing compounds other than proteins. A cell or biological fluid contains many such compounds. Free amino acids, some of the lipids, urea, creatine and creatinine, thiamine, heme, and many other compounds contain nitrogen. While all of them are present in relatively small amounts, it must nevertheless be remembered that they as well as protein contribute to the amount of nitrogen present in a food, and that a change in the concentration of one of these components will alter the amount of nitrogen present as surely as a change in protein.

The number of grams of protein in a food is often calculated by multiplying the number of grams of nitrogen by 6.25. This constant is derived from the assumption that proteins contain 16 per cent nitrogen and $100/16 = 6.25$. The assumption is not valid since all proteins do not contain exactly 16 per cent nitrogen. Protein is often reported as "crude protein" because of these two approximations: determination of nitrogen in compounds which are nonprotein and the "protein constant."

Kjeldahl Method

The usual method employed for the determination of nitrogen in foods is the *Kjeldahl method* and various modifications have been devised to improve its accuracy and speed. It is an oxidation of organic compounds by sulfuric acid to form carbon dioxide and water and the release of the nitrogen as ammonia. The ammonia exists in the sulfuric acid solution as ammonium sulfate, but the carbon dioxide and water are driven off. Sulfur dioxide is the reduction product of the sulfuric acid and it too is volatile.

$$\text{Organic compounds} + H_2SO_4 \rightarrow CO_2 + H_2O + (NH_4)_2SO_4 + SO_2$$

The digestion of the sample to form ammonium sulfate is the most difficult part of the operation. Although considerable effort has been expended to obtain accurate analyses, many factors cause incomplete formation of ammonia or loss of some of it from the digestion mixture. Attempts to eliminate or minimize these errors have centered around studies of: (1) chemical form of some of the nitrogen, (2) addition of salts to digest mixture, (3) use of oxidizing agents, (4) catalysts, (5) length of time of digestion, (6) use of reducing or hydrolyzing agents prior to digestion, (7) contamination with ammonium salts used in fractionation. Kirk concludes that the most important sources of error are connected with the catalyst, the time of heating, and the addition of reducing and oxidizing agents.

Many catalysts have been used to speed up the decomposition of the sample. Copper, mercury, and selenium are most widely used. Mercury appears to give better recovery of nitrogen than copper, but it must be precipitated before the ammonia is estimated since it forms a complex with the ammonia. Selenium shortens the clearing time of the sample, but it has often given a false sense of completion of digestion when clearing time and completion have been confused. Many workers have used 1.5 times the clearing time as the length needed for complete digestion, but it has been demonstrated that even this length of time is not always sufficient. Chibnall, Rees, Williams, and Jonnard have advised 8 to 16 hours of digestion after clearing in order to insure completion. With a long digestion time, the quantity of acid present must be carefully watched. If the amount falls too low, decomposition of NH_4HSO_4 occurs with the loss of ammonia from the mixture.

Combinations of catalysts have been used effectively. Copper, mercury, and selenium are used together as well as copper and selenium, and mercury and selenium.

Numerous oxidizing agents have been added at the end of the digestion period to complete the oxidation. Potassium permanganate was the first reagent used and in recent years persulfates, perchlorates, and hydrogen peroxide have been used. On various types of samples losses of nitrogen through the formation of amines or of free nitrogen have occasionally been reported with all of these reagents except hydrogen peroxide. However, it has not been tested with a wide number of different types of products. Hydrogen peroxide must be used cautiously since it is commonly preserved with acetanilide to depress the formation of oxygen and water, and the nitrogen in the preservative will act as a contaminant of the determination.

The chemical form of the nitrogen is important. Thus the terminal amino group on lysine is difficult to release as is the imidazole nitrogen of histidine and tryptophan. Sufficient length of time must be allowed for complete transformation into ammonia. At the same time the amount of acid must be controlled so that it does not fall sufficiently to allow the loss of ammonia from the mixture.

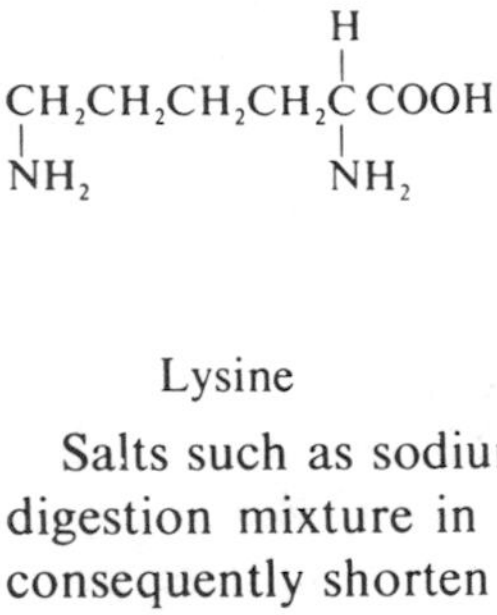

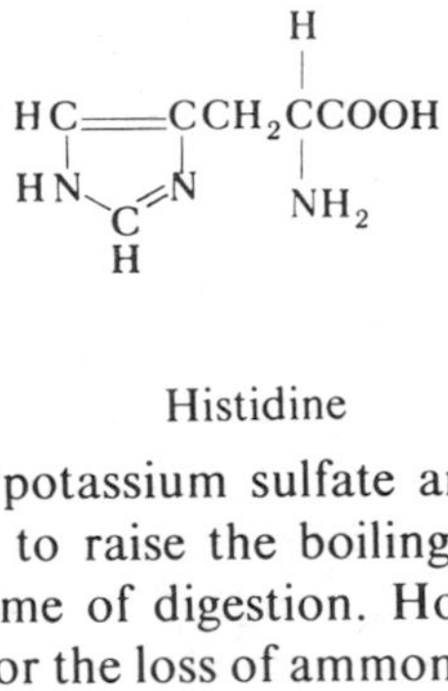

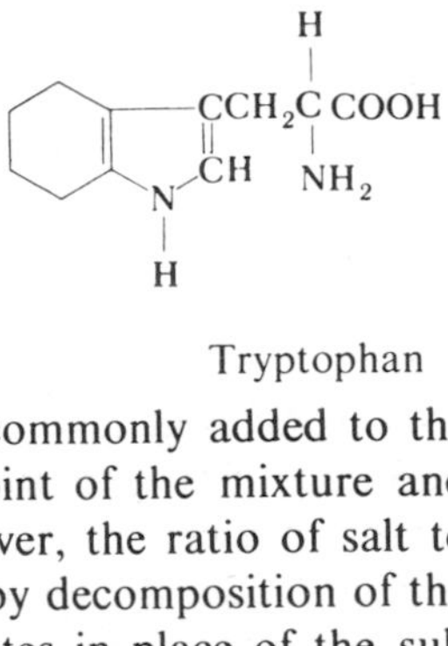

| Lysine | Histidine | Tryptophan |

Salts such as sodium or potassium sulfate are commonly added to the digestion mixture in order to raise the boiling point of the mixture and consequently shorten the time of digestion. However, the ratio of salt to acid must not rise too high or the loss of ammonia by decomposition of the NH_4HSO_4 will occur. Some workers use phosphates in place of the sulfates, with good results. Here, too, the ratio must not be too high.

Measurement of the ammonia after it is formed by digestion is carried out by a number of different methods, all of them quite accurate. In some the ammonia is distilled off after the addition of large quantities of alkali and trapped in a known quantity of standard acid. The acid is then back titrated to determine how much ammonia has distilled.

Another method uses the direct estimation of the ammonia by reaction with Nessler's reagent to form orange Nessler's salt.

$$2K_2HgI_4 + NH_3 + KOH \rightarrow NH_2Hg_2I_3 + 5KI + H_2O$$

The Nessler's salt is insoluble and precipitates; but if the amount is small, it can be produced in colloidal form and estimated colorimetrically with some degree of accuracy. Technically, the method is often difficult to carry out because of the tendency of the colloidal dispersion of Nessler's salt to become cloudy and then precipitate.

A method which has not been as widely used is oxidation of the ammonia with hypobromite and back titration of the hypobromite with iodine.

$$2NH_3 + 3NaBrO \rightarrow N_2 + 3H_2O + 3NaBr$$

Dumas Method

The *Dumas* method is a combustion method for the estimation of nitrogen in a sample, which has seldom been used on food samples. Comparisons of Kjeldahl methods with Dumas have occasionally been made. Often the Kjeldahl method gives nitrogen values a little lower than the Dumas. In the Dumas method the sample is heated in a combustion train in the presence of carbon dioxide. The gases are passed over hot copper gauze so that any nitrogen present as an oxide is reduced to free nitrogen. Carbon dioxide and water are absorbed and the nitrogen gas formed measured directly.

When a food sample is analyzed for nitrogen, the accuracy of the method of analysis, Kjeldahl or Dumas, varies with a number of factors, but often the error exceeds 1 or 2 per cent and sometimes even 5 per cent. Since the data are then multiplied by a factor that is approximate, the final answer cannot be exact and gives only the crude protein. Often the nitrogen-containing compounds other than protein are neither removed nor estimated, and, therefore, cannot be subtracted from the results. Most of our data have then this further approximation. Data on the protein amount of foods must, therefore, always be taken as *approximate* only, unless a careful check is made to assure that all approximation has been eliminated and that the protein constant actually applies.

Amino Acids

Amino acids are often determined in proteins. Although most proteins, and particularly those of animal origin, contain all 20 of the common amino acids, the amount in the protein is frequently of some interest. Since the essentiality of some amino acids have been recognized, a great number of studies have been made of the level of feeding one amino acid on the requirement for another. Although these data are in the province of nutrition, there is also a considerable interest in amino acid distribution in the field of food chemistry.

The problem of separating compounds as closely related chemically as the amino acids and estimating the concentrations of each is by no means small. In recent years the development of the technique of chromatography and the rapid increase in knowledge of microbiological assays have facilitated the solution of this problem.

Chromatography was studied in 1861 by Schoenbein and used in the separation of chlorophyll from other plant pigments by Tswett in 1906. The term "chromatography" is applied because Tswett separated colored compounds, mainly on a calcium carbonate column. When the method is applied to compounds which possess no color, it is still called "chromatography."

The technique depends on the differential adsorption of compounds, even those which are closely related chemically, by a solid adsorbent. The mixture of compounds can be dissolved in a suitable solvent and passed through a column of the adsorbent. As the solution trickles through, the compounds may move and be adsorbed at differential rates. Thus one compound which is strongly adsorbed will be held at the top of the column while one which is poorly adsorbed will be at the bottom. Sometimes all of the compounds are strongly held at the top of the column, but they can be separated by the use of a second solvent which detaches them preferentially. When the second solvent is poured over the column, the first fraction collected at the bottom contains the substance held least strongly and most readily dissolved in the second solvent, while subsequent fractions will contain other members of the mixture. The difficult part of the technique is finding those solvents which will effect separation, as well as a suitable adsorbent for the column. This is empirical and can be very time consuming.

Paper chromatography is an application of this technique to a filter paper strip and has frequently been applied to the separation of amino acids. A mixture of compounds is placed at one end of a strip of filter paper. The paper is suspended in a cylinder and dips into an organic solvent saturated with water. The atmosphere in the vessel is likewise saturated with the vapors of the two solvents. The filter paper has a stronger affinity for water than for the organic solvent and the components of the mixture move up the strip at different speeds. When the solvent front reaches the top of the paper, the paper is removed, dried, and sprayed with some compound that will react with the components to form colored products that can be seen. The components will be found at different places on the strip. This is ascending chromatography, but descending chromatograms can be made by dipping the top of the strip in a trough of solvent-solvent mixture and allowing the components to move down the strip. Two-dimensional separation can be achieved by running the mixture in one direction, then turning the paper 90° and running again. Better separation is sometimes possible by this device.

Chromatography can be used to separate even quite small samples of amino acids after hydrolysis of the parent protein. The quantities of amino acids present then can be estimated by various methods.

On acid hydrolysis of any protein, tryptophan is destroyed and humin is formed. Humin is a black or dark brown precipitate which always accompanies acid hydrolysis. In the presence of carbohydrate, it is especially abundant. Since most foods contain carbohydrate along with the protein, humin formation presents an error in quantitative estimations of amino acid content. The most satisfactory method is to remove carbohydrate. This is often very difficult to achieve without loss or change of the protein.

Quantitative estimation of some individual amino acids can be carried out by specific reactions to give colored compounds or precipitates. The Sakaguchi test, for example, is a test for arginine which can be applied either to proteins or to amino acid mixtures. It is carried out by treating the sample with α-naphthol and sodium hypochlorite. In the presence of arginine a deep red color develops. The method may be used quantitatively since the depth of the color is a function of the amount of arginine present. An example of a precipitating reagent is Reinecke salt—a complex salt of chromium, $NH_4[(NH_3)_2Cr(CNS)_4]$-ammonium diamminetetra-cyanatothiohromium III—which forms a precipitate with proline and hydroxyproline. Other reagents are discussed in textbooks of biochemistry or in special books on proteins and their analyses.

The *microbiological methods* are fairly recent developments which have been very valuable in analyzing mixtures of amino acids because of the speed and reproducibility of results obtained. The methods depend on preparing a nutrient medium which contains all of the compounds essential for the growth of a particular microorganism except one amino acid, which will be assayed. Addition of this amino acid results in growth of the microorganism in proportion to the amount of the amino acid added. Culture tubes are set up and graded amounts of the unknown are added to a series of tubes. Standards are set up at the same time with graded amounts of the pure amino acid. The unknown can be compared to the standards by measuring the rate of growth of the microorganism. Sometimes turbidity of the medium after a 24 hr period of growth in the incubator is used. With organisms which form acid such as some of the lactobacilli, titration of the acid formed can be used as a measure of the number of cells present. The only difficulty in carrying out these determinations is in finding an organism which requires a particular amino acid. *Streptococcus faecalis* requires nine amino acids and can be used in the estimation of these. Pure cultures must be used and all of the techniques necessary for handling microorganisms without contamination observed.

HEAT TREATMENT

For many years it has been recognized that the nutritive value of proteins changes with heat treatment, and a great many studies have been devoted to the problem. Unfortunately the "nutritive value" of a protein is still rather poorly defined because it is complicated by many factors. "Nutritive value" is usually measured by comparing the growth response in a species such as the albino rat to graded amounts of a purified protein or some food source. The growth response depends on numerous factors besides the level of protein consumed: (1) the amino acids present in the protein or proteins, (2) the quantity of each essential amino acid present, (3) the previous nutrional history of the test animal, (4) the extent of digestion of the protein, and (5) the rate of digestion of protein. Recently it has been shown that other factors such as mineral and vitamin levels in the diet, the presence of rancid fat, starvation, and hormones such as cortisone all influence the rate of growth and the response to specific levels of protein.

It is well established that heat treatment can alter the nutritive value of a protein when measured under constant conditions in a number of species. Many proteins show a decrease in value, but in a few there is an increase. Milk proteins, meat proteins, blood globulin, cereal proteins, fish, and coconut meal are among the numerous products which have been studied. In some studies the purified proteins are used and in others the whole food.

The chemical changes that take place on heating have been investigated in both isolated protein systems and in foods and have been found to consist of three types of reactions: (1) the reaction of protein with carbohydrate, resulting in the destruction of some amino acids and a change in the digestibility of the protein by proteolytic enzymes; (2) reaction of protein in the absence of carbohydrate which results in a decrease in the availability of an amino acid; and (3) heat inactivation of enzyme inhibitors such as ovomucoid.

When proteins are heated with carbohydrates, particularly the mono and disaccharides, interaction of the protein with the carbohydrate occurs. On continued heating, browning and destruction of amino acids occurs. In a simple system where only amino acids and glucose are used, there is a progressive loss of histidine, arginine, valine, and leucine in the order given, as shown by microbiological assay. The same type of reaction has been observed in numerous protein-carbohydrates systems as well as in many foods. For example, evaporated milk has a lower nutritive value when measured by rat growth, than fresh, whole milk. And if the evaporated milk has been subjected to heating before feeding, it suffers a further loss in value proportional to the extent and amount of heating. It has been demonstrated that the reaction occurs through the amino groups of the protein since if these groups are acetylated and rendered unavailable for reaction, the protein loses its power to bind carbohydrates.

Not only has it been shown in some studies that amino acids are destroyed, but the protein also becomes resistant to hydrolysis by some proteases. For example, heating casein or wheat gluten with glucose causes resistance to hydrolysis of the protein by trypsin or papain and to a less marked degree by pepsin, chymotrypsin, or pancreatin (which contains a mixture of proteases).

Heating protein in water or alone also affects the amount of amino acids and the resistance to enzymatic hydrolysis of the protein. Denaturation tends to make a protein more susceptible to enzymatic hydrolysis. In one study is was shown that if casein or ethanol-precipitated lactalbumin is heated, hydrolysis by enzymes is not as complete. Soluble lactalbumin did not show this behavior.

A number of foods have been studied in an attempt to determine what effect cookery and food processing has on the nutritive value of protein. In many foods carbohydrate is closely associated with protein and on heating there is a reaction between these compounds. In one study it was shown that on heating, sunflower meal, peanut meal, cotton seed meal, and corn are decreased in biological value, while beef (no carbohydrate present) is unchanged and linseed meal is improved. However, in another study, canning or heating beef decreased its biological value. Canning peas causes a small decrease in the amount of methionine and a slight increase in resistance to digestion.

Improvement in nutritive value occurs when enzyme inhibitors are destroyed. A number of enzyme inhibitors have now been found in foods. Ovomucoid, one of the proteins of egg white, was the first recognized. It inhibits the activity of trypsin. Although it is rather resistant to heat at the temperatures used for most cooking and food-processing operations, it can be destroyed by sufficient heat. It is significant that this anti-trypsin factor is apparently ineffective in the human, although it appears to interfere with protein utilization in the dog and the mouse. A trypsin inhibitor extracted from soybeans is found to depress growth of rats on diets in which the protein source is either casein or soybean protein. This inhibitor is destroyed by heating in steam.

Soybeans and their reaction to various types of treatment have been thoroughly studied. (See Almquist[1], Swanson and Clark[19], and Rice and Beuk[17].) The data of Melnick and Oser[15] indicate that methionine is liberated very slowly from heated soybeans although digestibility is not affected. They suggest that the low biological value may result from the tardiness of methionine to enter the metabolic pool while other amino acids are available in good supply. Other investigators who have measured the digestibility of heat-treated proteins as well as their biological value for growth promotion, favor the view that the molecules absorbed from the gut are not free amino acids but derivatives, unavailable for protein synthesis.

The significance of these observations in human food is questionable. Low biological value can only be demonstrated in animals on low protein diets. Since the diets eaten by man usually contain proteins from many sources, these proteins can supplement one another. Toasting bread may decrease the amount of some of the amino acids in the wheat protein; but when it is supplemented by egg or milk at the same meal, the small loss is insignificant. When men eat restricted diets during times of famine or in areas of poverty, the small loss could have significance.

PURE PROTEINS FROM SOME FOODS

Plant Proteins

The study of the proteins of plants has been plagued through the years by the same difficulties that the study of any protein entails. There is the difficulty of isolation of a protein from the complex mixture which always occurs in cells, without modifying the molecules. Many proteins occur in cells conjugated with lipid, carbohydrate, and other molecules, but only those conjugated proteins which are not readily dissociated are isolated intact. With plant proteins there is the usual trouble encountered in handling a large, relatively reactive molecule such as a protein. Most proteins are readily denatured by a wide variety of reagents and conditions, and the utmost care must be used to avoid reaction.[3, 13]

In view of the tremendous number of plants in the world, and the variations in cells with structure, the amount of knowledge concerning plant proteins that has accumulated is relatively meager. See Table 4.1. The spermatophytes (flowering plants) have received the greatest amount of attention while the gymnosperms (conifers, etc.) have been scarcely studied. The seeds commonly used for food have been investigated most frequently. See Table 4.2. The thallophytes, which include algae, fungi, actinomycetes, and bacteria, have had considerable study. This has occurred because of the interest of the brewing and baking industries in yeast and because of the rapid development of the field of antibiotics, which are produced by some of these plants. Much of the old work is incomplete and of question-

TABLE 4.1. AMINO ACID COMPOSITIONS[a] OF BULK PROTEINS IN SPERMATOPHYTES, PTERIDOPHYTES, A BRYOPHYTE, AN ALGA, FUNGI, AN ACTINOMYCETE, BACTERIA, AND PHYTOPATHOGENIC VIRUSES*

Amino Acid	Spermato-phytes[b]	Pterido-phytes[b]	Bryo-phyte[b]	Alga[b]	Thallophytes		Bacteria	Phyto-pathogenic Viruses
					Fungi	Actino-mycete		
Glycine	0.4	—	—	—	—	—	—	1.4- 2.1
Alanine	4.4- 5.1	—	—	—	—	—	—	4.5- 6.1
Valine	3.3- 4.5	—	—	—	2.2- 4.0	4.6	3.5- 5.1	4.5- 6.6
Leucine	7.1	—	—	—	2.7- 4.9	4.5	3.4- 5.1	5.8- 6.0
Isoleucine	3.6	—	—	—	1.1- 4.0	1.9	3.3- 4.7	3.0- 4.3
Phenylalanine	2.4- 2.6	—	—	—	1.1- 1.9	1.5	1.4- 2.2	2.8- 5.1
Cyst(e)ine	1.1- 1.6	1.1-1.2	1.3	2.2	—	—	1.5- 1.6	0- 0.5
Methionine	1.2- 1.6	1.5-1.7	1.4	1.2	0.4- 0.9	0.6	0.6- 1.5	0.0- 1.2
Tyrosine	2.3- 2.7	2.1-2.6	2.4	1.3	1.1- 2.3	—	0.8- 2.2	1.7- 3.1
Tryptophan	1.4- 1.9	1.1-1.4	1.7	1.0	0.6- 1.0	1.0	0.3- 0.6	0.4- 1.8
Arginine	12.4-14.0	15.3	—	—	4.9-13.3	10.3	4.8-12.2	17.5-20.5
Histidine	3.6- 4.0	—	—	—	1.4- 7.4	2.5	1.2- 4.4	0.0- 1.2
Lysine	5.0- 6.8	6.4	—	—	3.9- 9.1	4.6	5.7-10.6	1.6- 2.9
Aspartic acid	4.7- 5.4	—	—	—	—	—	—	8.0- 8.8
Glutamic acid	6.4- 7.8	—	—	—	4.9- 5.8	—	4.0- 7.3	3.7- 8.9
Proline	3.1	—	—	—	—	—	—	4.0- 4.3
Hydroxy proline	—	—	—	—	—	—	—	+?
Serine	—	—	—	—	—	—	—	4.6- 7.5
Threonine	3.0- 4.0	—	—	—	2.0- 3.7	3.1	2.4- 3.6	4.5- 6.7
Amide	4.7- 6.0	4.9-5.3	5.5	5.6	—	—	—	8.0

[a] As per cent protein nitrogen.
[b] Main photosynthesizing tissues.
*From Lugg, J. H., "Plant Proteins," Advances in Protein Chem., 5, 230 304 (1949).

able validity, in the light of recent developments but, nevertheless, of some interest.

Work so far seems to indicate that the bulk proteins of plants, particularly among the spermatophytes, are very similar in amino acid composition. Seeds, however, appear to possess proteins which are quite dissimilar in composition. A few thallophytes do not fall in line with the analyses of tissue proteins of the rest of the plant world, but these data need to be reinvestigated.

Milk Proteins[14]

Milk is one of the excellent sources of protein in man's diet. Its role as a source of protein has been emphasized frequently both in the scientific literature and in that for the general public. Its place as one of the Basic Seven in the recommended diet plan for Americans stems from its fine contribution of protein as well as calcium and other nutrients to the diet.

The proteins of cow's milk have been extensively studied, and those of human milk have received considerable attention. Very little is known of the proteins present in the milk of mammals other than that of the cow or human. In this book when the term "milk" is used, it refers to cow's milk.

In recent years with refinements in the techniques of separating proteins and in the development of physical measurements, electrophoresis, sedimentation, osmotic pressure, and other methods, it has been possible to extend earlier work and with more confidence describe the proteins of milk. These proteins are separated by precipitation at definite pH, salt fractionation, and occasionally heat coagulation. These protein fractions must be further purified in order to gain homogeneity.

Casein. Casein is the name assigned to the fraction precipitated by acidifying milk to a pH of 4.7. It is present in cow's milk to the extent of 3.0 to 3.5 per cent, in human milk 0.3 to 0.6 per cent. It may be further purified by redissolving and precipitating again. Since it is very readily obtained, it has been studied for many years. The casein produced by this method was shown many years ago to be a mixture. Many attempts have been made to prepare pure proteins from this mixture, and it appears at present to contain at least three proteins. These have been named α-, β-, and γ-casein and differ from one another in their molecular weights, their rate of migration in an electric field, and their phosphorus content. The amounts of some of the amino acids which have been determined likewise show that these are different proteins. See Table 4.3.

Casein is also precipitated from milk by the action of the enzyme rennin (the extract is called rennet) isolated from calves' stomachs; and the prob-

TABLE 4.2. APPROXIMATE PERCENTAGE OF AMINO ACIDS IN PLANT PROTEINS: PROTEINS OF SEEDS CALCULATED TO 16.0 g OF NITROGEN*

Amino Acids	Cotton Seed Meal	Linseed Meal	Peanut Flour	Soybean Meal	Oats	Rice
Arginine	7.4	6.9	9.9	5.8	6.0	7.2
Histidine	2.6	1.9	2.1	2.3	2.0	1.5
Lysine	2.7	2.0	3.0	5.8	3.3	3.2
Tyrosine	3.2	5.1	4.4	4.1	4.5	5.6
Tryptophan	1.3	1.6	1.0	1.6	1.3	1.3
Phenylalanine	6.8	5.8	5.4	5.7	6.9	6.3
Cystine	2.0	1.9	1.6	0.6 ± 1.4	1.8	1.4
Methionine	1.6	2.3	1.3	2.0	2.3	3.4
Threonine	3.0	4.5	1.5	4.0	3.5	3.9
Leucine	5.0	7.5 ± 2.8	5.5	6.6	8.0	9.0
Isoleucine	3.4	3.4 ± 0.3	3.4	4.7	5.3	5.1
Valine	3.7	5.8 ± 1.3	4.0	4.2	6.5	6.4
Glycine	5.3	—	5.6	—	—	—

*Adapted from Block, R. J. and Bolling, D., "Amino Acid Composition of Proteins and Foods: Analytical Methods and Results," 2nd ed., Charles C. Thomas, Springfield, Ill., 1951.

lem of discovering the difference between acid precipitated casein and rennin casein has been studied extensively. Conventional methods do not show any difference in casein precipitated by either acid or rennin. They do differ in the fact that rennin casein cannot be clotted a second time. This may indicate a difference in the structure of some fraction of the rennin casein. It has been found that the electrophoretic pattern of rennin-treated casein differs from that of the original casein in that the α-casein shows two peaks, indicating the presence of two types of molecules. Some investigators believe that rennin hastens the dissociation of α-casein into two components. Since rennin is incapable of clotting milk if the calcium ions are first precipitated, in the past it has been customary to consider the reaction with acid or with rennin as following quite different pathways. In America it has been customary to call the native protein *casein* and the soluble protein formed from rennin action, *paracasein*. The course of the reaction has been often written:

$$\text{Casein (soluble)} \xrightarrow{\text{rennin}} \text{Paracasein (soluble)} \xrightarrow{Ca^{++}} \text{Calcium paracaseinate}$$

However, writing this series of reactions in such a manner is an oversimplification. At present it is not possible to describe the course of the reaction.

Berridge[2] says that any theory of rennin action must account for the fact that there is a considerable time lag at low temperatures between the completion of the action of rennin and the clotting. It must also account for

TABLE 4.3. COMPOSITION OF CASEIN FRACTIONS*

Constituent	α-Casein (Per Cent)	β-Casein (Per Cent)	Probably γ Alcohol Soluble Casein (Per Cent)
Phosphorus	0.99	0.66	0.06
Total N	15.5	15.4	15.6
Amino N	0.99	0.73	—
Arginine	4.3	3.4	2.9
Histidine	2.9	3.1	2.3
Lysine	8.9	6.6	4.0
Tyrosine	8.1	3.2	2.5
Tryptophan	1.6	0.6	—

Mobility in Veronal buffer at pH 7.78 and ionic strength of 0.1

6.98 $\qquad$ 3.27 (cm^{2}volt^{-1}sec^{-1} × 10^{-5})

*From McMeekin, T. L. and Polis, B. D., "Milk Protein," *Advances in Protein Chem.*, **5**, 201–228 (1949).

the fact that the concentration of calcium ions has little effect at concentrations below 0.011 M and there is virtually complete precipitation of the casein at all concentrations above 0.0125 M. Neither the theory that the action of rennin splits α-casein into two types of molecules, nor the old reaction adequately explains the facts so far observed.

Whey Proteins. Milk contains approximately 0.6 to 0.7 per cent protein which is not precipitated on acidification to pH 4.7. This represents about 20 per cent of the protein contained in skim milk. These whey proteins were separated long ago into two fractions which were called globulin or lactglobulin and albumin or lactalbumin. The old names still persist; but in the light of modern technique and the demonstration that whey contains a number of proteins, not just two, it is desirable to modify them.

Classical globulin is separated from whey by saturation with magnesium sulfate. This fraction contains a number of compounds as shown by electrophoresis. It has been possible to separate two homogeneous fractions which are called euglobulin and pseudoglobulin.

The fraction which is soluble in saturated magnesium sulfate is the classical albumin. A crystalline component which appears to be homogeneous has been prepared from this fraction. It has been named β-lactoglobulin since it appears to be identical with the β-fraction from ultracentrifuge studies of milk serum.

Other fractions have been purified and more than two peaks occur in the electrophoretic pattern of whey. There is, therefore, the probability that these components occur in milk. They are still to be fully characterized.

Colostrum. Colostrum is the first secretion of the mammary gland on parturition and differs markedly from milk in its protein composition. During the first few days after the birth of the calf, the composition of the lacteal secretion gradually changes until that characteristic of milk appears. On birth of the calf the colostrum contains approximately 17.5 per cent protein with about 5 per cent of this casein. The most striking difference is in the globulin fraction that carries the antibodies which give protection against certain diseases. Since the health of the young animal often depends on protection against some of these diseases, the importance of colostrum feeding in both the calf and human infant has been studied. Euglobulin and pseudoglobulin are absent in the blood of the new born calf but when it feeds on colostrum they rise rapidly. Colostrum does not appear to be so essential in the feeding of human infants, since antibodies are more readily transferred by way of the placenta.

Egg Proteins[8]

The eggs of many species of birds have been eagerly sought and eaten by man, probably since the dim ages when he rummaged for food. The time of the domestication of the chicken is unknown, but it was in the prehistoric period. Whole egg is an excellent food because it is a very rich source not only of protein and lipid but also of most of the vitamins, except ascorbic acid, and many of the required minerals except calcium. In some countries the shells are used for food and are one of the major sources of calcium in the diet. The proteins of egg are rather numerous and together yield an excellent assortment of amino acids for animals. They are considered to be of the highest biological value.

The egg, of course, is produced for the nourishment of the embryo, which develops in the fertilized egg. Some of the proteins of egg white have unusual properties which give them the ability to protect the embryo from bacterial invasion. Others possess properties with perhaps a regulatory effect on the nutrition of the embryo. Lysozyme is antibiotic, ovomucoid is a trypsin inhibitor, while ovomucin is an inhibitor of hemagglutination. Avidin binds biotin, while conalbumin binds iron. The direct functions in the developing embryo have not as yet been demonstrated, but the properties of these proteins are surely suggestive of how they may function.

The data considered here are those accumulated for chicken eggs. The eggs of a few other species have had slight study; but when the term "egg" is used, it is meant to imply "chicken egg."

Egg White Proteins. These have been studied for many years, and hundreds of papers have been written about one or all of the proteins present. So far there is evidence for the occurrence of 8 different proteins in egg white. All have been isolated with a considerable degree of purity except globulin G_2 and G_3. Fevold[8] prepared the data shown in Table 4.4 from existing data of a number of investigators.

Ovalbumin is the most abundant of the egg white proteins and is also the protein studied most extensively. It is a phosphoprotein with a molecular weight of approximately 45,000 and a small amount of carbohydrate. Various workers have reported that from 1.8 to 2.8 per cent of the molecule consists of a polysaccharide composed of 2 glucosamines, 4 mannoses, and a nitrogen group. When the electrophoresis of crystalline ovalbumin is measured at various pH's, it is found that two peaks indicating two types of molecules occur. On storage of the ovalbumin the relative proportion of the two components changes. This has been determined on the basis of the phosphorus content of the two albumins. One type of molecule is believed to contain two phosphoric acid residues, while the other, only one. On standing, some of the diphosphate probably changes to monophosphate. Ovalbumin is readily denatured and precipitated by heat and a number of denaturation reagents.

Conalbumin was recognized as early as 1900 by Osborne and Campbell when it did not crystallize with ovalbumin. This is a protein with a molecular weight of approximately 70,000. It is able to bind iron ions. This property was discovered when egg white was found capable of rendering iron unavailable to a microorganism, *Shigelle dysenteriae*, which requires iron in the nutrient medium. It was then discovered that although the property was not lost on dialysis, it was lost on heating and eventually it was shown that conalbumin is the protein of egg white with this function. Ferric iron is bound to the protein molecule by coordination. Amino groups, carboxyl groups, guanido groups, and amides are essential for the activity of the protein. If these groups are blocked by reaction with suitable reagents, conalbumin loses its power to bind ferric iron. The signifi-

TABLE 4.4. PROTEIN COMPOSITION OF EGG WHITE*

Total protein	10-11 per cent wet basis
	82.8 per cent dry basis
Ovalbumin	70 per cent of total protein
Conalbumin	9
Ovomucoid	13
Lysozyme (G_1)	2.6
(G_2)	
	7
(G_3)	
Mucin	2
Avidin	0.06

*From Fevold, H. L., "Egg Proteins," *Advances in Protein Chem.*, **6**, 90 (1951).

cance of the function of conalbumin for the development of the chick embryo is as yet unknown.

Ovomucoid was early distinguished from albumin and conalbumin by the fact that it is not coagulated by heat. It has a molecular weight in the region of 27,000 to 29,000 and is a glycoprotein containing mannose and glucosamine. It acts as an antienzyme for trypsin, diminishing the protease activity of the enzyme. The manner in which it is able to exert its influence has been studied. It has been shown that the groups on the ovomucoid essential for antitrypsin activity are the carboxyl, guanidyl, and phenolic groups and that they react with the amino groups on the trypsin. Amino groups are not essential for the enzymatic activity of trypsin. Inhibition is consequently not the type where the antienzyme and substrate compete for a place on the enzyme, but instead inhibition represents some other change in the enzyme molecule.

The *globulins* of egg white consist of three different types of molecules according to electrophoretic data. *Lysozyme* has received the greatest amount of attention because of its antibiotic activity. In 1922 Fleming published a paper on an agent capable of lysing or dissolving bacteria, widely distributed, and which he called "lysozyme." He found activity in tears, saliva, plasma, numerous tissues, and a particularily active form in egg white. The activity of egg white has now been shown to reside in the G_1 fraction of the globulin. Since the antibiotic activity in this fraction is so interesting, the name "lysozyme" rather than "globulin G_1," is widely used. Lysozyme is a protein with a molecular weight of approximately 14,000 to 17,000. It is a basic protein with an unusually high percentage of histidine, lysine, and arginine. Lysozyme is quite stable to heat, cold, and many denaturation reagents. It is not stable in alkali. Some proteases such as trypsin and papain do not attack it.

Globulins G_2 and G_3 have received relatively little attention. Although G_2 has been separated and shown to be a euglobulin, G_3 is still almost unknown as a separate protein. Neither have been crystallized.

Ovomucin is likewise relatively unknown because it is the most difficult of the egg white proteins to handle, very insoluble, and consequently difficult to purify. It precipitates from egg white on dilution with water at pH 6.4. The molecular weight is still in some doubt, but sedimentation data give a value of 7,600,000. This interesting protein appears to be the one in egg white capable of inhibiting hemagglutination. The ability of influenza virus to agglutinate red blood cells, called "hemagglutination," is inhibited by egg white. It seems likely from studies of the activity of ovomucin fractions that the hemagglutination inhibitor is ovomucin:

Avidin is a protein present in low concentration in egg white, but it nevertheless has been crystallized because of the interest in its ability to bind biotin and render it unavailable to animals. If raw egg white is fed in the ration to rats, they develop a biotin deficiency. Avidin is a protein with a molecular weight of not more than 70,000 as determined by sedimentation data, although other methods indicate a slightly lower value. It was the first naturally occurring substance shown to possess antivitamin activity. The manner in which it renders biotin unavailable to an animal when raw egg white is fed in the ration has not been discovered. It has been shown that blocking most of the reactive groups on the protein molecule, amino, carboxyl, guanidyl and amide, and imidazole and phenolic, does not interfere with the activity. Some combination of protein and biotin must occur, of course, but the nature of the bonds which hold the biotin to the avidin are still obscure.

Egg Yolk. The proteins of egg yolk were first studied over 100 years ago. Then followed a long period in which little attention was devoted to them. It is only in recent years that they have been reinvestigated. When egg yolk is diluted with water, protein precipitates. When the yolk is heated, the proteins undergo heat denaturation and precipitate. Egg yolk appears to contain at least two lipoproteins, *lipovitellin* and *lipovitellenin.* Lipovitellin contains between 17 and 18 per cent lipid, while lipovitellenin contains 36 to 41 per cent lipid, mainly lecithins. The protein portions of these conjugated proteins are called *vitellin* and *vitellenin,* respectively. They can be prepared from the lipoproteins by exhaustive extraction with 80 per cent alcohol. They are phosphoproteins and they appear to be very similar except for the amount of phosphorus present. Vitellin contains approximately 1 per cent phosphorus, while vitellenin contains only 0.29 per cent.

Egg yolk also contains water soluble protein that does not precipitate on dilution of the yolk. This fraction is called *livetin.* It can be prepared by precipitation on half saturation with ammonium sulfate, followed by thorough extraction with alcohol-ether at $-51°C$ to remove lipids. Electrophoretically there appear to be three components present in the fraction. The enzymatic activity of the yolk is associated with this fraction.

The *membranes* within the shell and around the yolk are composed of proteins which belong to the class of either keratins or mucins. Fevold[8] describes how by staining techniques Moran and Hale indicate that the shell membrane is composed of an outer keratin layer, then two mucin layers followed by another membrane of a keratin and a mucin layer. The yolk membrane appears by their method to consist of a mucin, keratin, and mucin membrane.

REFERENCES

1. ALMQUIST, H. S., "Nutrition," *Ann. Rev. Biochem.*, **20**, 317–322 (1951).
2. BERRIDGE, N. J., "Rennin and the Clotting of Milk," *Advances in Enzymol.*, **15**, 423–448 (1954).
3. BLOCK, R. J. and ALEXANDER, P., eds., "Laboratory Manual of Analytical Methods of Protein Chemistry," Pergamon Press, New York, N. Y., 1960.
4. BLOCK, R. J. and BOLLING, D., "Amino Acid Composition of Proteins and Foods: Analytical Methods and Results," 2nd ed., Charles C. Thomas, Springfield, Ill., 1951.
5. BLOCK, R. J. and WEISS, K. W., "Amino Acid Handbook: Methods and Results of Protein Analysis," Charles C. Thomas, Springfield, Ill., 1956.
6. EDSALL, J. T. and WYMAN, J., "Biophysical Chemistry," Academic Press, Inc., New York, N. Y., 1958.
7. FERRY, J. D., "Protein Gels," *Advances in Protein Chem.*, **4**, 1–79 (1948).
8. FEVOLD, H. L., "Egg Proteins," *Advances in Protein Chem.*, **6**, 187–252 (1951).
9. FOX, S. W. and FOSTER, J. F., "Introduction to Protein Chemistry," John Wiley & Sons, New York, N. Y., 1957.
10. HILL, R. L., KIMMEL, J. R., and SMITH, E. I., "The Structure of Proteins," *Ann. Rev. Biochem.*, **28**, 97–144 (1959).
11. IDSON, B. and BRASWELL, E., "Gelatin," *Advances in Food Research*, **7**, 236–339 (1957).
12. KIRK, P. L., "The Chemical Determination of Protein," *Advances in Protein Chem.*, **3**, 139–167 (1947).
13. LUGG, J. H., "Plant Proteins," *Advances in Protein Chem.*, **5**, 229–304 (1949).
14. McMEEKIN, T. L. and POLIS, B. D., "Milk Protein," *Advances in Protein Chem.*, **5**, 201–228 (1949).
15. MELNICK, D. and OSER, B. L., "The Influence of Heat Processing on the Functional and Nutritional Properties of Protein," *Food Technol.*, **3**, 57–71 (1949).
16. NEURATH, H. and BAILEY, K., eds., "Proteins: Chemistry, Biological Activity and Methods," 2 vol., Academic Press, Inc., New York, N. Y. (1953, 1954).
17. RICE, E. E. and BEUK, J. F., "The Effect of Heat on the Nutritive Value of Protein," *Advances in Food Research*, **4**, 233–279 (1953).
18. SCATCHARD, G., ONCLEY, J. L., WILLIAMS, J. W., and BROWN, A., "Size Distribution in Gelatin Solutions," *J. Am. Chem. Soc.*, **66**, 1980–1981 (1944).
19. SWANSON, P. P. and CLARK, H. E., "The Metabolism of Proteins and Amino Acids," *Ann. Rev. Biochem.*, **19**, 235–260 (1950).
20. WAUGH, D., "Protein-protein Interactions," *Advances in Protein Chem.*, **9**, 326–439 (1954).

CHAPTER FIVE

The Flavor and
Aroma of Food

FLAVOR IS THE SUBTLE and complex sensation that is the source of much of the delight man finds in food. To both connoisseur and ordinary man, flavor is of utmost importance in regulating preferences. It is the difference between a cheap wine and the most expensive, the highest grade butter and the lowest, rancid cookies and fresh.

THE SENSATION OF FLAVOR

Flavor is a combination of taste, smell, and feel. In the mouth and pharynx are many taste buds capable of detecting sweet, sour, salty, and bitter. In the nose are olfactory endings that can detect a huge number of different odors. The "mouth feel" of a food is likewise part of its flavor—whether it is smooth or rough, tender or tough, unctuous so that it clings to the tongue and roof of the mouth or watery so that it slides down readily. "Aftertaste" is a quality for the most part that is the combination of all of these sensations after the particle of food has been swallowed. However, even in aftertaste the sensation of feel has specific importance since the stickiness or greasiness of the small amount remaining in the mouth and on the teeth contributes to the general aftertaste.

Taste

This sense is detected through the solution of soluble compounds in the saliva or in the food juices and the contact of those dissolved compounds with the taste buds. The commonly accepted theory (although not without challengers) of taste is that there are four primary tastes that can be detected: sour, sweet, salt, and bitter. The areas in which these tastes are detected overlap, but the sensation of sour is most readily detected on the

FIGURE 5.1. THE FLAVOR OF FOODS IS ONE OF THE DELIGHTS OF EATING. Orange, lemon, and vanilla are natural materials often added in small amounts to food for their flavor.

Courtesy of Food Materials Corp.

sides of the tongue, salt on the sides and tip, sweet on the tip, bitter at the back of the tongue and on the pharynx. The taste buds are present in greatest number in the vallate papillae, the tiny nipple-shaped elevations distributed in a V on the tongue. They are also present in papillae on the rest

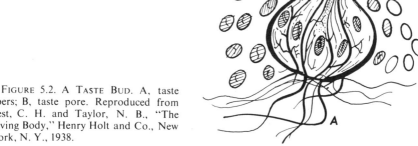

FIGURE 5.2. A TASTE BUD. A, taste fibers; B, taste pore. Reproduced from Best, C. H. and Taylor, N. B., "The Living Body," Henry Holt and Co., New York, N. Y., 1938.

of the tongue, the soft palate, pharynx, and epiglottis. The taste buds are composed of a number of cells arranged in a tiny well around a nerve ending.

A compound to be tasted must occur in solution or dissolve in the saliva. The solution seeps into the taste bud and the compound stimulates the nerve ending. An impulse is transmitted along a nerve to the brain and we recognize a taste. Then more saliva washes the solution out of the tiny well.

Taste thresholds are measures of the sensitivity to a given taste in terms of the detectable concentration. A common method for measuring the threshold of, say, salt, is to give a number of samples with varying concentrations of sodium chloride, a number of samples of pure water, and sometimes of other compounds with other tastes. The subject attempts to detect the taste present, rinsing the mouth after each attempt. The lowest concentration consistent for an individual is his taste threshold for a particular substance. The low levels at which tastes can be detected is truly amazing. In a group of students almost everyone could detect 0.087 per cent sodium chloride and 0.4 per cent sucrose. Considerable variation between individuals occurs—a low threshold for one taste not always accompanied by a low threshold for the others.

Influence of Chemical Constitution. Taste depends on a number of factors, the most important of which is chemical constitution. Saltiness is a

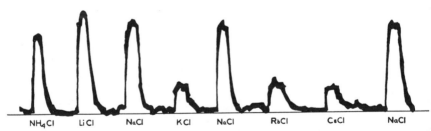

FIGURE 5.3. INTEGRATED ELECTRICAL RESPONSE OF CHORDA TYMPANI TO VARIOUS 0.1 M CHLORIDE SOLUTIONS FLOWED OVER TONGUE OF RAT. Reproduced from Beidler, L. M., "Techniques and Methods for Research in Flavors, Chemistry of Natural Food Flavors— A Symposium," Dept. of Army, Research and Development Command, Natl. Acad. Sci. and Natl. Research Council, Washington, D. C., 1957.

property of electrolytes, and the halides particularily. In order of their saltiness, the following ions are detectable: Cl , Br , I , SO_4 , NO_3 . The cations also influence the taste. Na^+ and Li^+ salts tend to have salty tastes while K^+, salts have considerable bitterness.

Sweetness is found in a number of organic compounds. Alcoholic hydroxyls tend to endow a compound with sweetness, but other groups are sometimes effective. Saccharin, for example is a totally different compound from sucrose, possessing no hydroxyl groups, but rather a sulfon-amide. Sucaryl with 30 times the sweetness of sucrose is also a sulfonamide.

$$\text{Saccharin}$$

Saccharin   Sucaryl Sodium

Generalizations concerning groups that make for sweetness in organic compounds are difficult.

Bitterness is a property of some organic and inorganic compounds. Some of the alkaloids such as quinine and brucine are exceedingly bitter, and NH_4^+, Mg^{++}, and Ca^{++} are also bitter.

Sourness is a property of the hydrogen ion; its concentration is of primary importance in determining whether or not the sensation of sour is detected. The acids that occur in foods are organic with relatively low ionizations and consequently relatively low hydrogen ion concentrations. Added to this is the common occurrence of a salt of the organic acid in the food, which further lowers the ionization by common ion effect. Some workers believe that aside from the effect on ionization, the anion likewise has an influence on the perception of sourness. Research upon rats concerning their nerve response to acids shows variation in response when hydrogen ion concentration is constant. (See Figure 5.4.)

Influence of Other Factors. Taste is also influenced by temperature, texture, and the presence of other compounds. Mackey and Valassi[14] measured the taste thresholds for the four primary tastes in water as well as in tomato juice and in egg-milk custard prepared as liquid, gel, and foam. They found that the primary tastes were harder to detect in gels than in liquids and that the foams were intermediate.

FIGURE 5.4. RESPONSE OF RAT TO VARIOUS ACIDS. The large peaks from left to right show response to hydrochloric, citric, formic, oxalic, acetic, and hydrochloric acids. Reproduced from Beidler, L. M., "Techniques and Methods for Research in Flavors, Chemistry of Natural Food Flavors—a Symposium," Adv. Board on Quartermaster Research and Development, Dept. of the Army, Washington, D. C., 1957.

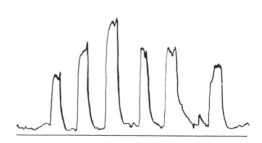

The effect of one primary taste on another is well known to everyone who cooks. Salt tends to decrease the sweetness of sucrose, so that candy to which salt is added has a richer, less sweet taste than that without salt. Sugar likewise can tone down and round out saltiness. Many a cook has salvaged the mashed potatoes that were too salty, by adding a small amount of sugar. The effect of sweet on sour, such as sugar on lemon juice or vinegar, is also important. Any food quite sour and yet quite sweet has a rich, full flavor that is lacking when either sourness or sweetness is present alone. Salt also tones down sourness of food acids. This toning down of sensation when two tastes are presented simultaneously is called *compensation*.

Successive contrast, on the other hand, tends to sharpen a sensation. Grapefruit seems unusually sour if eaten immediately after a sweet cereal. A plum tastes quite sour after candy but quite sweet after grapefruit.

An interesting phenomenon has been discovered in relation to the taste of a few compounds. Sodium benzoate tastes quite different to different individuals and the threshold at which it can be detected is very different. Phenyl thiocarbamide (PTC) has been extensively studied: "taste blindness" to this compound is inherited according to Mendelian ratio. If both parents are nontasters, the children do not taste PTC. Women tend to have a slightly lower threshold than men do for this compound. Taste blindness in women occurs in 22.2 per cent and in men 25.9 per cent.

Fatigue for taste does not occur rapidly and some physiologists do not believe it ever occurs. However, it is a common experience to notice a diminution in the taste of a food as one continues to eat it. Some experiments indicate that fatigue occurs for one taste but the others are not affected or are even enhanced.

Odor

We can detect differences in the odor of thousands of compounds impinging on the olfactory nerve endings. The sensation is experienced, of course, only when the nerve impulse is transmitted to the brain and recorded there. Our remarkable ability to remember odors and the ability of the brain to receive and detect differences in odors cannot be adequately explained at present. Many people can recognize odors that they have experienced only once, many years before. The author once heard a 70 year old woman proclaim that a noxious odor must be a stinkhorn mushroom since she remembered smelling one as a girl. The hoots of derision were quieted when a stinkhorn was found to be the source of the odor. Many individuals have experienced this remarkable recall for odors.

The aroma of food is recognized as a tantalizing and delectable prelude

FIGURE 5.5. OLFACTORY NERVES. Shown at bottom are cells of olfactory mucous membrane whose nerve fibers pierce skull and reach processes of olfactory bulb cells. Reproduced from Best, C. H. and Taylor, H. B., "The Living Body," Henry Holt and Co., New York, N. Y., 1938.

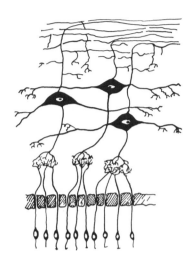

to eating but not always as an invaluable part of flavor. We have only to suffer loss of our ability to detect odor through congestion of the nose, to realize what a contribution it makes. Food tastes salty, sweet, sour, and bitter when we have a bad cold, but how flat and uninspired it is without odor. All the subtle nuances of the flavor complex are gone.

The olfactory mucous membrane is located in the upper part of the nasal cavity on portions of the turbinates and septum, and contains olfactory nerves imbedded in a special epithelium. Glands are present that secrete fluid. In order for a chemical compound to possess an odor it must vaporize and pass into the nasal cavity. It is not known whether the molecules dissolve in the fluid on the lining and the solution comes in contact with the olfactory cells or whether the hairs of the cells penetrate the mucous and come in contact with gas. The nerve then transmits an impulse to the brain. Allison and Warwick have shown that a rabbit has about 100,000,000 receptors, each with 6 to 12 hairs. This is a tremendous area for receiving odor stimuli. The detection of an odor during breathing occurs through the diffusion of the gas into the still air of the upper cavity. The olfactory endings are not located on the epithelium in the direct path of the inspired air. When we detect a flavor in a mouthful of food, the odorous compounds change to gases and diffuse through the pharynx into the nose. As we raise the food to our mouth, we inhale its odor.

The sense of smell varies considerably between individuals and is, of course, much less keen in man than in many other animals. Nevertheless, it is still remarkably keen for some compounds. Vanillin can be detected by most individuals at a concentration of one part in ten million.

Fatigue for odors occurs very quickly. We have all observed how strong an odor of a food such as coffee or meat will be when we first come into a building or room where it is being prepared and how quickly we loose that sensation. Students who do not work in chemistry, often complain of the odors of the laboratory while those of us who work there constantly, do not detect them unless the concentration suddenly increases or a new gas is formed. People have been overcome by hydrogen sulfide, which has an unusually strong odor, after fatigue of the olfactory sense was so depressed that they forgot the concentration was high.

Although fatigue for odors occurs rapidly, it is selective. We may not be able to smell the odor of a compound after a short period of time, but the sense of smell for another odor is unimpaired.

Testing Sense of Smell or Aroma. Elsberg[6] devised a method for testing acuity of smell. A flask is filled with cotton and between the layers is a toothpick, cotton wrapped, that has been dipped in a solution of some essential oil or odorant, often a pure chemical. Air is admitted to the flask in measured amounts and the vapors sniffed. The amount of air required to perceive the odor and also the concentration of the solution on the cotton give a measure of the acuity of the individual to that particular odor. Some individuals have high acuity of smell but cannot readily recognize odors. Thus, another part of the test often introduced is proper description of the odor perceived.

Some physiologists are now studying taste and odor responses in animals and attempting to obtain objective data that will make possible an understanding of these phenomena. Since the impulse that travels over a nerve is electrical, measurement of the potential of a nerve shows the effect of a stimulus. It is possible in a living animal to make measurements on the large nerve bundle that serves many taste buds, on an isolated fiber showing the response of several taste cells in one taste bud, or by inserting microelectrodes to make measurements on a single taste cell. Olfactory response is more difficult to measure although record of the activity of the olfactory nerve can be made. It is also possible to measure the change in potential in the presence of an odorant across olfactory tissue after it is excised from the animal. None of these studies is easy, but it is here that we can hope to find better understanding of the physiology of taste and odor.[15]

Feeling

Some compounds not only stimulate the olfactory ending but also that of the trigeminal nerve—a nerve with endings in the skin of the face, tongue and teeth, giving a general sensation. Ammonia, a common substance that stimulates this nerve, has an odor for us but also another sensation described as "tingling" or "sharp."

Other "feel" factors contributing to the over-all sensation of flavor are the hot, burning effect of some peppers and spices as well as the coolness of peppermint. The texture of the food is also important: grainy fudge tastes sweeter but not as rich as creamy fudge. Crispness and stickiness are other sensations of mouth feel. Some wines contribute to flavor by seeming to fill the mouth and cover the pharynx immediately after a small sip.

Blends

A blend of sensations occurs with most foods so that sorting out all of them is difficult. This is often highly important to the satisfaction that a superbly delectable dish affords. Many foods have a slightly sour note that most people do not separate as "sourness" but simply as part of the delightful flavor. Broccoli, for example, has a volatile acid which is part of its fine flavor and most honey contains appreciable amounts of acid. Occasionally clements of flavor appropriate and desirable in some foods are highly repugnant when present in others. For example beer has a sour, horsey aroma and some ripened cheeses have an element of rancidity in the odor but, because it is expected and blended, it is pleasing. In the Arthur Little Flavor Laboratory scientists have tried blends of many ingredients, pure compounds and extracts, and conclude that in blending various reactions may result. With two odorants: (1) some of the major notes of each odorant may be suppressed; (2) all of the major notes of each odorant may be suppressed; (3) some of the major notes of one odorant are suppressed but none of those of the other odorant; (4) a complete blend takes place with the formation of a new odor; or (5) a partial blend occurs with a new odor formed but some characteristics of each odorant retained. There is still much to learn about blending, masking, and enhancement when two or more odorous compounds are mixed.

Appropriateness is of great importance in our enjoyment of flavors in foods. An onion flavor may be delectable in a stew or soup but objectionable in a custard. We become conditioned to expect certain sensations from certain foods and while a slight variation is titillating, a completely unexpected taste is unacceptable. Most of us have a "sweet tooth" and relish anything sweet. But while we may find that chicken prepared with the addition of a small amount of thyme is exceedingly interesting because the flavor departs a little from that which we usually experience, chicken prepared with a strongly sweet sauce may be rejected as improperly prepared and unappetizing.

Aftertaste is part of the sensation of flavor of some foods. It is particularily common with foods that leave a residue in the mouth after swallowing. Sticky foods such as syrup or greasy foods such as meat fat tend to leave a residue coating the mouth for a short time after swallowing and all or part of the sensations which are part of flavor, continue to be experienced. The persistence of the flavor is highly desirable in some foods, although usually we do not enjoy the sensation if it persists too long. Many people do not eat raw onions because of the persistent aftertaste. The volatile disulfides present are absorbed by the blood and are then secreted in the saliva and exhaled in the breath, so that the odor may persist for many hours.

CONTROL OF FLAVOR AND AROMA IN PROCESSED FOOD

Control of flavor and aroma in processed food is of utmost importance in determining the quality and selling price of the finished product; this is likewise true of simple cookery operations. Many factors, all of which must be considered, influence the flavor. The quality of the ingredients has, of course, a considerable effect on the product. Off-flavor ingredients cannot produce an item with excellent flavor. Conditions of processing must be carefully controlled. Thus in roasting coffee or cocoa beans the temperature and length of time roasted must be controlled in order to insure flavor of the right quality. Avoidance of contamination by flavorful compounds during processing or storage must be watched. Thus the lacquer lining of cans cannot contain soluble flavored substances. Many foods such as fats, and baked goods, pick up odors of other foods readily and must be adequately protected from contact with these odors. Foods must also be protected from contamination by bacteria and molds and stored under conditions where these microorganisms cannot grow rapidly. Molds produce compounds that impart a musty or sour flavor which is completely unacceptable.

In many food industries, tasters are employed to grade the product and often the raw materials. There are usually no chemical tests that can differentiate the subtle differences between a quality product, an ordinary one and one that is off-flavored. With a little experience, some tasters develop remarkable ability to differentiate products. Some can readily detect flaws in the processing operation and spot the location of the trouble by taste. In many industries these tasters are of great value to the company since flavor is one of the bases on which the consumer buys the product.

Good flavor in a food product is desired by everyone who works in food production whether a housewife or the director of a nation-wide industry.

Caul[3] has analyzed the pattern of good flavor as the following sensations: "(1) an early impact of appropriate flavor; (2) rapid development of an impression of highly blended and usually full-bodied flavor; (3) pleasant mouth sensations; (4) absence of isolated unpleasant notes; and (5) anticipation of the next mouthful."

Measurement of Flavor

The sensations of flavor are such complex reactions that it is impossible by any simple chemical or physical test to measure them. An appreciation of flavor depends on a human being and his reactions to the actual act of tasting. Since the flavor of food products is so important in determining their commercial value, many methods for measuring flavor have developed. These have been reviewed at various times. A recent review of the methods and their implications is in a symposium[7] in *Food Technology* for 1957.

Expert tasters are employed in some industries, particularly for wine, whiskey, tea and coffee, and spices. The expert taster has usually grown up in one industry and is of value only to this industry. Through interest and opportunity he has developed a "discriminating palate" for the food of his company. There is no indication that an expert taster has a keener sense of taste than the normal individual, but rather that he has years of training.

Other methods use a panel of tasters; sometimes these are individuals with some training, sometimes they are workers in the industry with sufficient interest to be willing to serve on a panel, and occasionally where preference is desired, they are large numbers of consumers. The panel may indicate its preferences or judgment of quality by scoring a food on some well defined qualities. Often a numerical scale is used so that the scoring of the individuals can be added readily to give a composite score. Scoring has been widely used in meat, bakery, and milk products and in recent years with dehydrated foods.

Difference tests are sometimes used in an attempt to get a more precise and reproducible test of flavor in foods. A trained panel is required for these tests. Boggs and Hansen[1] have summarized the various methods of difference testing. Either a pair of samples or a triangle in which two of the samples are identical and the third is to be separated are presented to the judges. Sometimes in a triangle a standard is submitted and one of a pair is matched to it. In other tests the judges only know that two samples are identical.

The dilution test is a type of difference testing. A sample is presented to the judges and then other samples that may or may not contain the unknown at a definite level of dilution are offered. The method was first applied to the detection of dried egg in fresh egg. The higher the quality of the dried egg the greater the difficulty of detecting it. In order to obtain a high degree of agreement, it would be necessary to have a high per cent of dried egg present. A poor quality product would be detectable at a low concentration or a high dilution.

Ranking is used with a flavor panel and with either a numerical or alphabetical evaluation of one property. A series of samples are supplied to each member of the panel and he arranges them in the order of increasing or decreasing quality of the characteristic. A high degree of agreement between members of a panel occurs when the samples differ to a marked degree; but when differences are subtle, agreement is not as common.

The *flavor profile* is a method for evaluating a flavor by describing it either as a whole or by characteristics. The vocabulary by which these characteristics are described are comparative terms such as "rubbery" and "eggy;" or, when possible, they are referred to a compound—phenyl acetic acid for "horsey." Feeling effects are described on the basis of the mouth, throat or nose reaction, such as "throat burn," "puckering," and "cooling." The reactions are broken down into (1) character notes, (2) order of appearance, (3) aftertaste, and (4) amplitude. The character notes are the protruding sensations, often exceedingly difficult to distinguish in a blend. The order of appearance may appear at first sight to be of little importance or even absent. But close attention to flavor during tasting will reveal that many flavors are not perceived as a whole but rather as a series of sensations. These come very rapidly but can be separated. Coffee bitterness is not perceived until after the other flavors have been experienced. Usually an unpleasant note should not be either the first or last sensation of the flavor. Aftertaste has been mentioned before, and its importance in the taste complex is readily recognized. "Amplitude" is the term used to express the total effect of flavor and aroma. The fitness of the flavor, the broad aspects of all characteristics is included here. It requires some understanding and experience to judge amplitude. It is rated as very low, low, medium or high. Examples of amplitude can show the meaning of this term. Tomatoes picked green have low amplitude, those grown in a hot house are medium, and the vine-ripened are high. Potatoes, boiled but unseasoned, have low amplitude; salt added at the table improves the amplitude by blending the flavor but salt added in preparation gives high amplitude.

Any evaluation by a flavor panel is subject to error. Many of the precautions necessary to the satisfactory use of a panel have been explored and described. It is essential that the members work alone in quiet, restful surroundings. Usually air-conditioned booths are used so that even the

FIGURE 5.6. OLFACTORIUM FOR TESTING ODORS. Courtesy of Food Acceptance Branch, Quartermaster Food and Container Institute for Armed Forces, Chicago 9, Ill.

temperature and the humidity can be controlled. To avoid influencing one another, the panel members do not discuss their reactions with each other. Even the day of the week and the hour of the day has an effect. Mitchell[16] found that the sensitivity of a duo-trio test where two samples out of three are matched, is greatest on Tuesday with Friday also better than Monday, Wednesday, or Thursday. In an eight hour day accuracy of judgement was the best during the fourth, fifth, and sixth hours.

Since our appreciation of flavor and our judgement are influenced by many psychological factors, the environment must be carefully controlled. However, some of the psychological factors that cannot be controlled are the experience background of each tester and his personal problems and reactions. Food experience affects everyone. If you have frequently eaten food highly seasoned with garlic, you will usually like that flavor but it will have nothing of newness or surprise for you. A person adventurous in other life experiences will find pleasure in a new eating experience, while the more conservative will dislike anything unfamiliar.

The *consumer acceptance* or *preference* has been used in recent years for some products. If a new food is introduced, only one sample is offered to a large consumer panel, but if a food is modified, then two samples are submitted and a preference requested. Occasionally more than two samples are submitted but usually this leads to confusion in the subjects and the results are doubtful.

Flavor Intensifier: Monosodium Glutamate

Large quantities of monosodium glutamate have been used in the Orient for years and in the United States for approximately 2 decades as a flavor intensifier. In China and Japan fermented soybean curd is added to many foods and soy sauce is a common condiment. Soy sauce is rich in monosodium glutamate and interest led eventually to the production of relatively pure monosodium glutamate. It is claimed by some authors that this compound has little or no flavor itself but intensifies the flavor of meats and vegetables through a rounding or blending effect. It is widely used in stews, canned meats, soups, chowders, etc. It does not have an effect on fruits or fruit juices or sweet spicy foods. Cairncross and Sjöström[2] reported that it suppresses the sharpness of onion, the rawness of many vegetables, and the earthiness of potatoes. They also reported that the effect is noticeable in the pH range 3.5 to 7.2, the range in which most foods are eaten. They found that fats, oils, and high viscosity in foods modify the influence of sodium glutamate considerably.

The mode of action of sodium glutamate and its effect on flavor are still in dispute. Much of the work carried out has been an investigation of the preference of tasters for a food with and without added monosodium glutamate, and the underlying physiology and biochemistry of the effect.

Glutamic acid occurs in high concentration in numerous proteins. Wheat gluten for example contains approximately 36 per cent. The free glutamic acid hydrochloride is prepared commercially from wheat, corn or sugar beet wastes by acid hydrolysis of the protein. Glutamic acid is formed by dissolving the crude hydrochloride, adjusting the pH to 3.2, and allowing it to crystallize slowly. The product is neutralized with either sodium hydroxide or sodium carbonate, decolorized and crystallized. The product is relatively pure monosodium glutamate.

$$
\text{Protein} \xrightarrow{\text{HCl}}
\begin{array}{c} \text{COOH} \\ | \\ \text{CH}_2 \\ | \\ \text{CH}_2 \\ | \\ \text{CHNH}_2 \cdot \text{HCl} \\ | \\ \text{COOH} \end{array}
\xrightarrow{\text{NH}_4\text{OH}}
\begin{array}{c} \text{COOH} \\ | \\ \text{CH}_2 \\ | \\ \text{CH}_2 \\ | \\ \text{CHNH}_2 \\ | \\ \text{COOH} \end{array}
\xrightarrow{\text{NaOH}}
\begin{array}{c} \text{COONa} \\ | \\ \text{CH}_2 \\ | \\ \text{CH}_2 \\ | \\ \text{CHNH}_2 \\ | \\ \text{COOH} \end{array}
$$

Glutamic Acid Hydrochloride   Monosodium Glutamate

Very large amounts of this material are now used commercially in the production of many meat and vegetable dishes. Although it has been claimed that it is not detectable and although it is true that pure monosodium glutamate is only slightly sweet and salty, the author has noticed that the presence of this "intensifier" at the level usually used in foods can be recognized. It gives a sameness of flavor to foods as dissimilar as chicken or tuna pies and makes the product markedly different from the freshly prepared dish. On this basis, it is highly objectionable to her.

Flavoring Extracts

These are widely used in the food industry, and in the United States are quite well standardized. The Food and Drug Administration defines an extract as "a solution in ethyl alcohol of proper strength of the sapid and odorous principles derived from an aromatic plant, or parts of the plant, with or without its coloring matter, conforming in name to the plant used in its preparation." If artificial coloring or synthetic flavoring compounds are added, it is required that the label state this. The flavoring extracts are made either by adding the essential oil to alcohol or to water and alcohol, or by percolating the chopped plant or plant part with a mixture of water and alcohol. Thus lemon extract is prepared by adding the lemon oil expressed mechanically from lemons, to alcohol or diluted alcohol. The Food and Drug Administration's standards require 5 per cent by volume of lemon oil. Also acceptable is an extract prepared after the terpenes, which add little to the flavor, have been removed.

Vanilla extract, the most widely used extract in the United States, is prepared by percolating macerated vanilla beans with alcohol, frequently in the presence of glycerol or sucrose. The flavorful compounds extracted are numerous but the most abundant is vanillin. Since vanillin can be prepared synthetically very cheaply, synthetic vanilla is sold in large quantities. Its flavor is not quite as fully rounded as real vanilla extract but it is used widely in many foods selling at relatively low cost. Usually coumarin is

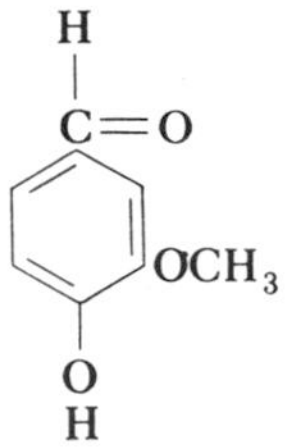

Vanillin

added to the vanillin in the synthetic extract to improve the flavor, and caramel or a food dye to simulate the brown color extracted from the vanilla
beans. Discussion of the methods of preparing flavoring extracts can be
found in a textbook of food technology.

Synthetic Flavoring Substances

These have been produced for many years. Originally esters were used in
small amounts for imitation flavors, particularly in cheap confections. Although an ester such as amyl acetate is reminiscent of banana flavor, simple
esters are crude, rough substitutes for natural flavor. Analysis of natural
flavors has developed slowly because of the difficulty of separating substances that occur in a concentration as low as that in most foods. New
methods of study (see below) are now opening the way for careful and detailed analysis of the flavoring compounds in many foods. During the past
50 years numerous compounds have been synthesized which are useful in
synthetic or imitation flavors. The common so-called *flavormatics* are listed
in a table prepared by the Food Additive Committee of the Flavoring Extract Manufacturers Association of the United States. (See Table 5.1.)
These are arranged in order of their importance. The type of flavor in
which they are usually used and the parts per million concentration in the
finished food are given.

Table 5.2 shows the composition of a few imitation flavors. It is interesting because it shows the complex nature of some of these imitation flavors.
The list of compounds isolated from natural flavors is far from complete.
Many more compounds have now been identified in strawberry, apple and
coffee flavors.

RECENT DEVELOPMENTS IN FLAVORING RESEARCH

Recent developments in flavoring research have followed three principal
paths: (1) careful identification of the composition of some flavors, particularly through the use of gas chromatography as a means of separating

(text continued on p. 165)

TABLE 5.1. THE PRIMARY FLAVORMATICS*

Aromatic Chemical	Flavor	Avg. Use (P.P.M.)	B.P. °C
Group I			
Vanillin	vanilla	31.5	81.5° MP
Ethyl Vanillin	vanilla	16.6	76.5° MP
Citral	lemon	17.6	229°
Benzaldehyde	cherry, almond	84.8	180°
Cinnamic Aldehyde	cinnamon, cola	110.7	252°
Methyl Salicylate	root beer wintergreen	129.7	221°
Amyl Acetate	banana, fruit flavors	78.4	142°
Amyl Butyrate	banana, fruit flavors	23.7	185°
Ethyl Acetate	fruit, rum	92.7	76°
Ethyl Butyrate	strawberry, fruit	31.1	120°
Methyl Anthranilate	grape	97.1	255°
Ethyl Oxy Hydrate	rum	472.7	
Anethole	root beer, anise	133.5	235.3°
Safrol	root beer, anise	16.9	236°
Menthol	mint	111.2	215°
Group II			
Heliotropine	vanilla, cherry	3.4	263°
Aldehyde C-18	coconut	17.6	263°
Diacetyl	butter	17.3	88°
Aldehyde C-16	strawberry	10.0	271°
Ethyl Oenanthate	grape	8.1	187°
Allyl Caproate	pineapple	11.7	187°
Aldehyde C-14	peach	8.0	297°
Eugenol	spice	48.8	253°
Amyl Valerinate	peach, fruit	9.6	200°
Ethyl Valerinate	fruit	14.4	145°
Carvone	spice, mint	190.3	230°
Anis Aldehyde	cherry, vanilla	3.6	247°
Group III			
Benzyl Acetate	strawberry, fruit	8.8	215°
Ethyl Formate	rum, strawberry	77.0	54°
Ionone	raspberry	0.9	147.5°
Ethyl Lactate	grape	42.1	154°
Tolyl Aldehyde	cherry	9.3	204°
Amyl Caproate	pineapple, apple	4.4	222.2°
Linalool	spice	30.2	199°
Citronellal	rose (general)	14.20	206°
Butyl Acetate	fruit	26.4	126.5°
Ethyl Pelargonate	grape, fruit	2.3	228°
Butyl Butyrate	fruit	17.9	166.4°
Aldehyde C-10	citrus	1.2	208°

TABLE 5.1 (*Continued*)

Aromatic Chemical	Flavor	Avg. Use (P.P.M.)	B.P. °C
Ethyl Heptoate	grape, pineapple	4.6	172.1°
Geranyl Acetate	fruit	8.4	245°
Methyl Phenyl Acetate	honey	2.4	222°
Group IV			
Aldehyde C-8	citrus	0.71	165°
Aldehyde C-9	citrus	2.81	191°
Cyclohexyl Butyrate	pineapple, fruit	8.3	212°
Ethyl Laurate	spice, fruit	6.9	269°
Linalyl Acetate	citrus	5.4	220°

*From Janovsky, H. L., *Food Technol.*, **9**, 500–502 (1955).

TABLE 5.2. SYNTHETICS*

	Aromatic Isolated	Sample Formula	%
Raspberry	beta ionone	beta ionone	10
	anisic aldehyde	iso-butyl acetate	23.49
	benzaldehyde	anisic aldehyde	1
	phenyl ethyl alcohol	phenyl ethyl alcohol	3
		phenyl ethyl iso-butyrate	4
	hexene	ethyl methyl p-tolyl glycidate	35
	n-hexanol	phenyl ethyl anthranilate	7
	iso-amyl alcohol	vanillin	3
	iso-butyl alcohol	hexyl butyrate	0.5
		iso-amyl acetate	4
		ethyl n-butyrate	5
		rose oil	4
		diallyl sulfide	0.01
Strawberry	borneol	ethyl methyl phenyl glycidate	35
	methyl caproate	amyl aldehyde	0.5
	amyl aldehyde	bornyl acetate	0.5
	ethyl caproate	ethyl caproate	2
		vanillin	3
		beta ionone	9
		ethyl methyl p-tolyl glycidate	12
		iso-butyl acetate	30
		ethyl butyrate	8
Cherry	geranyl butyrate	cinnamyl anthranilate	3
		iso-amyl acetate	12
	ethyl caprate	iso-butyl acetate	12
	terpenyl butyrate	cinnamic aldehyde dimethyl acetal	3
	ethyl caproate	benzyl alcohol	5.5
	benzaldehyde	geranyl butyrate	2
	geraniol	ethyl caprate	4

TABLE 5.2 (*Continued*)

	Aromatic Isolated	Sample Formula	%
cherry		terpenyl butyrate	0.5
(continued)		ethyl caproate	2
		benzaldehyde	8
		p-tolyl aldehyde	3
		p-methyl benzyl acetate	11
		ethyl methyl p-tolyl glycidate	16
		vanillin	7
		heliotropin	5
Peach	linalyl esters (as	linalyl formate	10
	valerate, acetate,	gamma undeca lactone	13
	formate)	linalyl butyrate	4
	furfural	heliotropin	14.5
		geranyl valerate	15
		furfural	1.5
		alpha-methyl furyl acrolein	6
		methyl cyclo pentenolone valerate	10
		benzaldehyde	5
		iso-amyl formate	15
		iso-butyl butyrate	6
Apple	geraniol	geranyl valerate	10
	acetaldehyde	geranyl n-butyrate	8
		geranyl propionate	8
		linalyl formate	10
		iso-amyl valerate	15
		vanillin	8
		allyl caprylate	6
		geranyl aldehyde (citral)	5
		acetaldehyde	6.5
		methyl cyclo pentenolone valerate	8
		alpha methyl furyl acrolein	2
		iso-amyl butyrate	13.5
Coffee	α-furfuryl mercaptan	α-furfuryl mercaptan	10
		ethyl vanillin	3
		solvent	87

*From Katz, A., *Food Technol.*, **9**, 636–638 (1955).

the large number of compounds that occur in low concentration in many natural foods, (2) production of natural essences by condensation of the vapors formed in vacuum or low temperature concentration of foods, and (3) the search for flavor precursors and their enzymes. As the food industry moves farther along the path of prepared or partially prepared foods, the problem of flavor in the final product becomes more trouble-

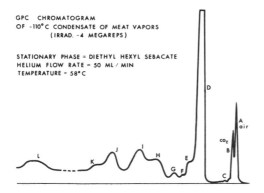

GPC CHROMATOGRAM
OF -110°C CONDENSATE OF MEAT VAPORS
(IRRAD. -4 MEGAREPS)

STATIONARY PHASE = DIETHYL HEXYL SEBACATE
HELIUM FLOW RATE = 50 ML / MIN
TEMPERATURE = 58°C

FIGURE 5.7. GAS CHROMATO-GRAM OF MEAT VAPORS. A is the pip for air, B for CO_2, and the subsequent rises are various compounds present in the vapor. Reproduced from Pilgrim and Schutz, "The Chemistry of Natural Food Flavor—a Symposium," Dept. of Army, Research and Development Command, Natl. Acad. Sci. and Natl. Research Council, Washington, D. C., 1957.

some. With the concentration of populations in cities and the rise in standard of living, a growing demand for quality food occurs at the same time as a need for food products stable enough to allow time for distribution. Maintenance of flavor is one of the chief problems in producing quality food.

The gas chromatograph is a fine instrument for the separation of the many compounds that make up the flavor complex in a natural food. A column is packed with finely divided particles of a solid, and wet with a liquid of low volatility. The nature of the liquid affects the separation achieved and as in all types of chromatography, the discovery of the suitable liquid is often one of the most time consuming parts of the research. A concentrate of the essence is made by one of a number of methods—for example, extraction or distillation—and this concentrate is placed on the column. The components are eluted by a gas such as helium that removes them from the column in fractions or as single compounds. The fractions are run on other columns and then separated into single compounds. Separations of very small quantities of material can be made very rapidly in a matter of minutes and the recoveries from the columns are almost quantitative. With large samples the gas chromatograph can be used as a preparative device.

After the components of a mixture are separated, the classical methods of identifying compounds can be used. In regular paper or column chromatography, the identity of a compound is often established by comparison of its behavior with that of a known compound. This technique may be applied to gas chromatography. Relative amounts of substances are recorded.

The flavors of many foods are now being studied by this method and in a fairly short time the chemistry of flavor will be on a firm foundation for the first time. An understanding of changes in flavor on processing or aging of a food such as coffee is at last possible with this technique. (See Tables 5.3. and 5.4.)

TABLE 5.3. COMPOUNDS FOUND IN THE GASEOUS EMANATION OF ONIONS

Compound	Relative Amount
Propionaldehyde Methyl Alcohol	very abundant
Propyl Mercaptan	abundant
Hydrogen Sulfide Acetaldehyde	small
Sulfur Dioxide Dipropyldisulfide Propyl Alcohol 4-Hexen-1-al 2-Hydroxy Propanthiol	trace

During the processing of a number of foods, particularly fruit juices, the loss of aroma during concentrating is appreciable. Many years ago vacuum evaporation was introduced with low temperature condensors to trap as much of the volatile compounds as possible. This was a great improvement in many cases over evaporation at atmospheric pressure. However, since air is present in the original product, venting the system is necessary; consequently, enough volatile material is lost so that the fresh

TABLE 5.4. VOLATILE COMPONENTS IN COFFEE AROMA

Formic Acid	Phenol
Acetic Acid	Resorcinol
Methyl Ethyl Acetic Acid	Cresols
n-Valeric Acid	Ammonia
iso-Valeric Acids	Methyl Amine
Higher Fatty Acids	Trimethylamine
Ethyl Alcohol	Pyrrole
Acetyl Methyl Carbinol	n-Methyl Pyrrole
Furfuryl Alcohol	Pyridine and Homologues
Acetaldehyde	Pyrazine
Methyl Ethyl Acetaldehyde	Guaiacol
Furfural	p-Vinyl Guaiacol
Acetone	Eugenol
Diethyl Ketone	Hydrogen Sulfide
Diacetyl	Methyl Mercaptan
Acetyl Propionyl	Dimethyl Sulfide
Hydroquinone	Furfuryl Mercaptan
Esters	Furane
Furfuryl Formate	Sylvestrine
Furfuryl Acetate	Vanillone
Methyl Alcohol	n-Heptacosane
2,3-Dioxyacetophenone	

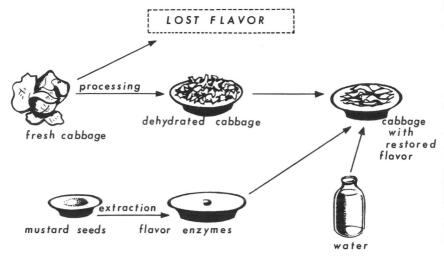

FIGURE 5.8. RESTORATION OF NATURAL FLAVOR TO DEHYDRATED CABBAGE. Reproduced from Hewitt, E. J., MacKay, D. A. M., Konigsbacher, K. S. and Hasselstrom, T., "Flavor Propagation Through Enzymatic Action, Chemistry of Natural Food Flavors—A symposium," Dept. of Army, Research and Development Command, Natl. Acad. Sci. Natl. Research Council, Washington, D. C., 1957.

flavor disappears. Improvements in design now allow the removal of most aroma in the first fraction by flash evaporation. The juice is heated rapidly in a matter of a few seconds by passing it with turbulent flow through a small pipe and then it is rapidly cooled. Flash evaporation is carried out at atmospheric pressure if the volatiles will take the high temperature, otherwise at low pressure. The condensate from the flash evaporation passes to a fractionating column and is separated into two fractions. The vent gases are scrubbed by a counter-current of cold column bottoms that trap the most volatile compounds by dissolving them. The essence is stripped from the column.

With the progress of biochemistry and the rapid rise in our understanding of the changes that occur in plant and animal cells, the time is ripe for some investigations of flavor precursors and attempts to utilize biochemical reactions for the maintenance and enhancement of fresh flavor. Studies are underway to find methods for the isolation or at least the concentration of flavor precursors and enzyme systems capable of developing fresh flavor. It is hoped that when these are obtained their addition to dried or concentrated foods will improve the flavor and acceptability even in the presence of off-flavors.

REFERENCES

1. BOGGS, M. M. and HANSEN, H. L., "Analysis of Foods by Sensory Difference Tests," *Advances in Food Research*, **2**, 220–253 (1949).
2. CAIRNCROSS, S. E. and SJÖSTRÖM, L. B., "What Glutamate Does in Food," *Food Indust.* **20**, 982–983, 1106–1107 (1948).
3. CAUL, J. F., "The Profile Method of Flavor Analysis," *Advances in Food Research*, **7**, 1–40 (1957).
4. CROCKER, E. C., SJÖSTRÖM, L. B., and TALLMAN, G. B., "Measurement of Food Acceptance," *Ind. Eng. Chem.*, **40**, 2254–2257 (1948).
5. DIMICK, K. P. and CORSE, J., "Gas Chromatography—a New Method for the Separation and Identification of Volatile Materials in Foods," *Food Technol.*, **10**, 360–364 (1956).
6. ELSBERG, C. A., *et al.*, "A New and Simple Method of Quantitative Olfactometry," *Bull. Neurological Inst.*, **4**, 1–4, 5–19, 26–30, 31–34 (1935).
7. "Flavor Symposium," *Food Technol.*, **11**, inserted between pp. 472 and 473 (1957).
   a. KIRKPATRICK, M. E., LAMB, J. C., DAWSON, E. H., and EISEN, J. M., "Selecting a Taste Panel for Evaluating the Quality of Processed Milk."
   b. PERYAM, D. R. and PILGRIM, F. J., "Hedonic Scale Method of Measuring Food Preference."
   c. SCHWARTZ, N. and FOSTER, D., "Methods of Rating Quality and Intensity of the Psychological Properties of Food."
   d. SIMONE, M. and PANGBORN, R. M., "Consumer Acceptance Methodology, One vs. Two Samples."
   e. SJÖSTRÖM, L. B., CAIRNCROSS, S. E., and CAUL, J. F., "Methodology of the Flavor Profile."
8. FOSTER, D., PRATT, C., and SCHWARTZ, N., "Variation in Flavor Judgements in a Group Situation," *Food Research*, **20**, 539–544 (1955).
9. HOWARD, L. B., "Flavor," *Food Technol.*, **13**, 8–16 (May 1959).
10. JANOVSKY, H. L., "Flavomatics in Foods," *Food Technol.*, **9**, 500–502 (1955).
11. KATZ, A., "Highlights in Newly Developed Flavoring Aromatics," *Food Technol.*, **9**, 636–638 (1955).
12. KONIGSBACHER, K. S., HEWITT, E. J. and EVANS, R. L., "Application of Flavor Enzymes to Processed Foods, I. Panel Studies," *Food Technol.*, **13**, 128–131 (1959).
13. LOCKHART, E. E., "Chemistry of Coffee," in "Chemistry of Natural Food Flavors," Natl. Acad. Sci. and Natl. Research Council, Washington, D. C., 1957.
14. MACKEY, A. O. and VALASSI, K., "The Discernment of Primary Tastes in the Presence of Different Food Textures," *Food Technol.*, **10**, 238–240 (1956).
15. MITCHELL, J. H., JR., LEINEN, N. S., MRAK, E. M., BAILEY, S. D., eds. "Chemistry of Natural Food Flavors, a Symposium," Dept. of Army, Research and Development Command, Natl. Acad. Sci. and Natl. Research Council, Washington, D. C., 1957.
16. MITCHELL, J. W., "Taste Difference Testing," *Food Technol.*, **12**, 476–477, 477–482 (1957).
17. PETTITT, L. A., "Informational Basis in Flavor Preference Testing," *Food Technol.*, **12**, 12–14 (1958).

18. RALLS, J. W., "Vegetable Flavors, Nonenzymatic Formation of Acetoin in Canned Vegetables," *J. Agr. Food Chem.*, **7**, 505–506 (1959).
19. STAHL, W. H., "Gas Chromatography and Mass Spectrometry in the Study of Flavor," in "Chemistry of Natural Food Flavors," Natl. Acad. Sci. and Natl. Res. Council Publ., Washington, D. C., 1957.
20. WALTERS, R. H. and ISKER, R. A., eds., "Monosodium Glutamate—a Second Symposium," Research and Development Assoc., Food and Container Institute, Inc., Chicago, 1955.

CHAPTER SIX

Meat and Meat Products

MEAT IS DEFINED by the Food and Drug Administration as follows: "Meat is the properly dressed flesh derived from cattle, from swine, from sheep or goats sufficiently mature and in good health at the time of slaughter, but is restricted to that part of the striated muscle which is skeletal or that which is found in the tongue, in the diaphragm, in the heart, or in the esophagus, and does not include that found in the lips, in the snout, or in the ears, with or without the accompanying and overlying fat, and the portions of bone, sinew, nerve, and blood vessels which normally accompany the flesh and which may not have been separated from it in the process of dressing it for sale.

"Flesh is any edible part of the striated muscle of an animal. The term animal as herein used, indicates a mammal, a fowl, a fish, a crustacean, a mollusk, or any other animal used as a source of food."

ANIMAL STRUCTURE

Muscle

There are three types of muscles in animals: striated or voluntary muscle which constitutes the meat in the above definition, smooth or involuntary muscle which for the most part is discarded when an animal is dressed, and heart muscle which is more or less intermediate between the other two in structure.

Striated muscle is composed of long cylindrical cells, called the muscle fibers, which lie parallel to one another and lengthwise of the muscle. Each cell contains a number of nuclei lying close to the outer edge near the membrane or sarcolemma. The cross striations that appear under the microscope have been the object of numerous studies but as yet a complete explanation of their appearance has not been established. The muscle cells are bound together by connective tissue and groups of them are associated and covered with more connective tissue (the *perimysium*) to form bundles. In some muscles the bundles are very noticeable when the muscle is cut crosswise, giving the grain to the meat. The bundles of muscle fibers are held together in a single muscle by a covering of connective tissue sheath called the *epimysium*.

The cells of the muscle fibers contain a complex mixture of proteins, lipids, carbohydrate, salts, and other compounds. The carbohydrate in living muscle is glycogen, but during slaughter and the ripening that occurs between the killing of the animal and the cooking of the meat, the glycogen is degraded and disappears from the tissues.

Muscle also contains lipid—phospholipid and cholesterol. However determining how much lipid is in the muscle cell and how much in the connective tissue is difficult. Striated muscle is believed to contain about 3 per cent lipid. The connective tissue contains large amounts of neutral fat that marbles the meat and is distributed between the bundles of muscle cells and is of great importance to meat quality at the table. It is difficult to remove lipids completely from the muscle cells.

The *proteins* of a muscle cell are numerous and comprise those important in contraction, in the functions of the nucleus, and in the enzymatic reactions of the cell. These latter may not be completely different from the others—for example, the protein myosin is necessary for contraction and also as an enzyme for ATP (adenosine triphosphate). The classical method of protein fractionation by extraction with salt solutions has been applied to muscle. This method often leads to fictitious separations and some of the older work is consequently open to question. Two proteins present in the cells have received much attention in recent years because they are responsible for the contraction of the muscle fiber. The two proteins are *actin* and *myosin* and the compound formed between the two, *actomyosin*. A protein fraction can be extracted from muscle with cold water. This fraction is called *myogen* and comprises a number of proteins including those enzymes essential for glycolysis. Many of the enzymes promoting the long series of reactions that occur when glycogen undergoes fragmentation have been isolated, purified, and some have even been crystallized. The pigment responsible for the pinkish red color of muscle, myoglobin, has also received some attention and is more completely discussed on p. 189 Another group, the proteins called nucleoproteins, is now being studied extensively.

The state of our knowledge of the proteins of a muscle cell is similar to that of our knowledge of any cell. Biochemistry is at the state where techniques for the separation of cellular compounds and an understanding of their role at the level of the cell are still developing. In the past it has been possible to fractionate tissues crudely, although separation of proteins into pure compounds has been uncertain. Now not only are techniques and knowledge growing to the point where criteria of purity for large molecules such as proteins are known, but also methods are available for separating parts of cells such as nuclei, mitochondria, and cell membranes.

The *salts* in muscle cells are the same as those in most cells of the body and are at the same concentration. Potassium is the most common cation, with magnesium and sodium following. The anions are acid phosphate, bicarbonate, and sulfate in the order of their concentration. The concentrations of these ions cannot be determined directly because the cells are in contact with extracellular (interstitial) fluid, and at present it is not possible to separate cells completely from extracellular fluid. In meat we cook relatively large pieces of tissue and consequently have both extracellular and intracellular fluid. The salts in the extracellular fluid have a concentration equal to that in the blood plasma. Sodium is the most important cation with potassium, magnesium, and calcium in small amounts. The most abundant anion is chloride, with fair amounts of bicarbonate and small amounts of acid phosphate and sulfate.

The organic compounds that can be separated from muscle by treatment with water and that are not lipid or protein are called the *extractives.* These substances appear in the water when meat is stewed, fricasseed, or in any way treated with moist heat and in the pan juice when meat is roasted or fried. Striated muscle has approximately 1 per cent organic extractives and 1 per cent salts. Glycogen is the most abundant extractive in resting muscle, but, as pointed out above, the amount decreases after slaughter. None can be detected in most cuts of meat on ripening. In the extract are products of glycolysis or intermediates from the tri-carboxylic acid cycle such as lactic acid and pyruvic acid and small concentrations of a number of low molecular weight nitrogen compounds. These latter are sometimes called as a group "nitrogen bases" since the nitrogen is present in an amino or imino group. The amino acids comprise some of the compounds; and there may be very small amounts of urea, creatine, and creatinine, which are undoubtedly intermediates and end products from muscle metabolism. Several other compounds, also present in small amounts, are anserine, carnosine, and carnitine whose structures are shown below. (Carnitine has been shown to have a role in the oxidation of fatty acids.)

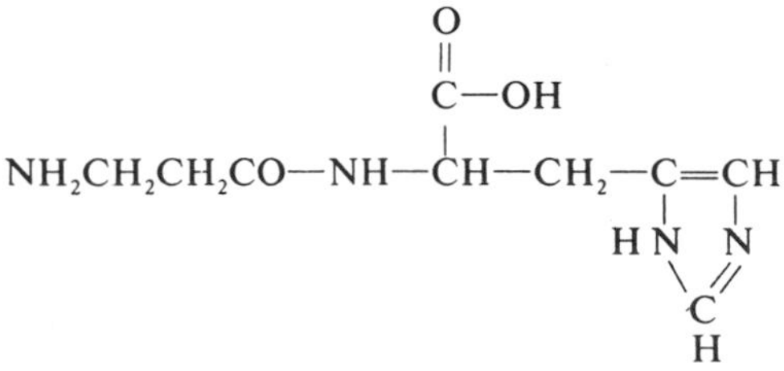

$$\text{NH}_2\text{CH}_2\text{CH}_2\text{CO}-\text{NH}-\underset{\overset{|}{\underset{\text{HN}\quad \text{N}}{\text{CH}}}}{\text{CH}}-\text{CH}_2-\text{C}=\text{CH}$$

Carnosine (β-alanylhistidine)

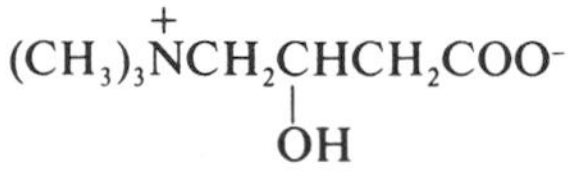

$$(\text{CH}_3)_3\overset{+}{\text{N}}\text{CH}_2\underset{\overset{|}{\text{OH}}}{\text{CH}}\text{CH}_2\text{COO}^-$$

Carnitine

$$\text{NH}_2\text{CH}_2\text{CH}_2\text{CO}-\text{NH}-\text{CHCH}_2-\text{C}=\text{CH}$$

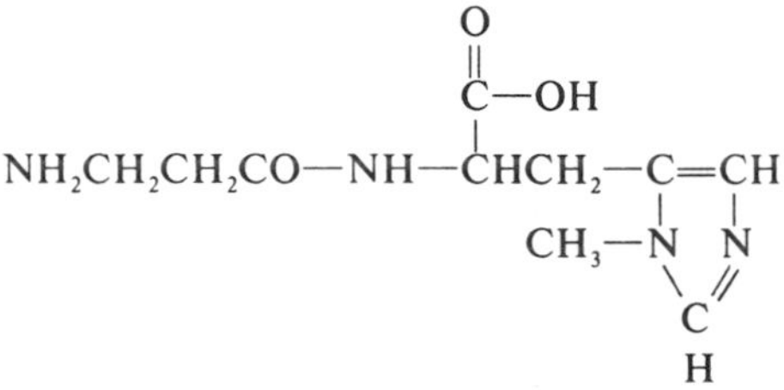

Anserine (β-alanyl 1-methyl histidine)

Water also extracts some proteins and derived proteins from muscle. Coagulable protein escapes in the meat juice and during cooking is coagulated. It forms the precipitate noticeable in the pan juice of roasted meat. The extract also contains gelatin which if not too dilute is noticeable through the increased viscosity or even jellying of the cold extract. Proteoses and peptones are present in small amounts. They originate from the hydrolysis of proteins that occurs as the meat ripens. The water soluble B-complex vitamins, particularly abundant in meat, are also extracted in water and juice.

Connective Tissue

This tissue is composed of a few rather large cells scattered through a matrix composed of fibers and amorphous ground substance. There are many types of connective tissue; but they all have this common structure of few cells and numerous fibers in ground substance. Connective tissue varies from extremely thin and fragile tissue found between organs to tough bands and membranes of great strength that form the capsule around most organs—the tendons, ligaments, and aponeuroses. A tendon is the tough band or cord that connects a muscle with some other structure such as a bone, while a ligament is a band which supports an organ or connects two bones or other parts. An aponeuroses is a type of muscle attachment to bone that forms a sheet. Connective tissue also forms the fat or adipose tissues as well as other specialized tissues such as bone and cartilage.

In connective tissue the size, shape, and number of cells varies enormously with the source, function, and type of tissue. Unlike most other tissues, connective tissue has relatively few cells and much intercellular material. Some connective tissues have few cells and these are separated or isolated. In others they appear more numerous and in some are arranged in definite patterns. In all connective tissue the two types of fibers, white collagenous and yellow elastic, and the ground substance are of great importance in determining the characteristics of the tissue.

The white collagenous fibers are arranged in bundles of wavy threadlike filaments and are coarse and strong but resistant to stretch. In tendons the collagenous fibers are arranged in parallel rows with rows of cells between them. Since the collagenous fibers are not elastic, tendons also lack elasticity. They are very strong however.

Yellow elastic fibers, particularly abundant in ligaments, are fine and branched and distributed in a random pattern to form a network. There are few collagenous fibers in ligaments. The tissue is elastic.

The collagenous fibers are composed principally of the protein, collagen, which on cooking hydrolyzes to form gelatin. If connective tissue containing a large per cent of collagen is cooked in moist heat for an extended period of time, much hydrolysis occurs, and the tissue almost disappears. The elastic fibers, however, are composed of the protein, elastin, which is very resistant to moist heat and indeed to most reagents. Cooking has little or no effect on these fibers.

Collagen. In 1952 Bear[4] presented a tentative model of the collagen fibril. See Figure 6.1. Under the microscope collagen fibers can be seen which have a diameter of approximatey $100-200\,\mu$ ($1\,\mu$ = .001mm) in tendon and $20-40\,\mu$ in the skin. Each fiber is composed of a number of *primitive fibers* approximately $2-10\,\mu$ in diameter. By mild mechanical or chemical means the primitive fibers can be divided into *fibrils* whose diameters are so small that they cannot be discerned with an ordinary microscope but must be detected by the electron microscope. Their diameter is measured in terms of hundreds of Angstrom units ($1A$ = $.0001\,\mu$) but even they are far above molecular dimensions. The smallest unit of organization above the molecule is the *protofibril*, a number of which constitute the fibril.

Data have been accumulating during the past decade with the use of X-ray diffraction methods and electron microscopy for collagen preparations subjected to various mild conditions. These data give some indications of the shape of the collagen molecule and the manner in which molecules are organized in the protofibril. Bear has suggested a tentative model which will be briefly described.

The fibril appears to be a very long thin body containing bands and

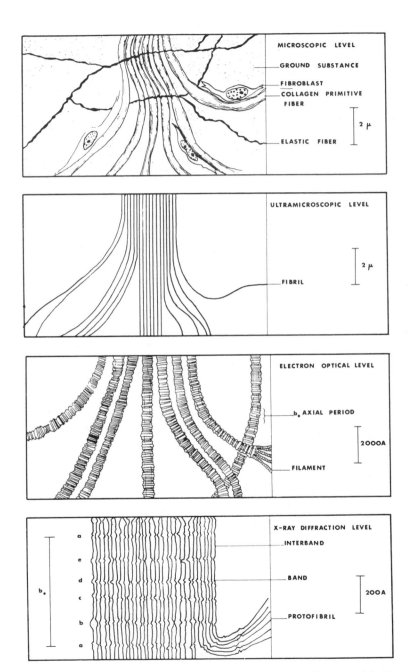

FIGURE 6.1. COLLAGEN FIBERS AND THEIR ARRANGEMENT. *Courtesy of R. S. Bear.*

interbands at definite distances. This series of bands reappears at 640Å along the length of the fibril and appears to have significance as far as the arrangement of the side chains on the collagen molecule is concerned.

The collagen molecule is a polypeptide chain with a very high percentage of glycine residues. Beef collagen, which has been studied most extensively, contains 19.9 per cent glycine. It contains a very high (64 per cent) percentage of amino acid residues which have nonpolar side chains (glycine, proline, alanine, the leucines, valine, phenylalanine, and methionine). Whether or not these amino acid residues are arranged in definite patterns has not been determined, but Bear believes that they are arranged in a definite pattern as far as type, with a section of nonpolar side chains followed by polar ones capable of hydrogen bonding and salt links. Polar side chains are those containing hydroxyl groups (hydroxyproline, threonine, serine, and tyrosine), those containing acidic groups (glutamic acid and aspartic acid or their amides), and those containing basic groups (lysine, arginine, histidine, and hydroxy-lysine). Hydroxyl and acidic groups are active in hydrogen bonding, while acidic and basic groups in close approximation form salt bridges. Collagens are unusual in their possession of fair amounts (from 5 to 13 per cent) of hydroxyproline.

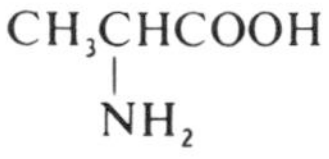

$$CH_3CHCOOH$$
$$|$$
$$NH_2$$

Alanine

$$CH_3CHCHCOOH$$
$$|\quad|$$
$$OH\,NH_2$$

Threonine

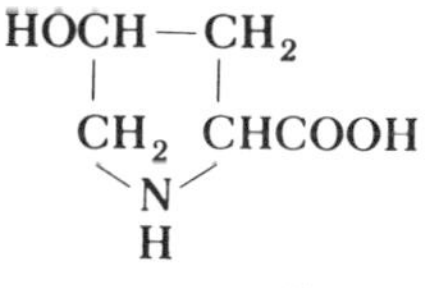

$$CH_3CHCH_2CHCOOH$$
$$|\qquad\quad|$$
$$CH_3\quad NH_2$$

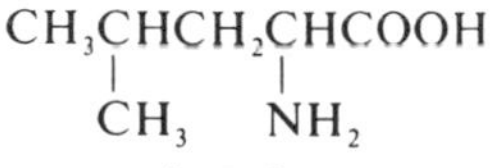

Leucine

$$HOCH-CH_2$$
$$|\qquad\quad|$$
$$CH_2\quad CHCOOH$$
$$\diagdown N\diagup$$
$$H$$

Hydroxyproline

The collagen molecule is coiled and the protofibril is made of parallel molecules coiled or twisted together to form a helix. The protofibril is formed of many molecules in length; and although the manner in which the end of one molecule abuts the end of another is unknown, Bear believes that the molecules do not overlap in the protofibril as they do in some textile fibers but instead lie parallel to one another along the length of the polypeptide chain.

The bands and interbands are believed to be caused by the side chains on adjacent molecules and the way in which they mesh or crowd one another. Bear suggests that the interbands are regions in which the relatively small nonpolar groups can mesh with one another in such a manner that the polypeptide chains are able to lie quite straight in the helix. The interbands are areas where the relatively large polar side chains stick out and react with one another, producing a kinking of the polypeptide chains. A short portion of the coil, straightened out, is shown in the illustration (Figure 6.1).

Collagens from animals in widely separated phyla show many similarities, but some differences. Collagens from mammals have been studied most extensively with fishes next; a few examples from other phyla have been examined. The preliminary work indicates that the collagens from closely related animals have marked resemblances, while the farther apart they are in the animal kingdom, the greater are the differences in collagen. The gross structure of the molecules of animals from different phyla appears to be similar, but the amino acid content is different.

Elastin. Elastin is the protein that forms the yellow elastic fibers. Unlike collagen, it is not hydrolyzed on boiling with water and consequently a tissue that contains a considerable number of yellow elastic fibers shows little softening or dissolving upon cooking. These fibers are abundant in tendons and ligaments which are often trimmed away before meat is cooked. The human Achilles tendon contains 20 times more elastin than

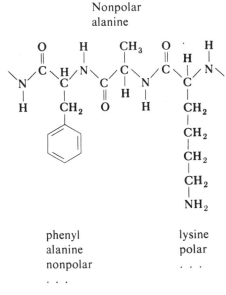

Polypeptide Chain

collagen and in the ligamentum nuchae, which extends down the neck of cattle and helps hold up the head, there is approximately 5 times as much. There has been some study of the structure of elastins by the methods applied to collagens. They appear to be linear proteins but of a different type than the collagens. Beef elastin differs from beef collagen in its amino acid composition. It is unusually rich in amino acids with nonpolar side chains (glycine, the leucines, valine, phenylalanine, and proline) with 78 per cent of the molecule made up of these amino acids. It is similar to collagen in its small content of histidine, cystine, tyrosine, and tryptophan.

The ground substance of connective tissue varies from a soft jellylike mass to a tough matrix. In cartilage and bone the ground substance is quite different from that in other connective tissue. It is principally composed of water in which salts, glucose and other molecules are dissolved and which achieves its jellylike characteristics from the presence of heteropolysaccharides, colloidally dispersed. In cartilage, chondroitin sulfates are most abundant while in soft connective tissue and skin, hyaluronic acid occurs. (See p. 80.)

Adipose Tissue. This is a specialized type of connective tissue in which the cells rather than the intercellular material are most abundant. Fat cells are found scattered through, or in groups in loose connective tissue, but the term "adipose tissue" is reserved for those tissues containing large deposits of fat. Adipose tissue varies in color from very light cream to dark yellow. It is often classed as white and brown adipose tissue. It occurs in the subcutaneous connective tissue, in the region of the kidney, in the omentum, and around and between some of the muscles and organs.

Each fat cell is filled with a large single vacuole of fat, and the cytoplasm and the nucleus are pressed by this vacuole close to the cell wall. The fat cells are arranged in groups or lobules with delicate connective tissue that is rich in fibers and many blood capillaries separating the groups.

Adipose tissue is by no means composed entirely of fat. It is true that the chief lipid is neutral fat, but the cells possess proteins of many kinds, water, salts, and other compounds commonly present in cells in small amounts. The neutral fat present in the adipose tissue is characteristic of the species; the assortment of fatty acids follows fairly closely a species pattern. But in the carcass some variation does occur in the composition of fat in different tissues. In general, the outer layers of adipose tissue contain fat with higher iodine numbers and lower melting points than the inner layers. Lipid occurs in the vacuoles in the liquid state, and the fat near the skin is at a slightly lower temperature than that in the interior of the living

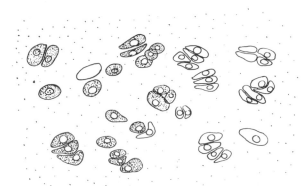

FIGURE 6.2. SECTION OF RIB CARTILAGE OF CALF. Matrix is slightly fibrous. × 240. (Redrawn after E. Sharpey-Schafer) Carleton, H. M. and Short, R. H. D., "Schafer's Essentials of Histology," Longmans, Green and Co., London, 1956.

body. It has been suggested that in the living animal the fatty acid content of the fat deposited is as saturated as it is possible for it to be and still remain in the liquid state. Some observations on the effect of external temperature appear to support this. The higher the skin temperature, the higher is the melting point of the fat deposited.

It is well known that the properties of lard produced from different tissues of a hog vary and this fact is utilized in commercial fat production. See p. 52 for a discussion of the production of fat from animal sources.

Cartilage. This is also a specialized type of connective tissue; like all other connective tissue cartilage (Figure 6.2) is composed of cells, fibers, and ground substance. The cells occur for the most part in islands called lacunae, surrounded by ground substance in which the fibers, both collagenous and elastic, are embedded. The ground substance, is a gel composed of chondroitin sulfate, chondromucoid, and albumoid in water.

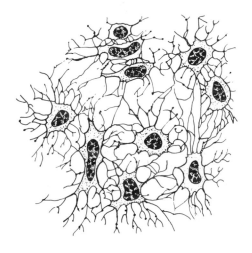

FIGURE 6.3. BONE CELLS AND THEIR CANALICULI IN THIN MEMBRANE BONE OF MOUSE. (After Maximow and Bloom) × 1040. *Courtesy of W. B. Saunders Co.*

TABLE 6.1. COMPOSITION OF BONES*

	Cattle bones
$CaCO_3$	7.07
$Mg_3(PO_4)_2$	2.09
$Ca_3(PO_4)_2$	58.30
CaF_2	1.96
Organic	30.58
Total	100.00

*Adapted from Edelmann, R. H., Mohler, J. R. and Eickhorn, A., "Textbook of Meat Hygiene," Lea & Febiger, Philadelphia, Pa., 1943.

Bone. Bone (Figure 6.3) is also a specialized connective tissue. Like cartilage, it is composed of cells located in lacunae, fibers, and ground substance in which tiny crystals of salts are deposited. In bone usually only one cell occurs in each lacuna, and it can communicate with others through a series of small canaliculi which traverse the matrix. The fibers are numerous and are collagenous. They are arranged in small bundles and run in definite patterns. Bones are traversed by numerous canals, some of which run perpendicular to the shaft (Haversian) and some of which run obliquely or at right angles (Volkman's). These canals make the bone porous and less dense. The ground substance is composed of the proteins ossomucoid and ossoalbuminoid surrounding small crystals of salts. The most abundant salt is calcium phosphate, but calcium carbonate, magnesium phosphate, and calcium fluoride are also present. See Table 6.1.

The proteins in bone are for the most part unidentified. Food gelatin is manufactured from fresh bones by boiling them in water, a process converting collagen to gelatins. In some processes the bones are boiled in very dilute hydrochloric acid, the phosphate is precipitated and the acid neutralized with lime water, after which the product is dried. The product is called "ossien" and on extraction with hot water yields gelatin. The terms "ossomucoid" and "ossoalbuminoid" have been assigned to the proteins in the living bone, but little is known about them.

POSTMORTEM CHANGES

When an animal dies, the skeletal muscles stiffen in *rigor mortis* and remain in this condition for a period after which they soften and become flexible again. The speed with which rigor develops and the length of time it persists is variable. It is widely believed that the onset of rigor is speeded up by high temperatures and delayed by low ones. But Bate-Smith[3] found that the temperature of the carcass was of little importance compared to other factors and Stewart, Lowe, Harrison, and McKeegan[47] confirmed this for chickens. They found the onset of rigor was gradual in the chickens studied and usually varied between 1 and 2 hours. However, one chicken developed rigor mortis in 5 minutes. Rigor mortis is important in meat products since muscles cooked while still in rigor are much tougher than if it is allowed to pass before cooking.

The stiffness that develops when muscles pass into rigor is the result of changes in the proteins. Living muscle fibers contain protein in a soft, pliable gel. During rigor this gel stiffens, but when rigor passes, the muscle again becomes soft and pliable. Finally during cooking, another change called *rigor caloris* occurs and the proteins stiffen again. The changes in the proteins of muscle are still incompletely understood although a number of theories to explain rigor have developed. While some knowledge has accumulated, there is still much ignorance concerning the chemical changes involved.

In a living muscle changes occur in the proteins actin and myosin when a muscle contracts. Szent-Györgyi[49] has presented a rather complex theory to explain the many reactions that lead to a change in actin and myosin, the formation of a complex, actomyosin, and the contraction of these molecules. It would take too much space to give the theory here, and it is constantly changing; but the fact that investigators have come so far in their study of muscles that the formation of a theory is possible is encouraging. The understanding of the chemistry of the muscle in life and after death of the animal will not be far behind. The energy for contraction in living muscle is supplied by adenosine triphosphate, ATP,

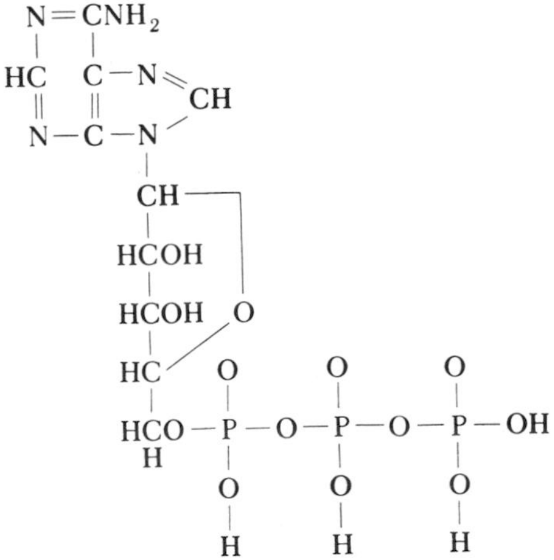

Adenosine Triphosphate

which possesses a high energy phosphate bond in the second and third phosphate.

The energy for contraction and its accompanying heat production comes principally from these high energy phosphate bonds. Resynthesis of ATP in the living cell is at the expense of glycogen which passes through a long series of enzymatic reactions to form pyruvic acid. The pyruvic acid is then oxidized by way of the citric acid cycle to carbon dioxide and water, and small bursts of energy are released as each carbon and hydrogen are oxidized. This series of reactions is completely discussed in any modern textbook on biochemistry. The research that has led to the elucidation of this series of reactions is a triumph for modern biochemistry. The over-all series is the following:

$$2(C_6H_{11}O_5)_n + 5[O] \longrightarrow 4n\ CH_3COCOOH + H_2O$$
$$\text{Glycogen} \qquad\qquad \text{Pyruvic Acid}$$

$$CH_3COCOOH + 5[O] \longrightarrow 3\ CO_2 + 2\ H_2O$$

The transformations of glycogen to pyruvic acid require a small amount of oxygen, but the citric acid cycle for the oxidation of pyruvic acid to carbon dioxide and water requires more (5 atoms of O for each $CH_3COCOOH$). The oxygen is carried to the muscle by the blood. When the demand for oxygen in a contracting muscle exceeds the supply, a condition called "oxygen debt" develops and some pyruvic acid is reduced to lactic acid to supply oxygen.

$$CH_3COCOOH + H_2O \rightarrow CH_3CHOHCOOH + (O)$$
$$\text{Pyruvic Acid} \qquad\qquad \text{Lactic Acid}$$

When an animal is killed and circulation of blood ceases, the degradation of glycogen continues, small amounts of intermediates accumulate, and with the influx of oxygen stopped, lactic acid is the most abundant product. Muscle contains some buffers that neutralize the first lactic acid formed. Later as more and more forms, the pH of the tissues begins to fall. The amount of glycogen stored in the muscle at the moment of death controls the amount of lactic acid formed and, consequently, the ultimate pH of the meat. A well-nourished muscle will have as much as 1 per cent glycogen, form 1.1 per cent lactic acid, and drop to a pH of 5.6. The pH never falls below 5.3 because some enzymes become inactive at this low pH.

Factors that influence the amount of glycogen in an animal have been the object of a number of studies since a low pH, we shall soon see, is usually desirable in meat from many standpoints. Whether or not animals are fed 24 or 48 hours before slaughter affects the amount of muscle glycogen and hence the resultant pH of the meat. It is only in the well-fed animal that muscle cells contain maximum glycogen. The amount of exercise an animal has just before slaughter is important. An animal that is chased or excited, or driven into the killing pen, will use up much of the muscle glycogen during the vigorous exercise, and without a rest period to allow for restoring the supply, will produce meat in which the development of a low pH is impossible.

The series of chemical reactions occurring after slaughter produces enough heat to cause a rise in the temperature of the meat. The average body temperature is 99.7° F in cattle, but shortly after death, the internal temperature of a round of beef may rise to 103° F. The fresh meat cools very slowly even in a refrigerator because of the continuing production of heat. This heat production, called *animal heat*, has been recognized by farmers and butchers for hundreds of years. It has been often considered something mysterious and mystical. Farmers will often say that meat must not be eaten until the animal heat has escaped, that it is not fitting to eat meat with animal heat, or make similar statements. Because of the recognition of animal heat, freshly killed meat is seldom eaten. A short period of hanging during which rigor develops and passes and during which some ripening may even begin improves the meat before cooking.

Both creatine phosphate and adenosine triphosphate are hydrolyzed as rigor begins to develop. Creatine phosphate forms creatine and phosphate. One of the muscle enzymes, a phosphatase capable of hydrolyzing adenosine triphosphate to inorganic phosphate and adenosine diphosphate, becomes more active as the pH drops to 6.5 and below. The adenosine triphosphate is then further decomposed to ribose, phosphate, ammonia, and hypoxanthine (from the adenine). Contraction is no longer possible in the muscle, and a tension develops which is incapable of relaxation until rigor passes.

Both Bate-Smith[3] and Szent-Györgyi[49] have proposed theories of rigor mortis that include reactions of the proteins of muscle and changes in them. Bate-Smith believes that the disappearance of adenosine triphosphate is of great importance in the development of the stiffening of rigor mortis. Szent-Györgyi points out that creatine phosphate recently has been shown essential for relaxation of contracted muscle, and he believes its disappearance may be of prime importance in the stiffening of rigor mortis.

The sarcolemma, the membrane around the muscle cell, shows a decrease at this time in electrical resistance, and a free diffusion of ions occurs. The change in the semipermeable membrane indicates that not only is the protein of the muscle fiber changing but the substances that make up the membrane, the sarcolemma, might be changing chemically. The electrical resistance of the muscle as a whole changes also.

The ultimate pH attained by the meat has important effects on other properties of the meat. Meat with a high pH is darker in color than normal, slimy to touch, and does not allow salt of curing pickle to penetrate readily. It is also difficult to express juice from meat with a high pH. The dark color that sometimes develops in beef carcasses has been studied by several groups because it is of considerable economic importance. Consumers are reluctant to buy beef that is darker than normal. The pigment of muscle, myoglobin (p. 189), is present in the same concentration in dark cutting beef as in normal; but the light passes into and is reflected from deeper layers of the muscle, giving a darker appearance. A second factor appears to be a difference in the openness (or closeness) of the grain of the meat. At high pH the swelling of the protein is greater, and because of this the ability of oxygen to diffuse through the tissue is probably decreased. Less oxygen causes a decrease in the proportion of oxymyoglobin to myoglobin, and a darker color. It has also been shown that the consumption of oxygen by dark meat is higher than by light meat, but the significance of this observation is unknown.

Meat that develops a relatively high pH is difficult to cure properly because the pickling salts do not penetrate the meat at a normal rate. Callow has studied the relation of the pH of meat to its electrical resistance and found that between pH 5.8 and 6.0 there is a rapid increase in resistance. This is also the range of pH's in which the grain of the meat shifts from open to closed. These observations have been interpreted to mean that swelling of the protein occurs in this range and that the passage of ions is thus impeded. It is thought that the increased resistance to the passage of an electric current reflects this difficulty of ion migration.

Watts[54] points out, however, that a relatively high pH is desirable in meat which is to be frozen since in this range of pH's discoloration and the rate of fat oxidation are inhibited and the drip losses are minimized.

Callow has also related the pH of the meat to the rate of growth of bacteria in Wiltshire cured hams. He showed that the occurrence of taint in the hams was correlated with the electric resistance and with the pH. Hams with low pH showed lower electrical resistance and a greater freedom from taint. He concluded that with lower pH the penetration of the pickle was more rapid and that the bacteria were not able to grow as readily in the more acid medium. Ingram showed that this was true by isolating organisms from spoiled hams and measuring the bacterial growth at various pH's. Growth was very low at pH 5, increased rapidly with rising pH, and reached a peak at near pH 8.*

*Data discussed in Reference 3.

Ripening or Aging

After the passing of rigor mortis, meat becomes progressively more tender, juicier, and more flavorful. The speed with which this ripening or aging occurs depends on the time the carcass is kept and the temperature. Changes occur quite rapidly at room temperature but more slowly at refrigerator temperatures. Only well-finished carcasses with a good layer of fat are satisfactory candidates for ripening since a growth of mold which must be trimmed away before the meat is cooked appears on the carcass. Poorly finished carcasses rapidly develop off-flavors and putrify. Lowe[32] discusses the work which Hoagland, et al. carried out many years ago on the ripening of beef stored just above the freezing point for 17 to 177 days. The organoleptic qualities of the meat were judged by a panel which considered the meat stored 15 to 30 days better in tenderness and flavor than the fresh meat. After 45 days the meat began to develop off or gamey flavors. A taste for well-ripened meat varies with different groups of people. Well ripened meat is preferred for the most part by the English but not by Americans.

Some aging occurs during the time which must inevitably elapse in normal commerce between slaughter and preparation of the meat for the table. Purposeful aging is costly because of the use of expensive refrigerator space, the additional trimming required to remove mold, and the shrinkage caused by evaporation. It is used to a limited extent and then only for prime and choice grades of meat.

A recent study[44] shows that aging beef at elevated temperatures with high humidity, with air velocity 5 to 20 lineal ft per min and with ultraviolet radiation to control microorganisms for 2 or 3 days produces beef equal in quality to that aged in a refrigerator 12 to 14 days. Quarters were judged on tenderness, flavor, aroma, and juiciness.

Although a number of studies of composition before and after aging have been completed, the numerous changes that occur on ripening and aging still are not fully known. The reactions are the result of autolytic action in the cells—reactions that fragment large molecules through the action of enzymes and reactions that result from changes in cell conditions, particularly the change in pH. Ginger, et al.[18] determined the content and distribution of nitrogen compounds and particularly the amino acids arginine, lysine, leucine, tyrosine, histidine, and glutamic acid in raw and cooked beef and in the drippings from the meat before and after aging. The cut used was rib steak. Aging caused an increase in the free amino acid nitrogen. The nitrogen in drippings of freshly slaughtered meat was present mostly as nonprotein nitrogen (NPN) with a large proportion of it as amino nitrogen. After the beef had aged for two weeks, drippings showed an increase in the amount of NPN. The amino acids histidine, leucine, tyrosine, glutamic acid, and lysine were present in a bound form. Histidine accounted for as much as 26 per cent of the NPN in the extracts. It was concluded that part of the bound histidine is present as carnosine.

$$HC = CCH_2CHNHCOCH_2CH_2NH_2$$

$$N \quad NH \quad COOH$$

$$C$$

$$H$$

Carnosine

(β-alanyl histidine)

Solov'ev and Piulskaya[46] studied the change in concentration of a few components of beef during ripening. In the first study the flesh of cows was stored at 0°C to 40°C for 144 hours and samples were taken and analyzed at intervals. Actomyosin showed a drop in solubility during the first 24 hours and then a slow rise. The abstract reports that the "activity" of the actomyosin increased from 6 per cent at 12 hours to 56 per cent at 48 hours and that there was some increase during resolution of rigor mortis. ATP is quickly destroyed and they found that at 12 hours 90 per cent of the ATP was gone. The first report likewise confirmed the fall in free glycogen and reported that there was also a decrease in "difficultly extractable glycogen" (quoted from abstract). In the second report[45] beef stored at 0°C, 8°-10°C, and 17°C was analyzed daily for some constituents as well as organoleptically. Optimum changes occurred in 10 days at 0°C, in 4 days at 8°-10°C, and in 3 days at 17°C. The chemical analysis paralleled these organoleptic changes with maximum increases in volatile fatty acids, volatile reducing substances, total nucleotide N, adenylic acid, free purine, and hypoxanthine. These observations serve to confirm in part the hazy idea that ripening is a period when autolytic changes occur and large molecules continue to form small ones although syntheses of the large ones that maintain the tissue during life have stopped.

Sizov[43] and Drozdov[9] present evidence that autolysis in rabbit and beef continues even in frozen meat. Sizov measured the myosin and its enzyme activity for adenosine triphosphate in rabbit meat stored at -14°C. He found that at the end of 5 days the amount of myosin had fallen to 58.5 per cent and eventually it disappeared. At the end of 22 days the enzyme activity was only 30.3 per cent of the fresh meat. Drozdov studied beef under a number of storage conditions and found evidence of autolysis in all samples. Some were stored from 2 to 4 months at -10°C and showed glycolysis, accumulation of lactic acid, and decomposition of phosphates.

Harrison, *et al.*[22] reported studies of four muscles removed from four steers 14 hours after slaughter. The aroma and flavor reached a maximum in 10 days and decreased after 30. Tenderness as judged by the shear test and the taste panel increased particularly during the first 10 days. Acidity rose during the first 2 hours.

Paul, *et al.*[39] compared tenderness in beef steaks and roasts held for varying lengths of time in cold storage. They used the *semitendinosis* and the *biceps femoris* of 2 prime, 2 good, and 2 commercial carcasses. The meat was hung 0, 5, 12, 24, 48 to 53, and 144 to 149 hours. Although steaks showed decreased tenderness up to 24 hours, with longer time tenderness increased; whereas roasts were least tender at 0 time and showed increasing tenderness on aging. The tissues were studied histologically and showed evidence of the occurrence of rigor in the roasts cooked at 0 time. As rigor passed, breaks and areas of granulation occurred in the muscle fibers.

COLOR OF MEAT

The principal pigment present in muscle cells is myoglobin, a red conjugated protein closely related to hemoglobin of the red blood cells. When meat is prepared for market, it is not completely bled and some red blood

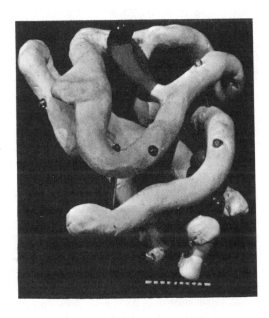

FIGURE 6.4. MODEL OF MYOGLOBIN MOLECULE. Constructed after mapping the globular molecule with X-ray diffraction studies. Polypeptide chains are white; the gray patch is the heme. *Courtesy of Dr. J. C. Kendrew.*

cells remain in numerous blood vessels which traverse the muscle tissue. These contain hemoglobin. Small amounts of colored enzymes which are heme pigments also occur in muscle cells. They are cytochromes and peroxidases. However, most of the color of meat is caused by the pigment in the muscle cells, myoglobin.

Myoglobin has been prepared from beef and pork,[17] as well as other muscles, but it has not been studied nearly as extensively as hemoglobin. Myoglobin is a globular protein composed of one heme moiety for each molecule of protein while hemoglobin contains four heme moieties and one of protein. The molecular weight of myoglobin is just one fourth that of hemoglobin, 17,000 versus 68,000. Likewise myoglobin shows a difference in amino acid composition and solubility. It can be saturated with oxygen at lower oxygen pressures than hemoglobin and its reaction with carbon monoxide and nitric oxide and the stability of the products are also distinct. The porphyrin, heme, in each molecule is the same, but the proteins are different and doubtless the manner in which they are combined is different. Mapping by X-ray[29] analysis suggests a structure for myoglobin shown in the picture. (Figure 6.4).

The porphyrins are a group of compounds that form the prosthetic groups of many colored conjugated proteins. In chlorophylls (Chapter 7) the porphyrins contain magnesium, while in some forms of life such as mollusks the porphyrins contain copper. In hemoglobin and myoglobin the metal held by the porphyrin is iron. The porphyrins are ring compounds composed of four pyrrole rings held together by CH or methene groups. A convenient shorthand formula for the basic porphyrin structure is shown below in which each of the corners is occupied by a carbon or a CH.

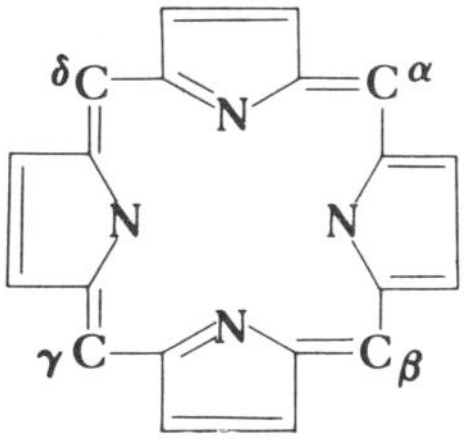

Porphyrin Nucleus

Further abbreviation of the formula occurs when it is written as a cross. Porphyrins differ from one another in the side chains commonly attached to the "corners of the cross."

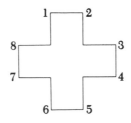

In hemoglobin and myoglobin the porphyrin "corners" hold methyl, vinyl, and propionic acid.

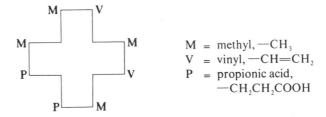

M = methyl, —CH₃
V = vinyl, —CH=CH₂
P = propionic acid,
    —CH₂CH₂COOH

It will be noticed that in the structure which shows the nitrogens, two of the nitrogen atoms have three bonds and two have only two bonds. The third valence of each of these latter nitrogen atoms is satisfied by combination with ferrous iron. The iron atom has the ability to coordinate or share electrons from nitrogen atoms, and in the hemoglobin molecule it may be thought of as being attached through an ordinary covalent bond to two of the nitrogen atoms and through a coordinate covalent bond to the other two. The whole structure is in resonance, and it does not matter which of the nitrogens is considered to be held by ordinary valence and which by coordination since in any case the bond consists of a pair of electrons.

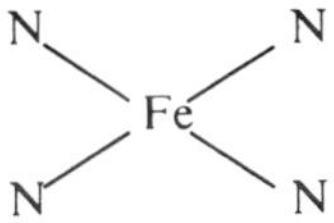

Iron has a coordination number of six.

Heme, ferroprotoporphyrin, of hemoglobin and myoglobin can combine with a number of organic nitrogen compounds. Among them are histidine, although it does not react with arginine and lysine. In hemoglobin a nitrogen (one of those circled) from the imidazole ring of histidine is believed to coordinate with the ferrous iron and hold the heme in place

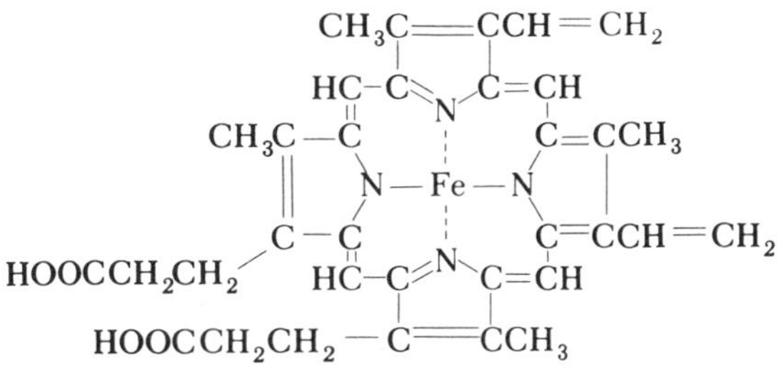

$$CH_3C = CCH = CH_2$$

Heme
Ferroprotoporphyrin

on the protein molecule. The sixth pair of electrons comes from the oxygen of a molecule of water in hemoglobin or a molecule of oxygen in oxyhemoglobin.

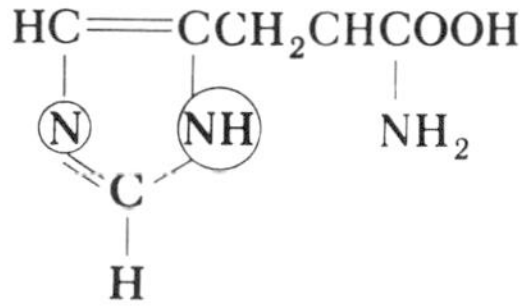

$$HC = CCH_2CHCOOH$$
$$NH_2$$

Histidine

The four nitrogens of the porphyrin are in one plane with a nitrogen of one imidazole ring above the iron, and H_2O or O_2 below. There is likewise a possibility that the carboxyl group of the propionic acid side chains may be linked with protein, but this has not been proved.

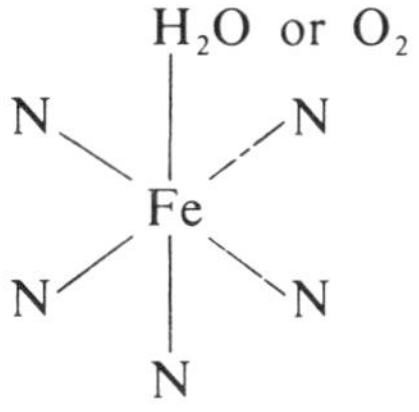

$$H_2O \text{ or } O_2$$

Imidazole

The unique property of hemoglobin and myoglobin is the capacity to bind a molecule of oxygen without a change in the oxidation of the iron. The reaction is readily reversible and is most important in the transfer of oxygen to the tissues in the living animal. In meat it accounts for the change in color that occurs when meat is cut. The myoglobin-oxymyoglobin shift occurs readily when myoglobin is exposed to oxygen. The same change in color occurs when meat is cut. Oxyhemoglobin and oxymyoglobin have a very bright red color, while hemoglobin and myoglobin are more purplish red. The bluish color of meat is caused by the large amounts of myoglobin (and some hemoglobin) present, but when the meat is cut and the surface exposed to air, the color becomes bright red as oxymyoglobin is formed. Other heme pigments of muscle, peroxidases and cytochromes, cannot combine in this fashion with oxygen.

When hemoglobin is treated with a strong acid or hydroxide, the protein and the porphyrin are separated.

$$\text{Hemoglobin} + \text{HCl} \rightarrow \text{Globin·HCl} + \text{Ferroprotoporphyrin (heme)}$$

The ferroprotoporphyrin, heme, is rapidly oxidized under most conditions to ferriprotoporphyrin, hemin, in which the iron exists in the ferric state. The crystallization of hemin is a common and delicate test for the detection of blood. Hill and Holden demonstrated that the globin formed by this reaction will dissolve in water near pH 7 and recombine with heme, ferroprotoporphyrin, to form hemoglobin, or with hemin, ferriprotoporphyrin, to form methemoglobin.

$$\text{Ferroprotoporphyrin (Heme)} + \text{Globin} \rightarrow \text{Hemoglobin}$$

$$\text{Ferriprotoporphyrin (Hemin)} + \text{Globin} \rightarrow \text{Methemoglobin}$$

If hemoglobin is treated with oxidizing agents, it is changed directly to brown methemoglobin.

$$\text{Hemoglobin} \xrightarrow{\text{oxid.}} \text{Methemoglobin}$$

Methemoglobin can be separated into ferriprotoporphyrin (hemin) and globin.

Interestingly enough the globin from different animals will combine with any heme, no matter what its source. The hemoglobin produced by this combination always resembles the hemoglobin of the species from which the globin was prepared. It is the protein portion of hemoglobins from different animals that is species specific. Edmundson and Hirs[12] have shown that sperm whale myoglobin contains the following proportion of amino acids: $Glu_{19}Asp_8Gly_{11}Ala_{17}Val_8Leu_{18}Ileu_9Ser_6Thr_5Met_2Pro_4Phe_6Tyr_3Try_2His_{12}Lys_{19-20}Arg_4$.

Methemoglobin is formed by the reaction of a number of oxidizing agents such as peroxides, ferricyanides, and quinones with hemoglobin. It has a brown color, as does metmyoglobin. Metmyoglobin is similarly formed by oxidation of myoglobin and is consequently believed to be the analogue of methemoglobin. In both of these conjugated proteins the iron exists in the ferric state. Methemoglobin is normally produced in small amounts in the living animal, but it is again reduced to hemoglobin. The reaction has been studied because it is important in some pathological conditions and in the preservation of whole blood. In meat the formation of brownish and gray pigments, or the dulling of the bright red of fresh meat as it ages, may be caused by metmyoglobin formation.

Browning of meat may sometimes occur through oxidation of heme after dissociation from the protein. Ferroprotoporphyrin, heme, is much more rapidly oxidized than hemoglobin and probably more rapidly than myoglobin.[54] Often when meat darkens or browns, conditions are right for the denaturation of the protein and the formation of free ferroprotoporphyrin, heme. Unless conditions are such that oxygen is excluded, and this usually is not true for meat, the ferroprotoporphyrin is quickly oxidized to ferriprotoporphyrin, hematin. When cooking occurs, heat coagulates or denatures the protein; when meat is marinated in acid such as vinegar, denaturation occurs. Some agents allow the heme to become disengaged from the protein and consequently be more susceptible to ferric oxidation. Freezing, salt, acid, or ultraviolet light are capable of denaturing protein and are known to affect the color of meat.

Occasionally after meat has been cut for some time, the color of the surface will deepen but it will remain bright red. In a study of packaging, Landrock and Wallace[30] found that this deepening is caused by dehydration of the surface of the meat and a consequent concentration of the pigment.

Green pigments occasionally occur on meats. The green pigments formed from the porphyrins that have been studied are all products in which the alpha methene carbon bridge is attacked. In the living animal ferroprotoporphyrin, heme, is normally changed into green pigments and in some animals they are discharged in the bile as the bile pigments. In other animals green pigments are changed to red and then excreted in the bile. Choleglobin is a compound in which the porphyrin is still conjugated with protein (and in which the iron remains as ferrous iron) but where the alpha methene bridge is partially oxidized but not broken. Verdoheme is the compound formed when oxidation of the alpha methene bridge is extensive enough to break it. The green pigments formed in meat may be similar compounds.

In meat, oxidation of the porphyrin ring can occur under a number of different situations. Jensen[27] has shown that the hydrogen sulfide produced by certain bacteria in the presence of oxygen can add directly to the porphyrin ring and form an analogue of choleglobin. Hydrogen peroxide-

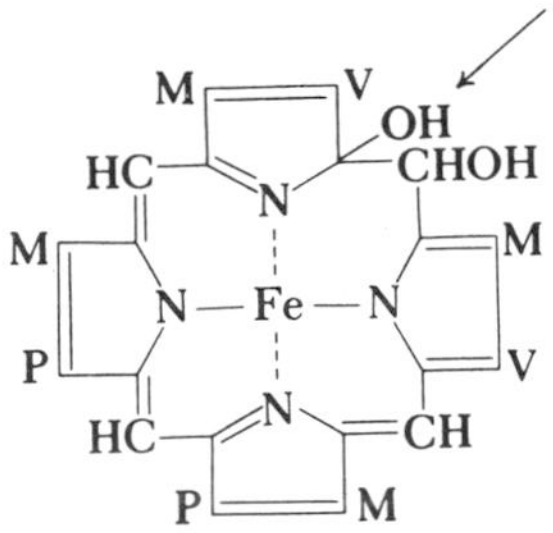

M = methyl, $-CH_3$
V = vinyl, $-CH=CH_2$
P = propionic acid,
$\qquad -CH_2CH_2COOH$

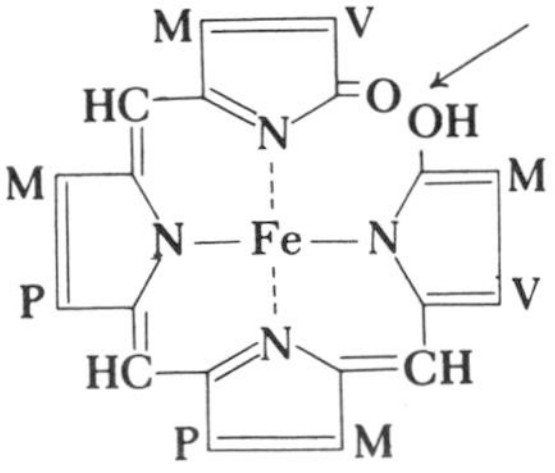

Choleglobin   Verdoheme

forming bacteria will bring about the same type of reaction with the formation of choleglobin itself. In fresh meat any hydrogen peroxide formed is rapidly destroyed by the action of the enzyme catalase but in cured meats, catalase is absent and greening of the meat occurs readily. Other compounds are able to bring about similar reactions. The one which has been most extensively studied is ascorbic acid which, of course, is normally present in fresh meat at low concentrations.

Both Jensen[26] and Urbain[51] say that the identity of green pigments is unknown. In 1936 they published work on the spectra of pigments obtained on exposure of nitric oxide hemoglobin to bacteria. Five different curves for green derivatives were obtained. They have shown that numerous non-pathogenic organisms can produce either oxidizing agents or hydrogen sulfide and that either type of product will react with nitric oxide hemoglobin to form green pigments.[27]

Iridescence or mother of pearl effect is sometimes observed in either cured or cooked meat. The play of color changes as the point of view or the source of light changes. Iridescence is commonly produced in glass, soap bubble films, or when white light passes through some sort of diffraction grating and is broken up into its components to form rainbows. In the case of iridescence of meat it is believed that light is broken up in this manner as it passes through a film of fat on the surface fibers. If the fat is removed, the iridescence disappears; if it is again added, the play of colors reappears.

The color of meat is also directly related to the development of rancidity in meat. It has been found that myoglobin and hemoglobin accelerate the onset of rancidity in the fat of meat. As oxidation of the fatty acid proceeds there is also an oxidation of the myoglobin to metmyoglobin.

In the cured meat nitric oxide hemoglobin and nitric oxide myoglobin are formed. Nitric oxide hemoglobin has been studied and found to be a compound very similar in structure to oxyhemoglobin. The nitric oxide,

Probable Reactions of Myoglobin*

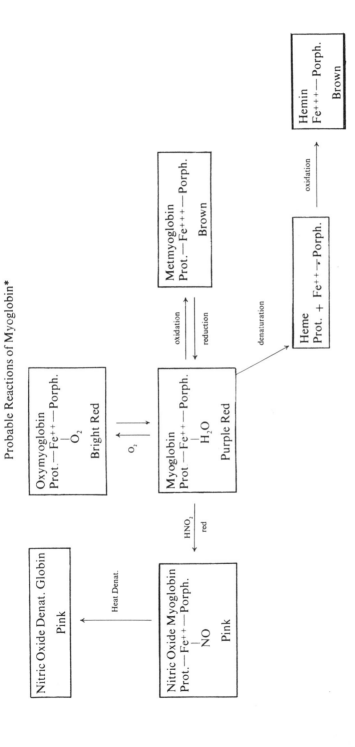

*From Watts, B. M., "Oxidative Rancidity and Discoloration in Meats," *Advances in Food Research,* **5,** 1–42 (1954).

NO, is held by coordinate valences by the ferrous iron in the same way that a molecule of oxygen is held in oxyhemoglobin. The two compounds have very similar bright red colors. When cured meat is cooked, the pink color persists because nitric oxide hemoglobin (or myoglobin) is denatured to the nitric oxide hemochromogen which is also pink.[3]

CURED AND SMOKED MEATS

Long ago when man began to cook his food and to live in some kind of shelter, he soon learned methods of storing meat. Probably he first discovered the method of smoking meat by hanging it near the roof of his cave or tent where it slowly dried and became smoked. Later he found that the addition of salt to the meat (called pickling) prevented its putrefaction and salt became one of man's most valued possessions. Salt routes developed and a man's wealth could be measured by his supply of salt. Jensen tells us that in Homer's time (about 1000 B.C.) both smoked and cured meats were generally consumed. Today with adequate refrigeration and rapid distribution, there is little need for curing and smoking as preserving measures. But our appetites have been developed for the particular flavors which smoking and curing develop, and these methods are still widely practiced. In early times spices were often added particularly to the comminuted meats, to disguise the flavor of "high" or slightly putrid meats. Today we relish these spices even though the meat must always be fresh and clean. During the middle ages in Europe each small region developed its own particular type of sausage. Here in America our cuisine is the result of food customs from all of Europe and to a limited extent from the rest of the world. The number of sausages on the consumer market today is therefore enormous, each with a subtle difference in its recipe. From a chemist's standpoint, they are all very similar.

Since the need for preservation of meat by curing and smoking is largely eliminated by our modern methods of refrigeration, the sanitary conditions of the meat packers, and the rapidity with which distribution occurs, it is no longer necessary to use as much salt in the pickle. Gradually the preference of American consumers has developed for mild cures—ones in which the salt content is kept very low. Meats that are mild-cured are not protected sufficiently by the salt to be stored at room temperature, but always require refrigeration.

There are four general methods of curing meat: (1) dry cure, (2) pickle cure, (3) injection, and (4) comminution and mixing. The Meat Inspection Division, FDA, allows five substances for curing: sodium chloride, sodium nitrate, sodium nitrite, sugar, and vinegar. In the dry cure a mixture of the first four compounds is rubbed on a cut of meat, and it is held under refrigeration for some days while the compounds penetrate the tissue. In the pickle method the compounds are dissolved in water, and the cuts of meat are submerged in the pickle for a period of time. Here the water hastens the penetration of the tissue but gives a watery product. The third method is now widely used in the United states, often in combination with one of the other methods. All meat contains arteries and veins and their branches—a complete vascular system. Pickle can be pumped in through one of these vessels and will be carried through the meat. In curing ham it is common practice to pump pickle in through the main artery (the iliac) of the leg. The needle is also inserted into the muscle and pickle forced into the muscular bed. In the fourth method the meat is simply chopped fine and the curing compounds mixed with it. Cure is rapidly effected because contact is immediate. Vinegar is not used in commercial pickle.

The term "corn" is an Anglo-Saxon word for a small grain, often applied by the people to the common cereal grain of the area—for example, in the United States corn is maize, in England, wheat. Since in early efforts to preserve meat the salt used was granular, the pickling or curing process came to be called "corning." The meat so cured is sometimes called "corned."

The action of salt, NaCl, in preserving meat is to inhibit the growth of bacteria. Jensen[26] reviews some of the rather sizeable body of research that covers the effect of salt on microorganisms. Salt shows a selective action and does not inhibit the growth of all microorganisms. Many yeasts, for example, grow in media containing as much as 15 per cent salt. It has been demonstrated that the effect of salt depends not only on its concentration but on the presence of other compounds in a nutrient medium. Some organisms will not grow in the pickle but will grow on the surface of the meat covered by the pickle since the nutritional environment is different there. Salt also has a detrimental effect on meat. It is well established that salt accelerates oxidative rancidity. When a dilute salt solution is brought in contact with fat, it may actually retard oxidation; but if the solution is partially evaporated so that a film of salt is deposited on the fat, rancidity is greatly accelerated. Gaddis[15] has shown that in bacon the greater the concentration of salt, the more rapid the development of rancidity. In uncured meat, salt also accelerates the formation of brown methemoglobin; but in cured meat, when nitrite is added, the opposite effect is noticed. Salt favors the development of the bright red color. The nature of the effect is not known but it may well be that it is through the denaturation of the globin. Salt is a denaturation agent.

The use of *nitrates* in curing meats originated from the discovery during the late Middle Ages that the color of meat becomes fixed when nitrate is added. The nitrate is reduced to nitrite by nonpathogenic organisms in the pickle. At the pH of the meat, 5.4 to 6.0, nitrite exists as nitrous acid, HNO_2. There are also reducing conditions in the meat because of the presence of readily oxidized compounds formed under conditions of oxygen debt following slaughter. The nitrous acid is reduced to nitric oxide, NO, which then reacts with myoglobin to form nitric oxide myoglobin.

(1) $NO_3 \rightarrow NO_2$ (bacterial action)
(2) $NO_2 + H^+ \rightarrow HNO_2$ (at pH of meat)
(3) $HNO_2 \rightarrow NO$ (reduction by compounds in meat)
(4) $NO +$ myoglobin $\rightarrow$ NO(myoglobin—bright red)

Saltpeter is the old name for potassium nitrate and Chile saltpeter for sodium nitrate. In meat packing houses either one is commonly called "saltpeter" although since the sodium salt is cheaper, it is usually used. The nitrate also has some bacteriostatic effect.

Watts[54] points out that the series of reactions is not quite as simple in many cases as that presented above. When nitrite comes in contact with oxyhemoglobin, the first reaction is the formation of methemoglobin, nitrate, and oxygen. Thus when ground pork sausage is mixed with curing salts, the color immediately turns gray as metmyoglobin is formed. This reaction is very rapid at the low pH of meat. When the meat is heated during smoking, the red color of the nitric oxide hemochromogen develops. Watts also believes that there is a good possibility that other intermediates are involved and that the reaction on heating is not as simple as the one given below.

$$NO—myoglobin \xrightarrow{heat} NO—hemochromogen$$

The development of the proper color during curing depends on the formation of nitrite, and this in turn depends on the presence of certain groups of bacteria which slowly reduce the nitrate to nitrite. When it was recognized that nitrite was the color-fixing agent, it became obvious that the problem with undercuring—poor color development—could be solved by the addition of nitrite as such to the pickle. In 1927 the use of sodium nitrite was allowed. Today the common practice is to use both nitrate and nitrite since nitrite alone often does not give good color. If nitrite is used alone, the concentration of nitrous acid is so great that it decomposes and N_2O_3 is lost from the pickle. The action is not prolonged or sustained during the whole of the time of pickling, so that a combination of both nitrate and nitrite is, in general, more successful.

Sugar not only adds a sweet flavor, delectable in cured meats, but also produces conditions during curing and storage that insure the best color and conserve protein. Ordinarily, sucrose is added to the pickle. In the presence of many types of bacteria, sucrose is hydrolyzed to glucose and fructose and then utilized by these organisms to form many different compounds, many of which are capable of reduction and some of which are acidic. The reducing conditions resulting from the organic compounds formed protect hemoglobin and myoglobin from irreversible oxidation of the iron to the ferric state and insure the formation of nitric oxide and nitric oxide hemoglobin. During storage these reducing conditions persist, and the chances that the nitric oxide hemoglobin will form methemoglobin are minimized. However, when exposed to oxygen, nitric oxide hemoglobin is oxidized to methemoglobin. Thus sliced bacon turns brown on standing for some time.

Since sucrose forms glucose and fructose, the effects of other mono- and disaccharides have been tested. For the most part they are not as successful in the pickle as sucrose because they are used by the bacteria so rapidly that the pH falls to a range where methemoglobin formation is speeded up and instead of protecting against browning these sugars can even hasten it. In short cures they are useful, and today glucose in the form of corn sugar or a mixture of glucose is sometimes used along with sucrose.

Smoking Meats

Hardwood smokes are known to produce superior preservation and flavor. Jensen[26] lists the effects of smoking: (1) drying effect, (2) imparting desirable organoleptic properties, (3) bringing out color inside the cured meat, (4) imparting antioxidants to the fat, (5) impregnating the outside portions of the meat with constituents of smoke that serve as antiseptics and germicides, (6) an adjuvant action of constituents of smoke and heat on microorganisms, if the process is carried out at a temperature above 49°C (120°F), (7) a tendering action from the high humidity of the smokehouse in combination with its high temperature, (8) a diminution of nitrite content probably by reason of the aliphatic diazo reaction with protein which occurs at higher temperatures, and (9) the imparting of a desirable finish or gloss on both the skin and the flesh sides of meat through the agency of aldehyde-phenol condensed resins from the smoke and the film of grease remaining on the piece.

Smoke contains many compounds produced by fracture of the large molecules present in the wood. In destructive distillation of wood, the wood is heated but not burned, and these compounds are distilled out and

TABLE 6.2. HARDWOOD SMOKE

	Parts Per Million
Formaldehyde, HCHO	25–40
Higher aldehydes	140–180
Formic acid, HCOOH	90–125
Acetic and higher acids	460–500
Phenols	20–30
Ketones	190–200
Resins	over 1000

separated. However, the conditions during smoking are more complicated because not only are these compounds formed and distilled out of the wood but, in the presence of fire and air, they may burn or react with one another. In the destructive distillation of wood sizeable quantities of phenols are formed, for example, but in smoke the concentrations are low because they react with formaldehyde and form resins. Jensen gives the following composition of hardwood smoke based largely on work of Pettett and Lane.[40] (See Table 6.2.)

The research laboratories of many meat packing houses have studied the bactericidal effect of smoking and have found reduction in numbers of bacteria. The concentration of formaldehyde in smoke is sufficiently great to account for this effect and some investigators believe that it is the active ingredient. There is the possibility that other compounds in the smoke such as some of the higher aldehydes may augment this effect. The temperature of the smokehouse is usually sufficiently high so that some of the bacteria are destroyed by heat. Smoked meat has a marked increase in keeping time over unsmoked meat. Small amounts of aldehydes appear to be retained on the surface of the meat and act as bactericides. However, since molds are much more tolerant of aldehydes than bacteria, smoking has little effect in preventing the growth of a number of types. Also many molds will grow in the presence of sizeable quantities of salt. Although molding is sometimes a problem with smoked meats, none of these molds produce disease in man.

The gloss formed on the surface of smoked meats is the result of two effects. Resins formed in smokes from the reactions of aldehydes—particularly formaldehyde and phenols—are deposited on the surface of the meat. Also at the temperature of the smokehouse some fat oozes out. Jensen reports that the aldehydes produce a glossy finish when the meat has oiled out a little.

During smoking there is a steady decrease in the amount of nitrite. Nitrous acid reacts with free amino groups to give the aliphatic diazo reaction and form hydroxyl groups.

$$-NH_2 + HONO \rightarrow -OH + N_2 + H_2O$$

When meat is smoked, the onset of rancidity is delayed. The protection is primarily on the outside of the smoked piece since there is little penetration of the smoke compounds to the center of a ham or bacon. When the piece is sliced, it is for the most part unprotected. The identity of the compounds that act as antioxidants has not been established; but since small amounts of phenols are in the smoke and since many antioxidants are phenols, it has been assumed that these are the active agents. Watts[54] reports her work with Faulkner on the effect of liquid smoke as an antioxidant. Liquid smoke is produced by the thermal decomposition of hardwoods and a number of these products are on the market. They are not allowed by the Bureau of Animal Industry in commercial production of smoked meat. Watts and Faulkner found considerable difference in the antioxidant power of these liquid smokes, ranging from no effect at all to a pronounced inhibition of oxidation.

When cured meat is smoked, the color brightens and becomes more stable. The reaction is not completely known, but it is believed that nitric oxide myoglobin is denatured and forms the corresponding porphorin and denatured protein. The solubility is diminished on coagulation and the increased stability or decreased reactivity may be the result of this change.

In the moist heat of the smokehouse the muscle proteins are greatly affected. Denaturation of the protein occurs at the higher temperatures. At the beginning of the smoking process when the hams are warmed, autolytic enzymes in the cells may be active and cause some tendering. As the internal temperature of the piece of meat increases, enzymes will also be denatured and cease to function. This will be particularly true of meat smoked to a high temperature or held in the smokehouse for a relatively long time. During smoking the changes that occur with cooking get under way; and when the time of smoking is long or the temperature high, as for example in the preparation of ready-to-eat ham, these changes go to completion.

CHANGES IN MEAT ON COOKING

Cooking causes many changes to occur in meat which develop appetite-stimulating flavors. Who has not smacked his lips over the thought of a rich brown steak dripping with juicy goodness? Yet the sight and odor of that same steak before it is cooked inspires none of the same response in us. Undoubtedly, cooking markedly increases those properties which are called organoleptic.

Probably as far as our well being is concerned, the most important change in cooking is the destruction of microorganisms. All cooking processes decrease the number of bacteria, yeasts, and molds, and if the meat is well done may even render it sterile. Spore-forming organisms are destroyed and the spores killed if the meat is cooked to a sufficiently high temperature. In canning meat this is important since the meat must be completely sterile so that it will keep. Pork is often infected with a parasite, *Trichinella spiralis*, which is dangerous to man. This organism is killed at 132°F; Jensen points out that the Bureau of Animal Industry demands that pork be cooked to an internal temperature of 137°F to insure, with the 5° margin of safety, that all organisms are destroyed.

The chemical and physical changes which occur on cooking are numerous and although some studies have been made on this problem the reactions are not all known or understood. They are: (1) denaturation of the protein, (2) hydrolysis of collagen to gelatin, (3) color change, (4) formation of drip, (5) development of brown flavor, (6) rupture of fat cells and dispersion of fat through the meat, and (7) decrease in vitamins and possibly decrease in the nutritive value of the protein.

Denaturation of the Protein

This is evidenced by the change in the physical condition of the meat. The jellylike structure stiffens and often toughens. There is a shrinkage in volume and an increase in density as cooking proceeds. Ramsbottom and Strandine[41] studied the change in tenderness of various muscles from wholesale cuts of beef of three young heifers, good grade, by the shear test. This is a test in which tenderness is determined by measuring the force necessary to cut through a standard sample. They found that tenderness decreased in 35 of 50 samples cooked in lard to an internal temperature of 170°F. Lowe[32] discovered in the case of beef rib roasts that raw meat is tenderest, rare is intermediate, and well done, least tender. She used both the shear test and a penetrometer. Other workers have found similar results, but some have found divergent ones. Undoubtedly the difference in results is caused by the fact that a measure of tenderness in meat does not measure the change in tenderness of the protein. Instead, a change in the tenderness of meat is the result of numerous changes, only one of which is this change in protein. Heat denaturation of protein is a familiar experience in food cookery, but it is not always accompanied by a toughening of the protein.

Denaturation of a protein molecule is believed to be a series of reactions in which the protein molecule is altered by splitting of some of the links, particularly some of the hydrogen bonds and sulfur-sulfur links which help maintain the three-dimensional shape of the molecule. The molecular weight of the molecule may not be very much changed in the first steps of the reactions. In heat denaturation, as in other types of denaturation, it is believed that these changes in the molecule account for the alteration in such properties as solubility and density.

The Hydrolysis of Collagen

This is the reaction by which commercial gelatin is prepared. The extent to which it occurs in meat cookery has been the object of a number of researches. If a piece of meat is cooked for a long time, the loose connective tissue disappears and the gravy gels on cooling. This is particularly noticeable in moist cooking such as braising. However, even in dry cooking such as roasting and frying, the connective tissue and the collagen fibers are in contact with water from the tissue. Connective tissue itself has a fair amount of water. So there is always the possibility that hydrolysis of the large collagen molecules to the relatively small gelatin molecules can occur.

Winegarden, *et al.*[57] carried out careful tests of the connective tissue from nine beeves. They used the ligamentum nuchae in which most of the fibers are elastin, the deep flexor tendon in which most of the fibers are collagen, and strips from the aponeurotic sheet which is intermediate in collagen and elastin content between the first two tissues. Collagen and total nitrogen were determined on all the samples. The strips were heated in distilled water for periods of time of 1, 2, 4, 16, and 64 minutes and at temperatures ranging from 60°C to 95°C. Changes in shear force, weight, length, width, and thickness were measured. The tissues were also studied histologically. Heating for 64 minutes at 95°C caused some changes in the collagenous fibers, so that some were fused or merged, some straightened, and all were less distinct. The elastin fibers did not show these changes. All tissues showed softening and a decrease in shear force, even the ligamentum nuchae, on heating at the higher temperatures. There was little change at 60°C but the extent of change increased with increased temperature and increased time; 60° C or 65° C is the internal temperature to which meat is cooked when it is rare. Thus the study indicates that although little change in collagen can be expected for rare meat, collagen hydrolysis increases with doneness. The study also demonstrated that strips from meat aged 35 days had slightly lower shear forces than those aged 10 days; finally the older animals for the most part required more force for shearing the tissue even though tissues from the oldest animal, an 8-year-old cow, did not have the highest shear-force readings.

Griswold[19] studied a number of factors in beef rounds, commercial and prime grades, cooked by various methods. Collagen was significantly higher in the raw meat for the commercial grade over the prime but not for the top round over bottom round. The loss of collagen on cooking occurred with all methods and increased with an increase in the internal temperature to which the round was cooked. When pressure cooking was compared, it was found that meat cooked at 10 lbs or 15 lbs pressure showed greater losses of collagen than that cooked at 5 lbs pressure. It was higher in meat roasted at 250°F than in that roasted at 300°F or braised. Marinating in vinegar caused a greater loss; but other methods such as scoring, pounding, or using enzyme preparations had little effect.

Color Change

The color change from red or purplish red to brown or gray on cooking has already been discussed under the color of meat. When meat is cooked, oxyhemoglobin and oxymyoglobin (red) and hemoglobin and myoglobin (purplish red) are denatured. The ferrous iron in the free porphyrin formed is rapidly oxidized to ferric iron of hemin (brown). Some transformation of oxyhemoglobin, oxymyoglobin, hemoglobin, and myoglobin to methemoglobin and metmyoglobin (brown) can also occur.

Drip Formation

This accounts for some of the change in weight that occurs in a piece of meat during cooking. The drip is composed of water carrying a number of soluble compounds as well as some coagulable protein and fat. As cooking proceeds the coagulable protein is denatured by the heat and forms a curd in the pan gravy. Numerous compounds have been isolated from the drip, but it is questionable whether they account for the delicious flavor of the mixture. Most of the compounds are small molecules formed either during autolysis while the meat is ripening or by heat fragmentation of larger molecules. Creatine, creatinine, salts, particularly sodium chloride, small amounts of amino acids, amino acid derivatives such as amines, purines, and pyrimidines can be detected. These molecules are individually flavorless or almost so aside from the saltiness of sodium chloride and other salts.

Fat oozing always occurs but the extent to which fat runs out of the meat depends on many factors. The cut of meat and the amount of fattiness, the method of cooking, the extent to which the piece is cooked all obviously affect the amount of fat in the drip. There may also be fat degradation products in the drip. When steak is broiled and some burning of the fat occurs, small amounts of acrolein can be detected in the drippings. Free fatty acids are sometimes present. Griswold[19] found that the amount of soluble nitrogen compounds increased in the drippings during cooking but did not find that free amino nitrogen showed an increase. She cooked round steak by various methods.

Meat Aroma

The change in flavor that occurs with cooking is one of the most apparent changes to the man at the table. Particularly where the meat has browned, a strong flavor, sometimes called the "brown flavor," is evident. Crocker in a review[6] points out that raw meat, whether beef, pork, lamb, or chicken has little flavor. It is slightly salty and slightly sweet. Cooked meat likewise has this slightly salty, slightly sweet taste but it also possesses an aroma that adds greatly to the flavor. The aroma is composed of low molecular weight, volatile compounds such as amines, ammonia, hydrogen sulfide, and organic acids. These compounds probably arise from the cracking of amino acids during heating. The reactions may be decarboxylation, deamination, or desulfuring in which either the free amino acid or polypeptides react. The compounds formed vary according to the species, so that the flavor of lamb is recognizable and different from the flavor of beef.

Dispersion of Fat

Fat cells rupture and fat disperses through the meat on cooking. The proteins in fat cells of the adipose tissues undergo denaturation on cooking just as all other proteins in the meat. There is a change in the permeability of the cell walls and fat flows out. Wang, et al.[53] studied this change carefully in beef steaks. They used the *longissimus dorsi* (ribeye) and the *semi tendinosus*, eye of the round, and broiled them to an internal temperature of 150°F. Sections made of the cooked steak were stained with Sudan IV or Nile Blue to show the fat. It could be seen that fat had diffused out of the fat cells of the perimysium. Fat droplets seemed to be dispersed along the path of degraded collagen and mixed with it. The size of the droplets decreased from the dispersion center to the periphery. Wang, et al. interpreted this to mean that the degraded collagen emulsifies the fat. Photographs of their slides are shown in Figure 6.5.

Decrease in Vitamins

The decrease in vitamins that occurs during cooking of meat has been extensively studied. The B complex vitamins are all more or less sensitive to heating and thiamin and pantothenic acid are particularly labile. If cooking is prolonged and if the temperature is high, the destruction of the vitamins may be appreciable. There is always some loss. Niacin and riboflavin are

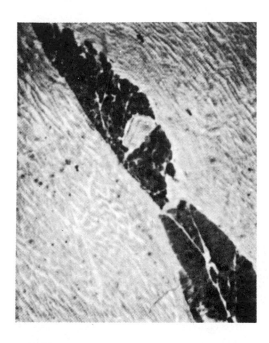

1. Raw fresh light good ribeye control, showing parts of two perimysial fat islands with cells intact and loaded with fat, × 20.

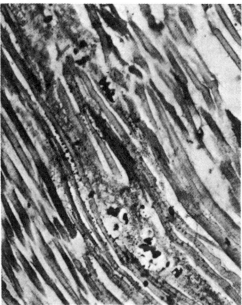

2. Section of above sample after it was cooked, showing partly depleted fat islands surrounded by partly hydrolyzed collagen. Three empty fat cells with distinct walls and nuclei are seen in upper fat island, whereas more than half of the cells in the lower island have released their fat. × 55.

FIGURE 6.5. CHANGES IN MEAT FAT ON COOKING. Fat is stained and appears dark.

3. Another cooked ribeye, showing typical fat dispersion center.
Dispersed fat has followed a
course approximately coincid-
ing with that of hydrolyzed
collagen with little or no fat in
endomysial spaces. × 55.

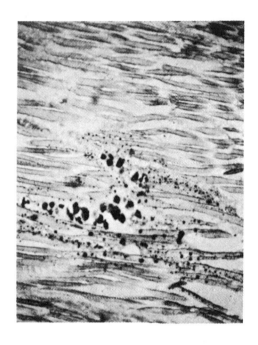

4. Another cooked
sample showing
extensive disper-
sion from fat
cells not shown,
but above upper
right.

FIGURE 6.5 (Continued)

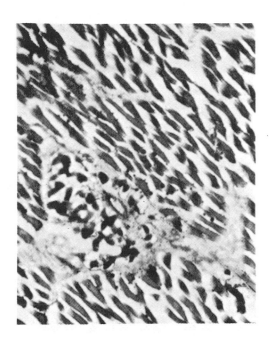

5. Cooked 14-day-aged prime rib-eye, with dispersion of fat into neighboring perimysial spaces which separate muscle bundles. × 55.

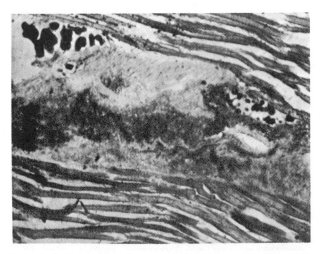

6. Cooked fresh prime ribeye perimysium contains thoroughly mixed fat and hydrolyzed collagen. Note relatively intact collagen fibers between 2 fat islands are free of fat. × 55.

FIGURE 6.5 (Continued)

7. Cooked unaged cow round. With intact (right) and hydrolyzed (left) collagen. Fat island at lower right. × 110. Reproduced from Wang, H., Rasch, E., Bates, U., Beard, F. J., Pierce, J. C. and Hankins, O. C., *Food Research*, **19,** 314–22 (1954). *Courtesy of American Meat Institute Foundation.*

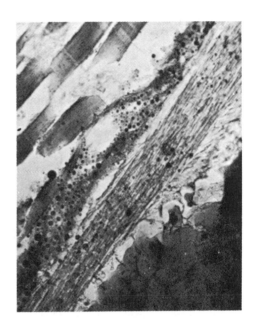

FIGURE 6.5 (Concluded)

more stable to heat and do not disappear as rapidly during the cooking of meat. However, the folic acid group is very sensitive to heat, and as much as 90 per cent may disappear. Ascorbic acid begins to decrease as soon as an animal is slaughtered, and during cooking any that is still present will suffer further diminution. Cooked meat cannot be counted on to supply more than very small amounts of ascorbic acid to our diet.

Surface Reddening

Occasionally a red surface is produced when meat is boiled, broiled, or roasted and sometimes it is seen in canned meat. We do not expect meat to get red when cooked, but instead anticipate a brown or gray color and, in cured meats, a pink. Jensen discusses the conditions under which it is possible for the reddening to occur, but points out that sometimes it is produced under conditions where as yet there is no explanation. The reddening does not make the meat harmful or unpalatable. A red color is produced whenever the meat comes in contact with nitrites, nitrates, carbon monoxide, or sulfites during preparation so that hemoglobin and myoglobin react. This occasionally occurs when nitrites are present in water supplies in very small amounts. Even as little as a few parts per million of nitrite can cause surface reddening as nitric oxide hemoglobin or the porphyrin is formed. Vegetables also sometimes contain small amounts of nitrites as well as nitrates. If meat is cooked with these vegetables, some leaching of the ions can occur and reddening of the meat surface will form. Carbon monoxide might come in contact with roasting meat through a leak in a gas pipe carrying coal gas (natural gas has little or no carbon monoxide). While carbon monoxide is a very poisonous gas, neither carbon monoxide hemoglobin nor carbon monoxide myoglobin are harmful. Sulfites can also produce reddening, but they are not likely to be present in either the meat or vegetables. Sulfites are prohibited in meat and do not occur in plant material.

Overcooking

This causes an excessive loss of drip and a toughening of the meat. The meat becomes stringy as the amount of connective tissue falls and the fat drips out. Not only are the tenderness and juiciness of the meat diminished, but the flavor as well decreases. Volatile, odorous compounds that contribute to the flavor are driven off during the excessive cooking period, and the meat becomes less and less flavorful. A decrease in the nutritive value of the protein may also occur (p. 136). Overcooking can be the result either of too long a time or too high a temperature.

TENDERNESS

Tenderness is one of the most important attributes for rating meat. As we bite into a piece of meat, tenderness is one of the first sensations perceived. Like many other characteristics of meat, tenderness is incompletely understood. However, we are fairly certain that it is the result of a complex of factors. Most investigators find that tenderness of meat decreases on cooking. Indeed, some restaurants print on the menu that tenderness of steaks cannot be guaranteed unless they are served rare. But since cooking includes a whole series of changes, these observations do not explain the fundamental cause of tenderness or its absence.

Tenderness is measured by one of three methods or a combination of them: (1) the force necessary to penetrate through a piece of meat, (2) the shear test which measures the force necessary to cut across fibers, or (3) a panel of trained judges. The correlation between methods is sometimes excellent and sometimes poor.

The factors that may be significant in explaining tenderness are (1) the amounts of connective tissue in the muscle, (2) the amount of hydration of the muscle proteins, (3) the nature of the protein molecules, and (4) the amount of fat between muscle fibers. These factors will be discussed and evaluated.

Connective Tissue

The connective tissue that forms a fine network through muscle tissue has been credited traditionally with tenderness or lack of tenderness. However finding methods by which connective tissue can be estimated without doubt and separated or differentiated from other types of tissue has been exceedingly difficult. Investigators have for the most part used estimates of collagen, the protein present as fibers in the ground substance, to measure connective tissue. It has been claimed that collagen is completely hydrolyzed to gelatin by autoclaving and that it is the only protein in a meat sample capable of this reaction. A number of determinations of the amount of nitrogen in the gelatin formed have been presented as an estimate of the connective tissue present. Some years ago it was realized that the assumption that only collagen would be hydrolyzed is problematic, and the method was then refined by preparing the meat sample by extraction before autoclaving. In this modification it is assumed that the proteins capable of some hydrolysis on autoclaving are first removed and that none of the collagen is removed. Miller and Kastelic[36] have critically studied some of the methods used to evaluate connective tissue in bovine muscle and find that the amount of protein extracted by alkali increases with time of exposure of the sample and with the concentration of the alkali. (Sodium hydroxide is usually used.) The protein extracted probably is from the ground substance and not collagen since the protein is free of hydroxyproline. Their work shows that nitrogen determinations on insoluble substances do not give valid measures of collagen or elastin.

The hydroxyproline method for estimating either collagen or elastin depends on the presence of unusually large amounts of hydroxyproline in these proteins. However, the proteins used in setting up the method are those present in tendons, hide, and ligaments. When the method is applied to meat, collagen is calculated from the hydroxyproline determination on the gelatin formed by autoclaving, while elastin is calculated from a determination of hydroxyproline in the portion insoluble in alkali and after autoclaving. Elastin is assumed to be resistant to hydrolysis by mild alkali or by water on autoclaving. Sometimes the residue remaining has been measured by estimating the nitrogen in it and calculating this as the elastin portion.

Other methods of estimating connective tissue make use of histological means. Samples of meat are prepared on slides and stained so that the connective tissue can be seen and evaluated. This is very difficult to do precisely. Harrison, *et al.*[22] demonstrated that the most tender piece of beef (aged 1, 2, 5, 10, 20, and 30 days) had the smallest amount of connective tissue fibers, both collagen and elastin, as detected by staining sections. Deatherage's group[24] estimated the amount of collagen and elastin by determining the alkali insoluble fraction. Their beef, short loin, aged 3 days, showed no correlation between this fraction and tenderness but when the beef was aged 15 days, there was a definite correlation. In another series on beeves representing a wide variety of market grades, they[25] found excellent correlation between alkali insoluble protein and tenderness. These animals had all been aged 14 days at 3.5°C.

So far no simple and reliable method has been developed for estimating connective tissue. The connective tissue in muscle has not been separated, a difficult job, and the amount of collagen and elastin determined. The relation of connective tissue to tenderness is consequently still not clear. Whatever method is used, the amount of connective tissue can not be correlated with tenderness in all cases.

Hydration

The degree of hydration of muscle proteins may influence tenderness. A migration of ions occurs[2] as the meat ages. Calcium, sodium, and magnesium migrate out of the muscle cell while potassium ions move in. The total exchange is such that the over-all change is an increase of ions in the protein. These ions are held so firmly that they not only do not appear in the juice formed when the meat is cooked, but they also are not extracted with water. Deatherage concludes that the migration of ions allows greater hydration of the protein and consequently an increase in tenderness. This is a logical conclusion.

Water of hydration is held within a protein molecule by hydrogen bonding as the slightly positive hydrogen is attracted to relatively negative atoms of oxygen or nitrogen. Water molecules are held between the folds or coils as well as on the surface of the large protein molecules. Hydration is affected by pH and is lowest at the isoelectric point of the protein. Ions also influence the ability of the protein to hydrate—hydration tending to increase with increase in ionization. During rigor mortis, the pH of a carcass of a well nourished animal falls to approximately 5.6, very close to the isoelectric point of actomyosin; and the amount of hydration decreases. This is demonstrated by the increase in the amount of juice formed when meat in rigor is cooked. As rigor passes the pH rises slightly, so that the protein begins to form negative ions rather than the internal zwitter ions of the isoelectric point. The migration of ions allows salt formation and the protein may become more highly ionized and more able to hold water molecules by hydrogen bonding. Deatherage believes it is reasonable to expect a highly hydrated protein to be more tender than one which is not so highly hydrated. The protein molecules will separate from one another more readily if a layer of water lies between them.

Nature of Protein Molecules

There are some observations which suggest that changes in muscle proteins influence tenderness. Harrison, *et al.*[22] studied the histological appearance of sections of the psoas (tenderloin) muscle in beeves of various tenderness. In the most tender specimen there was an absence of longitudinal striations on the muscle fibers; whereas, in the least tender—a specimen from a cow 8 years old and of cutter grade—the fibers were twisted and gnarled, less uniform in diameter, and the cross striations less uniform in spacing. This observation indicates that protoplasm protein, the contractile material of the cell, is involved.

Two proteins, myoglobin and actomyosin, have been studied by Deatherage's group. Myoglobin content gave a positive correlation with tenderness in rib steaks which had been aged 15 days. They[56] found no correlation of actomyosin content with tenderness. As rigor occurs and the muscle becomes rigid and tough, actin and myosin combine to form actomyosin. As rigor passes and the meat begins to become tender, the content of actin and myosin continues to fall. The change in the muscle is not, therefore, a simple reversal of actomyosin formation. Likewise water extractible nitrogen which includes polypeptides and peptones, the fragments of proteins, does not increase in amount but instead falls.

Deatherage and his group have studied a number of other cell components and have shown that there is no correlation of tenderness with total nitrogen, trichloracetic acid-soluble nitrogen, nonprotein nitrogen, water-soluble nitrogen, pH, lactic acid, moisture or inorganic phosphate.

Fat

A number of investigators believe that good marbling contributes to tenderness of meat. Harrison *et al.*[22] noticed one or two rows of fat cells between the fibers in their most tender specimens and Husaini found a correlation, although of a low order, between tenderness and intermuscular fat. The effect of marbling on tenderness is still open to question.

JUICINESS

The amount and the quality of the juice formed when meat is chewed is another of the most important factors in judging the total quality of a piece of cooked meat. But it is a factor very difficult to define. Everyone has experienced the extremes of a juicy bite of meat or a very dry one, but the problem of exactly defining juiciness or setting up some device that can objectively measure it has been extremely difficult. Several instruments for measuring press fluid—the amount of juice that can be pressed out of a meat sample under high pressure—have been devised and used, but the correlation of the results of these instruments with the judgements of a panel have not been good. In general it has been considered that marbling of the meat with fat, is one of the most important factors in producing the sensation of juiciness. In one extensive study, Gaddis, Hankins, and Hiner[16] found a low but significant correlation between the amount of press fluid and the juiciness scores in 97 beef ribs, but none in 115 lambs and sheep (legs studied) or 11 goats. With beef, increase in percentage of fat in the press juice caused a decrease in the amount of juice yet an increase in the juiciness scores by the panel up to 2 per cent fat. They concluded that the sensation of juiciness is probably the result of numerous factors, only one of which is the juice in the piece of meat. They believe that fat adds flavor and stimulates the flow of saliva in the mouth. It also coats the mouth and prolongs the sensation of moistness and richness during the chewing and swallowing. These combined effects give the sensation of juiciness.

PHOSPHATES IN CANNED MEAT

The effect of phosphates on the ability of meat to bind water has been investigated in recent years. Some results appear to indicate that the effect is specific for phosphate, others that it depends on the increased pH which occurs. Hamm and Grau[21] found that the addition of one per cent phosphate to the sodium chloride that is put in meat has a definite effect on the amount of bound water and decreases the loss of drip. Bendall[5] studied the effect of orthophosphate, pyrophosphate, metaphosphate, Calgon (a commercial polyphosphate), and phosphate glass on rabbit meat. Although all influenced meat's capacity to bind water, the pyrophosphate had the greatest effect. Bendall considers that the binding of water depends on the ionic strength present except in the case of pyrophosphate.

Commercially, phosphate is now added to canned meat as well as to shrimp. The increased binding of water decreases the shrinkage during processing and improves tenderness.

REFERENCES

1. ALEXANDER, J. C. and ELVEHJEM, C. A., "Isolation and Identification of Nitrogenous Compounds of Meat," *J. Agr. Food Chem.,* **4,** 708–711 (1956).
2. ARNOLD, N., WIERBICKI, E., and DEATHERAGE, F. E., "Postmortem Change in the Interactions of Cations and Proteins of Beef and Their Relation to Sex and Diethylstilbesterol Treatment," *Food Technol.,* **10,** 245–250 (1956).

3. BATE-SMITH, E. C., "Physiology and Chemistry of Rigor Mortis," *Advances in Food Research*, **1**, 1–34 (1948).
4. BEAR, RICHARD S., "The Structure of Collagen Fibrils," *Advances in Protein Chem.*, **7**, 69–161 (1952).
5. BENDALL, J. R., "Swelling Effect of Polyphosphates on Lean Meat," *J. Sci. Food Agr.*, **5**, 468–475 (1954); *C.A.*, **49**, 1985d (1955).
6. CROCKER, E. C., "Flavor of Meat," *Food Research*, **13**, 179–183 (1948).
7. DAY, J., "The Mode of Reaction of Interstitial Connective Tissue with Water," *J. Physiol. (London)*, **109**, 380–791 (1949).
8. DRAUDT, H. N. and DEATHERAGE, F. E., "Studies in the Chemistry of Cured Meat Pigment Fading," *Food Research*, **21**, 122–132 (1956).
9. DROZDOV, N. S., "Muscle Tissue Autolysis During Cold Storage, Deep Freezing and Defrosting," *Biokhimiya*, **20**, 403–407 (1955); *C.A.*, **50**, 2087h (1956).
10. EASTOE, J. B., "Amino Acid Content of Fish Collagen and Gelatin," *Biochem. J.*, **65**, 363–368 (1957).
11. EASTOE, J. B. and EASTOE, B., "Organic Constituents of Mammalian Compact Bone," *Biochem. J.*, **57**, 453–459 (1954).
12. EDMUNDSON, A. B. and HIRS, C. H. W., "Sperm Whale Myoglobin," *Fed. Abst.*, **86** (1959).
13. ERDMAN, A. M. and WATTS, B. M., "Effect of Storage Conditions on Loss of Color and Free Sulfhydryl Groups in Cured Meat," *Food Technol.*, **11**, 183–185 (1957).
14. FENDER, A. E. and WOOD, T., "Meat Extractive," *Food Manuf.*, **31**, 223–227 (1956).
15. GADDIS, A. M., "Effect of Pure Salt on the Oxidation of Bacon in Freezer Storage," *Food Technol.*, **6**, 294–298 (1952).
16. GADDIS, A. M., HANKINS, O. G., and HINER, R. L., "Relationship Between the Amount and Composition of Press Fluid, Palatability and Other Factors of Meat," *Food Technol.*, **4**, 498–503 (1950).
17. GINGER, I. D. and SCHWEIGERT, B. S., "Chemical Studies with Purified Metmyoglobin," *J. Agr. Food Chem.*, **2**, 1037–1040 (1954).
18. GINGER, I. D., WACHTER, J. P., DOTY, D. M., SCHWEIGERT, B. S., BEARD, F. J., PIERCE, J. C., and HANKINS, J. O., "The Effect of Aging and Cooking on the Distribution of Certain Amino Acids and Nitrogen in Beef Muscle," *Food Research*, **19**, 410–416 (1954).
19. GRISWOLD, R. M., "The Effect of Different Methods of Cooking Beef Round of Commercial and Prime Grades: I. Palatability and Sheer, II. Collagen, Fat and Nitrogen Content," *Food Research*, **20**, 160–179 (1955).
20. GUSTAVSON, K. H., "The Chemistry and Reactivity of Collagen," Academic Press, Inc., New York, N. Y., 1956.
21. HAMM, R., and GRAU, R., "Effect of Phosphates on Bound Water of Meat," *Deut. Lebensm. Rundschau*, **51**, 106–111 (1955); *C.A.*, **49**, 12744d (1955).
22. HARRISON, D. L., LOWE, B., MCCLURG, B. R., and SHEARER, P. S., "Physical, Organoleptic, and Histological Changes in Beef During Aging," *Food Technol.*, **3**, 284–288 (1949).
23. HUGGINS, M. L., "The Structure of Collagen," *Proc. Natl. Acad. Sci.*, **43**, 209 (1957).
24. HUSAINI, S. A DEATHERAGE, F. E., and KUNKLE, L. E., "II Observations on the Relation of Biochemical Factors to Changes in Tenderness," *Food Technol.*, **4**, 366–369 (1950).

25. HUSAINI, S. A., DEATHERAGE, F. E., KUNKLE, L. E., and DRAUDT, H. N., "Biochemistry of Beef as Related to Tenderness of Meat," *Food Technol.*, **4**, 313–316 (1950).
26. JENSEN, L. B., "Meat and Meat Foods," Ronald Press Co., New York, N. Y., 1949.
27. JENSEN, L. B. and URBAIN, W. M., "Bacteriology of Green Discoloration in Meats and Spectrophotometric Characteristics of the Pigments Involved," *Food Research*, **1**, 263–273 (1936).
28. KELLEY, G. C. and WATTS, B. M., "Effect of Reducing Agents in Cured Meat Color," *Food Technol.*, **11**, 114–116 (1957).
29. KENDREW, S. C., BODO, G., DINTZIS, H. M., PARRISH, R. G., and WYCKOFF, H., "A Three Dimensional Model of the Myoglobin Molecule Obtained by X-ray Analysis," *Nature*, **181**, 662–666 (1958).
30. LANDROCK, A. H. and WALLACE, G. A., "Discoloration of Fresh Red Meat and its Relationship to Film Oxygen Permeability," *Food Technol.*, **9**, 194–196 (1955).
31. LEMBERG, R. and LEGGE, J. W., "Hematin Compounds and Bile Pigments," Interscience Publishers, Inc., New York, N. Y., 1949.
32. LOWE, B., "Experimental Cookery," 4th ed., John Wiley & Sons, Inc., New York, N. Y., 1955.
33. LOWE, B., "Factors Affecting the Palatability of Poultry with Emphasis on Histological Postmortem Changes," *Advances in Food Research*, **1**, 203–256 (1948).
34. LOWRY, O. H., GILLIGAN, D. R., and KATERSKY, E. M., "Determination of Collagen and Elastin in Tissue," *J. Biol. Chem.*, **139**, 795–804 (1941).
35. MEYER, K., LINKER, A., DAVIDSON, E. A., and WEISSMANN, B., "Mucopolysaccharides of Bovine Cornea," *J. Biol. Chem.*, **205**, 611–616 (1953).
36. MILLER, M. and KASTELIC, J., "Meat Tenderness Factors: Chemical Response of Connective Tissue of Bovine Skeletal Muscle, *J. Agr. Food Chem.*, **4**, 537–542 (1956).
37. MOHLER, K. and KIERMEIER, F., "The Action of Inorganic Phosphate on Animal Protein V. The Influence of pH Value on Binding of Water by Coagulated Meat Roast," *Z. Lebensm. Untersuch u. Forsch.*, **100**, 260–266 (1955); *C.A.*, **49**, 8519b (1955).
38. NEUMAN, R. E. and LOGAN, M. A., "Hydroxyproline Determination," *J. Biol. Chem.*, **184**, 299–306 (1950).
39. PAUL, P., BRATZLER, L. G., FARWELL, E. D., and KNIGHT, K., "Studies on Tenderness of Beef I Rate of Meat Penetration," *Food Research*, **17**, 504–510 (1952).
40. PETTET, A. E. and LANE, F. G., "The Chemical Composition of Hardwood Smokes," *J. Soc. Chem. Ind.*, **59**, 114–119 (1940).
41. RAMSBOTTOM, J. M. and STRANDINE, E. J., "Comparative Tenderness and Identification of Muscles in Wholesale Beef Cuts," *Food Research*, **13**, 315–330 (1948).
42. RICE, E. E., and BEUK, J. F., "The Effect of Heat on Nutritive Value of Protein," *Advances in Food Research*, **4**, 233–280 (1953).
43. SIZOV, M. I., "The Effect of Temperature and Time of Storage on the Physicochemical Properties of the Salt Soluble Proteins," *Biokhimiya*, **21**, 317–321 (1956); *C.A.*, **50**, 16903f (1956).

44. SLEETH, R. B., KELLEY, G. G. and BRADY, D. E., "Beef Aging under Control Condition," *Food Technol.*, **12**, 86–89 (1958).
45. SOLOV'EV, V. I., "Biochemical Processes and Protection of Ripening Meat," *Myasnaya Ind. S.S.S.R.*, **23**, 43–9 (1952); *C.A.*, **46**, 7246i (1952).
46. SOLOV'EV, V. I. and PIULSKAYA, V., "New Data on Ripening Meat," *ibid.*, **22**, 4, 74–81 (1951); *C.A.*, **45**, 10423h (1951).
47. STEWART, G. F., LOWE, B., HARRISON, D. L., and McKEEGAN, M., "Palatability Studies on Fowl. The Organoleptic, Rigor, and Histological Changes in Fowl after Slaughter," Special Rept. Iowa State College Library, discussed in Lowe B. (32).
48. SWIFT, C. E. and BERMAN, M. D. "Factors Affecting Water Retention of Beef, I Variations in Composition and Properties Among Eight Muscles, *Food Technol.*, **13**, 365–370 (1959).
49. SZENT-GYÖRGYI, A., "Chemical Physiology of Contraction in Body and Heart Muscles," Academic Press, Inc., New York, N. Y., 1953.
50. TSEN, C. C. and TOPPEL, A. L., "Meat Tenderization III Hydrolysis of Actomysin, Actin, and Collagen by Papain," *Food Research*, **24**, 362–364 (1959).
51. URBAIN, W. M., "Meat and Meat Products," in JACOBS, M. B., ed., "Chemistry and Technology of Food and Food Products," Interscience Pub., Inc., New York, N. Y., 1951.
52. WANDERSTOCK, J. J. and MILLER, J. I., "Quality and Palatability as Affected by Method of Feeding and Carcass Grade Beef," *Food Research*, **13**, 291–303 (1948).
53. WANG, H., RASCH, E., BATES, V., BEARD, F. J., PIERCE, J. C., and HANKINS, O. G., "Histological Observations on Fat Loci and Distribution in Cooked Beef," *Food Research*, **19**, 314–322 (1954).
54. WATTS, B. M., "Oxidative Rancidity and Discoloration in Meats," *Advances in Food Research*, **5**, 1–42 (1954).
55. WHITAKER, J. R., "Chemical Changes Associated with Aging of Meat with Emphasis on the Proteins," *Advances in Food Research*, **9** (1959).
56. WIERBICKI, E., KUNKLE, L. E., CAHILL, V. R., and DEATHERAGE, F. E., "The Relation of Tenderness to Protein Alteration During Postmortem Aging," *Food Technol.*, **8**, 506–511 (1954); **10**, 80–86 (1956).
57. WINEGARDEN, M. W., LOWE, B., KASTELIC, J., KLINE, E. A., PLAGGE, A. R., and SHEARER, P. S., "Physical Changes of Connective Tissue of Beef During Heating," *Food Research*, **17**, 172–184 (1952).
58. WYCKOFF, H., "A Three Dimensional Model of the Myoglobin Molecule Obtained by X-ray Analysis," *Nature*, **181**, 662–666 (1958).

CHAPTER SEVEN

Vegetables and Fruits

WEBSTER'S INTERNATIONAL UNABRIDGED DICTIONARY says, "There is no well-drawn distinction between vegetables and fruits in the popular sense; but it has been held by the courts that all those which, like potatoes, cabbage, carrots, peas, celery, lettuce, tomatoes, etc. are eaten (whether cooked or raw) with the principal part of the meal are to be regarded as vegetables; while those used only for desserts are fruits." Botanically, fruits are considered the ripened seeds and the adjacent tissues which contain them and this definition is often carried over into foods. So apples, peaches, bananas, cherries, and berries are always called fruits while tomatoes and melons are sometimes classed one way and sometimes another. Cucumbers, peppers, squash, and eggplant are usually classed as vegetables, although they are botanically the "fruit of the plant."

Fruits and vegetables add enormously to the interest and variety of our diet by their great range of color and texture and by their complex aromas, giving each variety and even each individual a slightly different flavor. Nutritionally, fruits and vegetables are important because they contain large amounts of certain vitamins and minerals. Cooking often brings about a change in the aroma and even in the relative sourness and sweetness. So through the use of fresh and cooked fruits and vegetables, the possible variations in menu pattern are immense.

Although fruits and vegetables vary greatly in their chemical composition, some generalizations are possible. All, with the exception of nuts and dates, are high in water with a range from approximately 70 per cent for pears, bananas, figs, etc. to 98 per cent for vegetable marrow. All, with the exception of legumes and nuts, are relatively low in protein. Although the quantity of protein in vegetables and fruits is sometimes neglected, no cellular material ever exists without a certain amount. Protein varies from approximately 0.3 per cent in apples to 4.4 per cent in brussels sprouts.

All vegetables and fruits contain some carbohydrate. Part of the carbohydrate in fresh fruits is present as cellulose and pectic substances in the cell walls, but these compounds are indigestible and not available to the human body. Starch is present in almost all fruit and vegetables although it may disappear on ripening. Glucose, fructose, and sucrose are widely distributed, and sweet taste is dependent on their occurrence. Glucose, fructose, sucrose, and starches constitute the "available carbohydrate" of fruits and vegetables, and the caloric value of the food depends in large measure on the concentration of these components. In practical dietetics fruits and vegetables are frequently classified by the approximate amount of available carbohydrate since all those in the same class have approximately the same number of calories. The common classifications are: 5, 10, 15 and 20 per cent available carbohydrate. The amount of lipids in fruits and vegetables is usually very small. Nuts are the general exception, while a few vegetables such as avocados are a rich source of fat.

VEGETABLES

1. *Leaf vegetables* (lettuce, mustard greens, chard, spinach, water cress, parsley, cabbage) are high in water and cellulose and low in calories and protein. They add valuable amounts of minerals and vitamins to the diet although they do not contain large amounts of most of these nutrients. They are usually rich in iron and provitamin A, and often in the B complex. The raw leaves often contain appreciable amounts of ascorbic acid.

2. *Flowers, buds, and stems* (broccoli, cauliflower, artichoke, asparagus, celery, kohlrabi) are relatively high in water and cellulose, but low in protein. They have moderate amounts of calcium and some are moderately rich in provitamin A. They all contribute small amounts of vitamins and minerals. A few have moderate amounts of ascorbic acid and riboflavin.

3. *Bulbs, roots, and tubers* (potatoes, beets, turnips, carrots, rutabaga) are high in water, moderate in cellulose, and contain an appreciable amount of available carbohydrate. The available carbohydrates are starches, glucose, and some sucrose. The amounts of the vitamins and minerals are not high but are valuable adjuncts to the diet. The amount of ascorbic acid in potatoes has often made the difference between gross scurvy and its absence in large numbers of Europeans.

4. *Seeds* (legumes, corn, rice) are relatively low in water and cellulose, containing a fair amount of protein and a large amount of starch. They are notable sources of the B complex vitamins and iron.

5. *Vegetable fruits* (cucumbers, peppers, melons, tomatoes, squash, pumpkin, eggplant, okra) are relatively high in water and cellulose but low in calories and protein. Many contain valuable amounts of vitamins and

small amounts of minerals. Some, such as the tomato, are notable for ascorbic acid; others, such as green peppers and squash, for provitamin A; and some, such as the tomato, for thiamine.

STRUCTURE OF FRUITS AND VEGETABLES

A fresh fruit or vegetable is a group of living cells, still undergoing metabolic reactions. Although when harvested it is cut off from tissues of the plant supplying water and other nutrients, it is nevertheless still living. The chief type of cell in the edible portion of most fruits and vegetables is the parenchyma cell. See Figure 7.1. It makes up the bulk of the cells in leaves, fruits, and even in young edible stems. A parenchyma cell is rather thin walled; it may be polygonal or cubical in shape, but all are about the same size. The parenchyma cells do not fit tightly together but are often separated by air spaces that contribute to the slightly chalky appearance of a fresh fruit or vegetable. The walls of parenchyma cells in young plants are composed almost entirely of fibrils of cellulose. The cells are held together by cementing substances, which in the young plant are composed of pectic substances. As the plant grows older the nature of these cementing substances often changes, lignins and other compounds are deposited, and the cellulose layer of the cell wall thickens. Such an old plant is not considered desirable for food since it is woody and tough even when cooked.

The material within the cell wall is protoplasm composed of a very large number of different molecules that form either a viscous fluid or a gel. Some of the molecules are dissolved in water in the protoplasm, but many such as proteins are colloidally dispersed. The protoplasm is not uniform but differentiated into various regions and cell parts, the most distinct of which is the nucleus. It is believed that much of the activity of a cell is directed by the nucleus and that a cell cannot survive long without it. Within the cytoplasm are numerous small bodies called plastids. The parenchyma cells of leaves and occasionally other tissues contain granular green plastids called chloroplasts—the site of the chlorophyll so important in photosynthesis by the plant. If the plastid is any other color because of the occurrence of other pigments in it, it is called chromoplast. The leucoplasts are colorless bodies containing starch granules.

Beside the plastids there are often large vacuoles in cells, made up of droplets of solutions with strands of cytoplasm around them. They contain salts, sugars, and other soluble material dissolved in water. This solution is sometimes spoken of as the "cell sap." In young cells the vacuoles are small and numerous. As the cell grows the total size of the vacuoles increases much more rapidly than does the amount of protoplasm, through the imbibition of water and other small molecules. The vacuoles coalesce

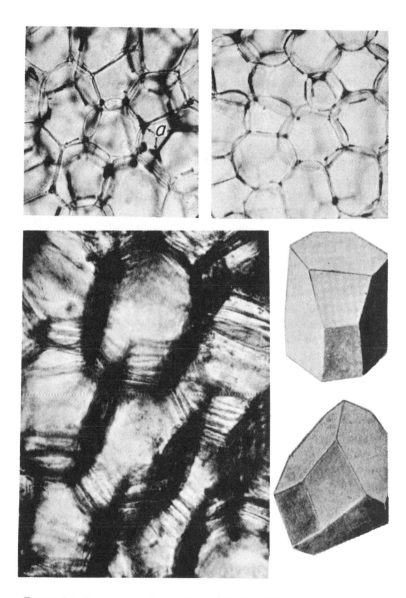

FIGURE 7.1. PARENCHYMA CELLS. *Upper left:* From living onion; dark regions, (a) are intercellular spaces filled with air. Currier, 1946. *Upper right:* From blanched onion; intercellular spaces are filled with water. Currier, 1946. *Lower left:* Nonliving, with much intercellular air. *Lower right:* Two diagrams, showing three-dimensional shape. Marvin, 1939.

Courtesy of Elliot T. Weier.

and become much larger in size and smaller in number with often only one large vacuole in a mature cell. Cells with a high fat content may also have an oil vacuole. Some cells contain crystals embedded in the cytoplasm.

Other types of plant cells besides parenchyma cells are the conducting cells, the supporting cells, and the protective cells. See Figure 7.2. The conducting cells are composed of long tubes through which water and salts or foodstuffs are distributed through the plant. These are of two types called xylem and phloem. The walls of the xylem are composed primarily of cellulose thickened at intervals in definite patterns with lignin. The walls of the phloem contain little lignin. Fibers composed of cellulose may occur associated with the phloem. When numerous or large, such fibers are objectionable since they are largely unchanged on cooking and produce stringiness and toughness.

The supporting tissues are not numerous in young plants or in the young parts of plants, desirable for foods. They are composed of long pointed cells whose cell walls of cellulose thicken as the plant ages and become encrusted with lignin. Some plants have another type of supporting cell in which the cell wall is composed of cellulose and pectic substances in place of cellulose and lignin.

The protective tissue is composed of specialized parenchyma cells that secrete cutin or contain suberin. Sometimes these cells are thick and corky, in other plants they are thin. But they are closely pressed together and are usually quite tough. Usually the epidermis of a fruit or leaf contains stomata, minute valves through which exchange of gases can occur when they are open. The cutin or suberin of the epidermal cells make them impervious to water and also protect them from injury. Often this layer of cells forms a skin or peel which may be removed in the preparation of the leaf or fruit for eating. The layer of epidermal tissue not only protects the organ from mechanical injury, but also prevents inroads by insects, fungi, and microorganisms. Everyone has observed that after the skin is broken the keeping time of a fruit is relatively short, whether it has been harvested or not.

The cells of a harvested fruit or vegetable are still *living*. It is only when food is cooked that cells are killed. If the food is frozen or kept for a long time, death of cells usually occurs. See Figure 7.4.

The composition of the cutin or suberin of plants is for the most part unknown. A number of studies have been made of the composition of waxes such as carnauba or japan wax which are produced commercially; and some investigation has also been directed to discovering the composition of the cutin of other plants, but not necessarily those which are used for food. The cutin of a few food plants, particularly apples, has been

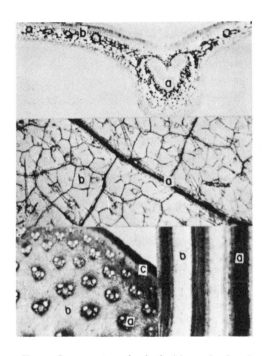

FIGURE 7.2. TYPES OF PLANT CELLS. *Upper: Cross section of a leaf* with conducting tissue and some strengthening tissue forming the main vein **a**, and showing the leaf blade **b** formed by parenchyma cells and small veins. *Middle: Surface view of a whole leaf*—dark lines **a** are veins composed of conducting and strengthening cells, and parenchyma cells **b** form the main body of the leaf blade. *Lower left: Cross section of an old (inedible) asparagus stem*—conducting and strengthening tissues **a** are grouped in discrete bundles and surrounded by parenchyma cells **b**, fibers have formed just under the epidermis of the stem **c**. (In a young asparagus stem there would be no fibers and the bundles would be in a rudimentary stage of development. In many stems the conducting cells form a continuous cylinder within the stem.) *Lower right: Longitudinal section of old asparagus stem* showing the conducting and strengthening bundles **a** embedded in parenchyma tissue **b**.

Courtesy of Elliot T. Weier.

fractionated and the components at least partly identified. The composition of most waxy coatings is complex but the compounds fall into a few general classes. The components of natural waxes have been classified in the following categories: (1) hydrocarbons, (2) fatty and wax acids, (3) ketones, (4) fatty and wax alcohols, (5) esters, (6) ethers, (7) pseudo esters, and (8) aromatic compounds. These compounds usually have long chains of between 18 and 22 carbons or more for the principal chain. The principal fatty acids, palmitic, stearic, oleic, linoleic, and linolenic, are found in many of the waxes as esters, as well as free acids. Some waxes contain fatty acids with a higher number of carbons in the chain than the more common fatty acids. Thus it is reported that carnauba wax contains 75 per cent myricyl cerotate.

$$C_{25}H_{51}COOC_{30}H_{61}$$
Myricyl Cerotate

Cerotic acid is the straight chain fatty acid with 26 carbons. A summary is presented in Table 7.1.

TABLE 7.1. COMPOUNDS REPORTED IN PLANT WAXES

No. of Carbons		
	Alcohols and Ketones	
12	1-12 dodecandiol	$CH_2OH(CH_2)_{10}CH_2OH$
25	Pentacosan-8-ol (or -one)	$CH_3(CH_2)_{16}CHOH(CH_2)_6CH_3$
26	Hexacosanol	$C_{26}H_{53}OH$
27	Heptacosan-9-ol (or -one)	$CH_3(CH_2)_{17}CHOH(CH_2)_7CH_3$
28	Octacosanol	$C_{28}H_{57}OH$
29	Nonacosan-10-ol (or -one)	$CH_3(CH_2)_{18}CHOH(CH_2)_8CH_3$
29	Nonacosan-15-one	$CH_3(CH_2)_{13}CO(CH_2)_{13}CH_3$
30	Triacontanol	$C_{30}H_{61}OH$
31	Hentriacontan-16-ol (or -one)	$CH_3(CH_2)_{14}CHOH(CH_2)_{14}CH_3$
33	Tritriacontan-16-ol (or -one)	$CH_3(CH_2)_{16}CHOH(CH_2)_{14}CH_3$
	Acids	
8	Suberic Acid	$COOH(CH_2)_6COOH$
12	Sabinic Acid	$CH_2OH(CH_2)_{10}COOH$
20	Eicosandicarboxylic Acid	$C_{18}H_{36}(COOH)_2$
22	Phellonic Acid	$CH_2OH(CH_2)_{20}COOH$
26	Cerotic Acid	$C_{25}H_{51}COOH$
26	Juniperic Acid	$CH_3(CH_2)_9CHOH(CH_2)_{14}COOH$
30	Ursolic Acid	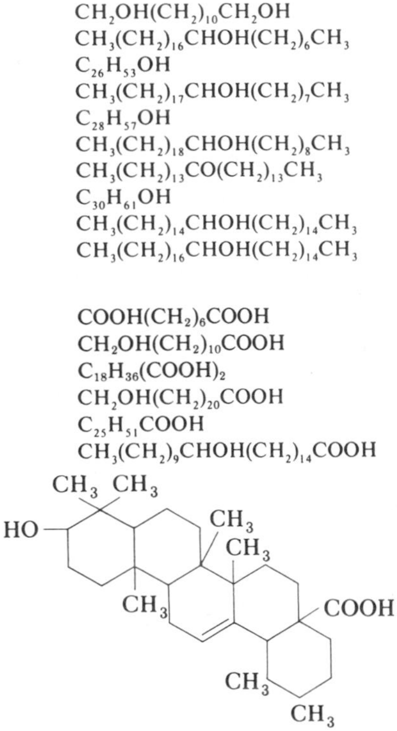

	Hydrocarbons	
27	Heptacosane	$C_{27}H_{56}$
29	Nonacosane	$C_{29}H_{60}$

Scott[67] has reported histological studies on the cells of numerous plants but none that are used for foods. She found that the walls of all intercellular air spaces in leaves, stems, fruit, and flowers are lined with a substance which does not dissolve in 80 per cent sulfuric acid or in chromic acid when the tissue section is flooded with the reagent. She calls the substance or mixture "suberin" and indicates that it is waxy. She also found the material in a thin layer on the inside of cells next to the protoplasmic membrane in root hairs and along vessel segments. The degree to which this material occurs varies with the type of plant and increases on ageing. The material can be stained. Usually the term "suberin" is only applied to corky tissues.

Thus we see that many plant tissues are covered with a water-impervious layer of waxy material called "cutin." The cutin limits the amount of water lost by transpiration, as well as the amount of water-soluble substances lost by the leaching of rain. The correlation between the thickness of the whole cuticle layer and the rate at which water is lost is often poor. This is because other factors such as the number of stomata and their opening and closing influence water loss.

TEXTURE OF FRUITS AND VEGETABLES

The texture of a fruit or vegetable depends on the turgor of the living cells as well as on the occurrence of supporting tissues and the cohesiveness of the cells. Turgor is that pressure of the cell contents on the partially elastic wall of a cell, tending to produce rigidity. It is produced by a delicate balance of forces which maintains the cell at a normal volume yet allows the exchange of substances. When the cell volume diminishes, the cell becomes soft and flaccid; but if the volume increases beyond the point that can be accommodated by the elasticity of the cell walls, the cell ruptures, the cell contents flow out, and rigidity is lost. The substance chiefly responsible for changes in volume is water. When a plant wilts, water has been so extensively lost from the cells that they no longer have normal turgor but are soft and flabby.

One of the best known forces affecting cell volume is osmosis. Plant cell walls are for the most part composed of cellulose permeable to many types and sizes of molecules. Inside the cell protoplasm may be stretched around a large storage vacuole. The protoplasm and cell wall act as a semipermeable membrane, allowing water and some other small molecules to pass through it. Water diffuses in greater amounts from a region in which it is high in concentration (dilute solution or pure water) to one in which water is low in concentration (a relatively concentrated solution). The vacuole contains soluble compounds as well as colloidal substances; and if the intercellular fluid is composed only of water or of a dilute solution, water will move into the cell. While the fruit or vegetable is still part of the plant, a fine balance is achieved and the volume of the cell is maintained at a certain normal turgor. The effect of a concentrated solution on cell turgor is readily demonstrated by placing cucumber slices in a concentrated salt solution or sprinkling sugar on strawberries. Both soon become limp as water rapidly passes from the cells into the solution.

Cell turgor depends on a number of factors:

(1) The concentration of osmotically active substances in the vacuole, in both true solution and colloidally dispersed.

(2) The permeability of the protoplasm.

(3) Elasticity of the cell walls. If the walls are relatively high in elasticity, a considerable increase in volume of the cell can occur before the cell ruptures. It will simply become more turgid as water flows into the cell. However, in the face of a shrinking volume a highly elastic cell wall will cause a rapid loss of turgor for the tissues as a whole even as it adapts itself to the small volume of the cell. A stronger but more rigid wall will maintain a firm texture even when the cell volume decreases.

When a fruit or vegetable is cooked, the protein is denatured, the cells die, and the vacuoles are no longer covered by a living membrane of protoplasm. The protein usually precipitates and the permeability is markedly affected. Often solutes and water stream out of the vacuoles into the intercellular spaces or even out of the fruit or vegetable and the food becomes very soft. However, when starch granules fill the vacuoles, they are such large molecules that they cannot escape unless the cell walls are broken. During cooking the starch granules swell, become gelatinized and retain moisture; and if there are sufficient of them and if the process continues, they may hold the water in the vacuole and maintain a firm texture. However cooked starch granules are never able to maintain crispness; this is noticeable in a sweet potato and in peas and beans.

The rigidity of the structural tissues and of the cell walls is an important factor in the texture of a fruit and vegetable. The changes in the compounds which strengthen the structure of the plant with cooking will be considered in the section on cooking changes. See p. 265.

PIGMENTS IN FRUITS AND VEGETABLES

The color of fruits and vegetables is exceedingly important to our pleasure at the table. It is not only the subtle variety of flavors that makes fruits and vegetables so pleasing, but also their variety of delicate and bright colors.

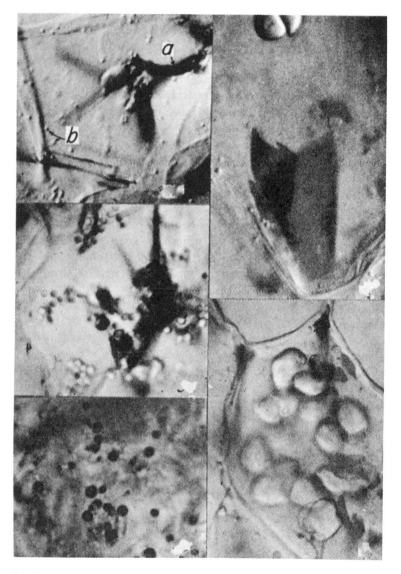

FIGURE 7.3. PIGMENTS IN PLASTIDS. *Upper left:* Chromoplasts **a** and oil droplets **b** in living parenchyma cell from carrot root. *Middle left:* Destruction of chromoplasts upon death of cell. *Lower left:* Carotene in solution in oil droplets in dead cell from carrot root. *Upper right:* Living carrot cell showing plate of carotene. *Lower right:* Living carrot cell containing starch grain with associated carotene plates.

Courtesy of Elliot T. Weier.

Most of the pigments occur in plastids, specialized bodies lying in the protoplasm of the cell. See Figure 7.3. For example, the chlorophylls occur in the chloroplasts, which under the microscope can be seen next to the cell wall as distinct bodies containing flecks of green pigment. Occasionally a pigment is present in the protoplasm as a crystal. Thus in carrot cells, platelets of carotene can be observed and in tomatoes, needles or platelets of lycopene. Sometimes the water soluble pigments are dissolved in the vacuoles, and not generally distributed through the cell.

In the early nineteenth century the only colored materials available for dyeing were natural products. Much of the early research was directed to an understanding of these natural dyes. When synthetic dyes were produced, interest in natural pigments declined, although it never disappeared. During the past thirty years there has been a tremendous interest in biochemistry, and a great variety of problems have been studied vigorously. Among them is research in the pigments of plant and animal tissues. The number of pigments whose structure and reactions have been completely or partially elucidated is enormous. Only a relatively small number of plant pigments will be considered here and only a very brief introduction to the field attempted. The number of studies on the changes in pigments in foods is relatively small. But the results of the chemists can be used to understand some color changes which occur when food is cooked or processed.

The chief pigments of fruits and vegetables can be classified as (1) the carotenoids, (2) the chlorophylls, (3) the anthoxanthins, and (4) the anthocyanins. Each group of pigments will be discussed and a brief introduction to their structure, stability, and the changes which occur during cooking considered. Tannins often account for the formation of off-colors during processing and they, too, will be considered.

The Carotenoids[34]

The carotenoids are a group of yellow, orange, and orange-red fat soluble pigments widely distributed in nature. In green leaves they occur in the chloroplasts, small bodies close to the cell walls of the palisade cells. These cells are next to the epidermal cells on the upper side of the leaf. The carotenoids are present in the lipid material along with the chlorophylls. The green color of the chlorophyll masks the yellow to red color of the carotenes except in very young leaves while the amount of chlorophyll is small. The bright, fresh yellow-green color of spring leaves is the result of carotenoids and small amounts of chlorophylls. These pigments are also present in a wide variety of fruits—peaches, banana skins, tomatoes, red peppers, paprika, rose hips, squash, etc.—as well as other parts of plants—

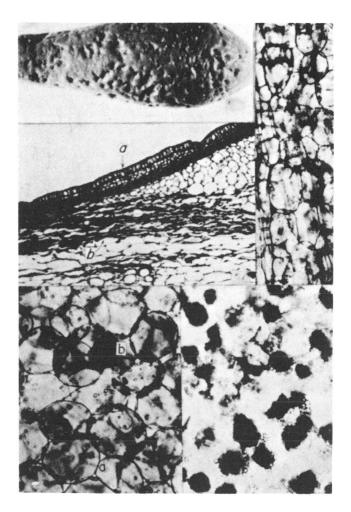

FIGURE 7.4. COMPARISON OF LIVING AND NONLIVING PLANT TISSUES AND THE EFFECTS OF DYES. *Upper left:* Whole summer squash showing pitting injury due to 16-day storage at 0°C (32°F) and one day at 21°C (70°F). Mann and Morris, 1947. *Upper right:* Parenchyma cells from cold storage summer squash, mounted in dilute neutral red. Cells are dead and therefore do not accumulate dye. Mann and Morris, 1947. *Middle left:* Showing section through an injury pit of cold storage summer squash. a, epidermis, b, collapse of dead cells forming the pits. Mann and Morris, 1947. *Lower left:* Living parenchyma cells, mounted in neutral red. Note accumulation of dye, b. Mann and Morris, 1947. *Lower right:* Living parenchyma cells, mounted in neutral red and strong sugar solution. Note plasmolysed cells. Mann and Morris, 1947. *Courtesy of Elliot T. Weier.*

in carrots, sweet potatoes, and in most yellow, orange, and red flowers. When they are consumed by animals, they tend to concentrate in lipids and hence are found in blood, milk, egg yolk, and depot fat.

The name carotenoid is applied to all pigments chemically related to the carotenes, which were the first isolated. In 1831 Wackenroder extracted a pigment from carrots and called the fraction carotene. We now know that this "carotene" is a mixture of three isomers, α-, β-, and γ-carotene.

The carotenoids are either hydrocarbons or derivatives of hydrocarbons and are composed of isoprene units. Isoprene is a diene,

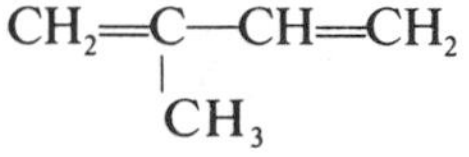

$$CH_2{=}C{-}CH{=}CH_2$$
$$\underset{\displaystyle CH_3}{\vert}$$

and this molecule is the unit out of which the carotenoids are constructed. It contains five carbon atoms while many, although not all, of the carotenoids contain 40 carbon atoms or eight isoprene units. Some of the molecules contain a long unsaturated hydrocarbon chain with a ring at one or both ends of the chain. Some of the molecules are symmetrical, so that if folded in half the left half would be the mirror image of the right half. β-carotene and lycopene are examples of symmetrical molecules.

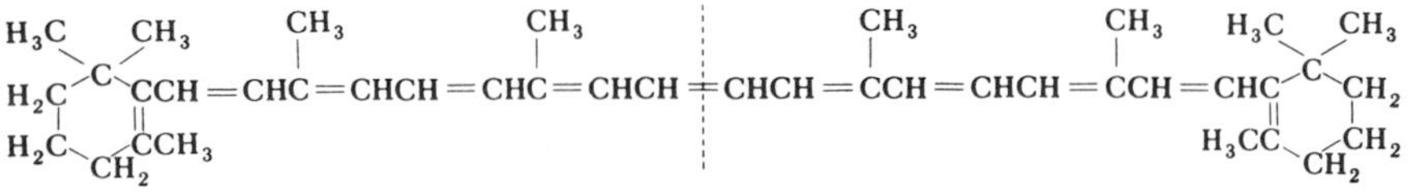

β-carotene

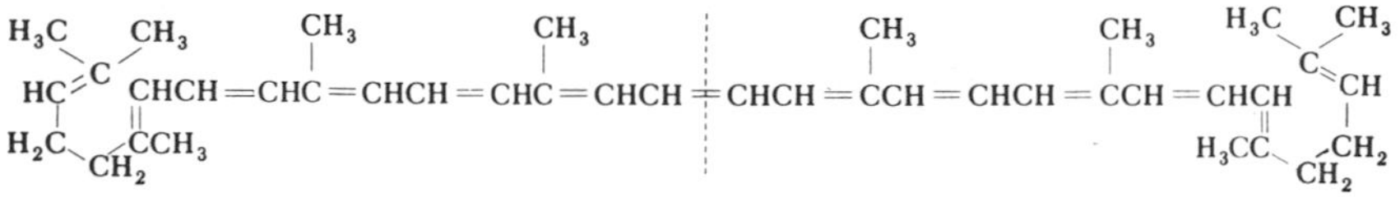

Lycopene

Notice that β-carotene and lycopene differ only in the cyclization of the end carbons of lycopene to form the rings of β-carotene. β-Carotene is widely distributed in plant materials. It is readily prepared from carrots, sorb apples, or paprika. It sometimes occurs free, although it is often accompanied by small amounts of α- and γ-carotene (see Table 7.2). Lycopene is the orange-red pigment of the tomato, but it is also found in rose hips, watermelon, apricot, and many other plants. It also is usually accompanied by other carotenoids.

Some carotenoids have two terminal rings or groups which are different. Examples are α-carotene and γ-carotene.

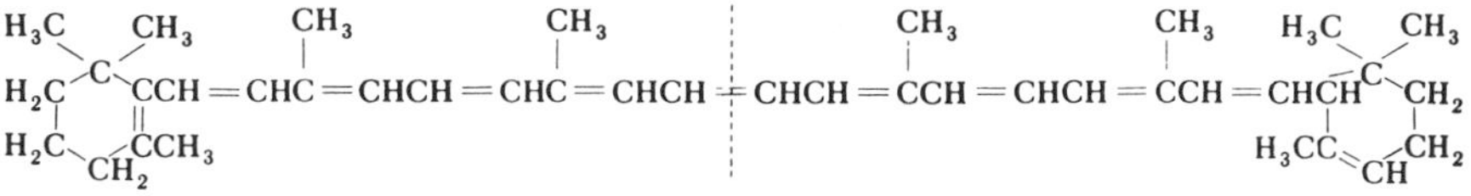

H₃C, CH₃, CH₃, CH₃, CH₃, CH₃, H₃C, CH₃ structures...

$$\text{H}_3\text{C} \quad \text{CH}_3 \qquad \text{CH}_3 \qquad\qquad \text{CH}_3 \qquad\qquad\qquad \text{CH}_3 \qquad\qquad \text{CH}_3 \quad \text{H}_3\text{C} \quad \text{CH}_3$$

$$\text{H}_2\text{C}\overset{\text{C}}{\diagdown}\text{CCH}=\text{CHC}=\text{CHCH}=\text{CHC}=\text{CHCH}\!+\!\text{CHCH}=\text{CCH}=\text{CHCH}=\text{CCH}=\text{CHCH}\overset{\text{C}}{\diagup}\text{CH}_2$$

$$\text{H}_2\text{C}\diagdown_{\text{CH}_2}\diagup\text{CCH}_3 \qquad\qquad\qquad\qquad\qquad\qquad\qquad\qquad\qquad \text{H}_3\text{CC}\diagdown_{\text{CH}}\diagup\text{CH}_2$$

α-Carotene

$$\text{H}_3\text{C} \quad \text{CH}_3 \qquad \text{CH}_3 \qquad\qquad \text{CH}_3 \qquad\qquad\qquad \text{CH}_3 \qquad\qquad \text{CH}_3 \quad \text{H}_3\text{C} \quad \text{CH}_3$$

$$\text{H}_2\text{C}\overset{\text{C}}{\diagdown}\text{CCH}=\text{CHC}=\text{CHCH}=\text{CHC}=\text{CHCH}\!+\!\text{CHCH}=\text{CCH}=\text{CHCH}=\text{CCH}=\text{CHCH}\overset{\text{C}}{\diagup}\text{CH}$$

$$\text{H}_2\text{C}\diagdown_{\text{CH}_2}\diagup\text{CCH}_3 \qquad\qquad\qquad\qquad\qquad\qquad\qquad\qquad\qquad \text{H}_3\text{CC}\diagdown_{\text{CH}_2}\diagup\text{CH}_2$$

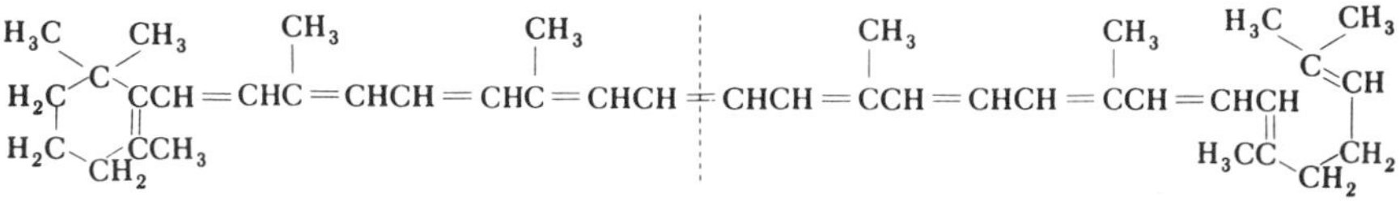

γ-Carotene

The difference between α- and β-carotene is in the placement of the double bond in ring 2. Unlike these pigments γ-carotene has only one ring; half of its molecule is like lycopene and half like β-carotene. Both α- and γ-carotene occur in nature associated with β-carotene, although usually in smaller amounts. In a very few plant products α-carotene predominates. For example, in red palm oil it accounts for approximately 30 to 40 per cent of the carotenes present.

Carotenoids which contain hydroxyl groups are called xanthophylls. A number of xanthophylls have been isolated from plants and their structures determined partially or completely. They are often associated with carotenes; leaves contain not only the hydrocarbon carotenes as their yellow pigments but also the closely related xanthophylls. Cryptoxanthin is an example of one of the xanthophylls.

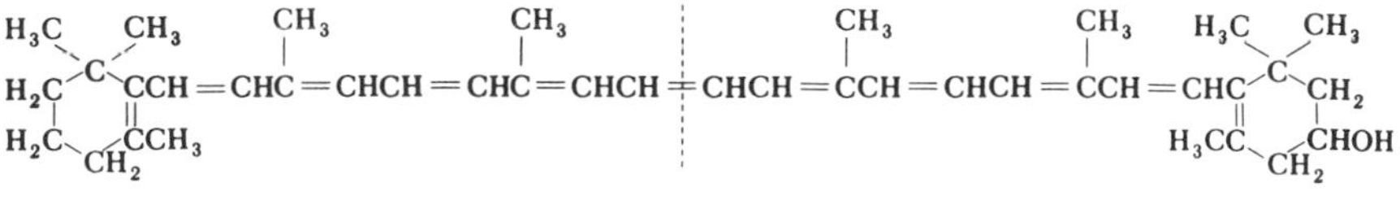

$$\text{H}_3\text{C} \quad \text{CH}_3 \qquad \text{CH}_3 \qquad\qquad \text{CH}_3 \qquad\qquad\qquad \text{CH}_3 \qquad\qquad \text{CH}_3 \quad \text{H}_3\text{C} \quad \text{CH}_3$$

$$\text{H}_2\text{C}\overset{\text{C}}{\diagdown}\text{CCH}=\text{CHC}=\text{CHCH}=\text{CHC}=\text{CHCH}\!+\!\text{CHCH}=\text{CCH}=\text{CHCH}=\text{CCH}=\text{CHC}\overset{\text{C}}{\diagup}\text{CH}_2$$

$$\text{H}_2\text{C}\diagdown_{\text{CH}_2}\diagup\text{CCH}_3 \qquad\qquad\qquad\qquad\qquad\qquad\qquad\qquad\qquad \text{H}_3\text{CC}\diagdown_{\text{CH}_2}\diagup\text{CHOH}$$

Cryptoxanthin

Notice that it is very similar to the carotenes and that it also is composed of isoprene units with the same system of conjugated (alternating) double bonds. Indeed, it differs from β-carotene only in the presence of the hydroxyl group. It often occurs as an ester of fatty acids, or it can be found in the free form. It is one of the chief pigments of yellow corn, paprika, papaya, and the mandarin orange.

There are also carotenoids which contain oxygens in other groups. Some are ketones, others are carboxylic acids, or hydroxycarboxylic acids. The number of carbons is not always 40 as in the carotenes and xanthophylls but may be smaller.

An example of a shorter chain pigment is crocetin which occurs as a glycoside, crocin, in the spice, saffron. The deep orange-yellow color of saffron is not the result of crocetin alone since lycopene, β-carotene, γ-carotene, and zeaxanthine are also present. Crocetin is:

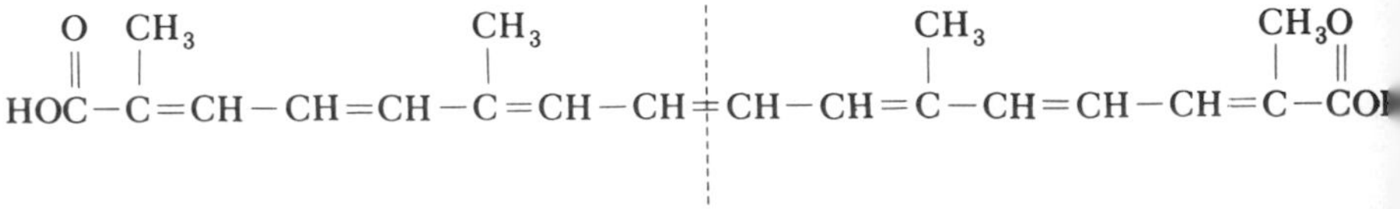

$$\underset{HOC}{\overset{O}{||}}-\underset{C}{\overset{CH_3}{|}}=CH-CH=CH-\underset{C}{\overset{CH_3}{|}}=CH-CH \vdots CH-CH=\underset{C}{\overset{CH_3}{|}}-CH=CH-CH=\underset{C}{\overset{CH_3O}{\overset{|}{\underset{||}{C}}}}-COH$$

Crocetin

In crocin the carboxyl groups are esterified by carbohydrates (gentiobiose).

The number of carotenoids that occur in nature is probably very large. Along with those described, a number have been isolated and their structures either completely or partially elucidated. Information on other carotenoids can be found in some advanced texts in organic chemistry and in the original literature.

Aside from the beauty that plant pigments impart to fruits and vegetables, some of the carotenoids are important in nutrition as precursors for the synthesis of vitamin A in the body. Vitamin A_1 is identical with one-half of a β-carotene molecule plus a hydroxyl group, and in animals, one molecule of β-carotene is converted to two molecules of vitamin A by hydrolysis.

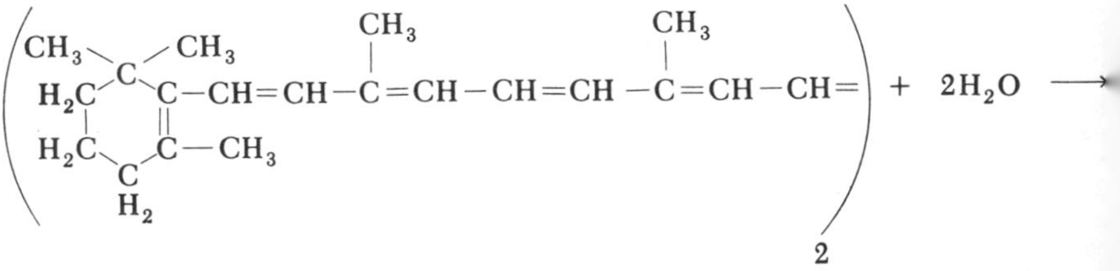

$$\left(\begin{array}{c} \underset{H_2C}{\overset{CH_3}{\diagdown}} \overset{\diagup CH_3}{\underset{C}{\diagup}} \\ \underset{H_2C}{\overset{|}{\diagdown}} \underset{C}{\diagup} C-CH=CH-\overset{CH_3}{\underset{|}{C}}=CH-CH=CH-\overset{CH_3}{\underset{|}{C}}=CH-CH= \\ \underset{C}{\overset{|}{\diagup}}-CH_3 \\ \underset{H_2}{} \end{array} \right)_2 + \quad 2H_2O \longrightarrow$$

β-Carotene

$$2 \quad \begin{array}{c} \underset{H_2C}{\overset{CH_3}{\diagdown}} \overset{\diagup \dot{C}H_3}{\underset{C}{\diagup}} \\ \underset{H_2C}{\overset{|}{\diagdown}} \underset{CH_2}{\diagup} C-CH=CH-\overset{CH_3}{\underset{|}{C}}=CH-CH=CH-\overset{CH_3}{\underset{|}{C}}=CH-CH_2OH \\ \underset{CH_2}{} \end{array}$$

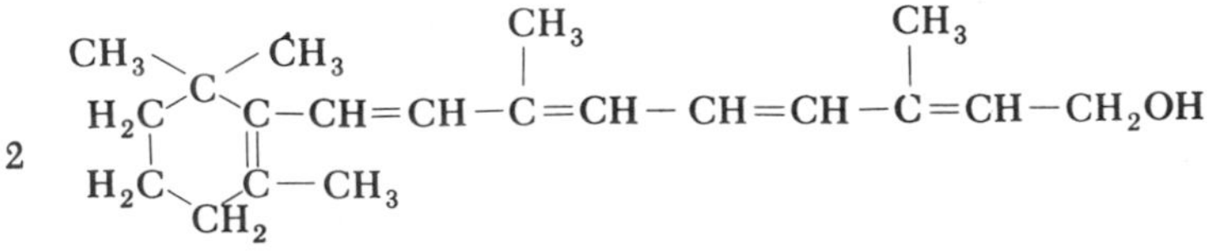

Vitamin A

The efficiency of the conversion has been the subject of extensive study. It appears to vary with (1) saturation of the subject, (2) the species of animal used, and (3) the solvent for the carotene fed. In animals deficient in vitamin A the conversion may be as great as 70 to 80 per cent; but if an animal has a large store of vitamin A in its tissues, if, in other words, it is saturated, then little carotene is absorbed and converted to vitamin A. Much is then excreted in the feces. Rats show a great capacity for the conversion of the carotenes to vitamin A, while dogs have only a small capacity and cats none. If the carotene is fed in vegetable oil, it is readily absorbed and converted to vitamin A; but if it is dissolved in mineral oil, it escapes in the feces. In American diets the quantity of carotenoid precursors is of great importance in total vitamin A nutrition.

$$
\begin{array}{c}
\mathrm{CH_3}\diagdown \quad \diagup \mathrm{CH_3} \qquad\qquad \overset{\displaystyle \mathrm{CH_3}}{\overset{\displaystyle |}{}} \qquad\qquad \overset{\displaystyle \mathrm{CH_3}}{\overset{\displaystyle |}{}} \\
\mathrm{H_2C}^{\diagup\mathrm{C}\diagdown}\mathrm{C-CH{=}CH-C{=}CH-CH{=}CH-C{=}CH-CH_2OH} \\
\mathrm{HC} \diagdown\;\diagup \mathrm{CCH_3} \\
\mathrm{C} \\
\mathrm{H}
\end{array}
$$

Vitamin A$_2$

The other carotenoids which are vitamin A precursors are those which have the same terminal ring as that in vitamin A$_1$: α-carotene, γ-carotene, and cryptoxanthin have one ring to each molecule. They are not as valuable

TABLE 7.2. DISTRIBUTION OF CAROTENOIDS IN SOME FOODS

Carrot (Daucus Carota)	α-carotene, β-carotene, γ-carotene, xanthophyll, two hydrocarbons of unknown composition.
Wheat Germ	Xanthophyll, carotene (?).
Corn (Zea mays)	Zeaxanthin, cryptoxanthin, xanthophyll, α-carotene, β-carotene, γ-carotene, κ-carotene (?), neo-cryptoxanthin, hydroxy-α-carotene.
Apricot (Prunus armeniaca)	β-carotene, γ-carotene, lycopene.
Peach (Prunus persica)	β-carotene, cryptoxanthin, xanthophyll, zeaxanthin, unknown carotenoid.
Soya Bean (Glycine max)	α-carotene, β-carotene.
Cowpea (Vigna sinensis)	β-carotene, xanthophyll.
Orange (Citrus aurantium)	β-carotene, lycopene, cryptoxanthin, xanthophyll, violaxanthin, zeaxanthin, β-citraurin, citraxanthin.
Grapefruit (Citrus grandis)	β-carotene, lycopene.
Squash (Cucurbita maxima)	α-carotene, β-carotene, xanthophyll, violaxanthin.
Red pepper or Chili (Capsicum fructescens)	Capsanthin, α-carotene, β-carotene.

as β-carotene in the synthesis of vitamin A, since they are capable of forming only one molecule of vitamin A for each molecule of pigment, but they are nevertheless significant. The other carotenoids do not contain this important ring and are therefore not precursors of vitamin A. The depth of color of a fruit or vegetable has been suggested as a rough index of its vitamin A value. But it can be seen that such an index could lead to gross errors since so many carotenoid pigments have no activity at all. For example, the yolk of the hen egg contains some vitamin A and is brightly colored because of the presence of two carotenoids, lutein and zeaxanthine, which are the dihydroxy analogs of α- and β-carotene, respectively. But these pigments have no vitamin A activity and the depth of the orange color of the yolk is no measure of its vitamin A value. Table 7.2 shows the distribution in some foods.

The carotenoids are insoluble in water but soluble in lipids and in lipid solvents. In processing fruits and vegetables, loss of these pigments into cookery or canning water is very slight. They do undergo oxidation when exposed to air, so that in drying fruits or vegetables which contain these pigments a problem is sometimes encountered. For example, carrots and apricots show loss of pigment on drying. These pigments do not undergo hydrolysis except when they occur in plant tissues as esters, and they are not affected by changes in pH.

In blanched carrots, or those blanched and then dehydrated, the loss of pigment is rapid. Carrots diced small and stored at 62°C in moist air after blanching show a complete loss of pigment in 20 hours. Weier and Stocking[88] have studied the change and find that antioxidants partially protect the pigment from deterioration in carrots ground to 48 to 16 mesh but have little effect on diced carrots. They were led to try antioxidants because carotene occurs in processed carrots (but not in the living cell) in fat droplets; and they reasoned that the degradation of the pigment might be associated with oxidative changes in the fat. In leaves the enzyme lipoxidase is important.

Numerous other studies of the disappearance of carotenes have also been made. Usually the loss, if it occurs, is small—5 or 10 per cent. Altogether the carotenes present few problems; they are bright and attractive and their color during food processing is easily maintained.

In a discussion of changes in fruits and vegetables on processing, one other aspect that should be mentioned is the distribution of the pigment after processing. Weier and Stocking discuss the work of Weier and of Reeve on the carrot. The carotenoids in carrots are present in the chromoplasts. In some cells starch is present and the granules may more or less surround the carotene crystals. In the outer cells of the root the carotene concentration is highest, there is little or no starch, and the carotene is present in crystals as needles, tubes, flakes, or spirals. These cells also contain tiny fat droplets. When the cell is killed by blanching, drying, or chemical reagents, the chromoplasts disintegrate and the carotene dissolves in oil droplets. The oil droplets are sometimes originally present in the living cell or they may separate from the protoplasm after it is killed.

Chlorophylls

The green pigments of leaves and stems are usually held close to the cell wall in small bodies, the chloroplasts, along with some carotenes and xanthophylls. Two chlorophylls have been isolated, chlorophyll a and chlorophyll b; and they occur in plants in the ratio of approximately 3a:1b. Chemically, they are very similar. They belong to that group of important biological pigments the porphyrins, which includes hemoglobin. They are fairly large molecules composed of four pyrrole rings held together by methene carbons ($-CH=$) to form a large flat molecule. In chlorophyll, a magnesium atom is held by the nitrogen on two of the rings by ordinary covalent bonds. The other two nitrogens share two electrons with the magnesium to form a coordinate covalent bond (indicated by dotted lines). The formulas for chlorophyll a and chlorophyll b are given below:

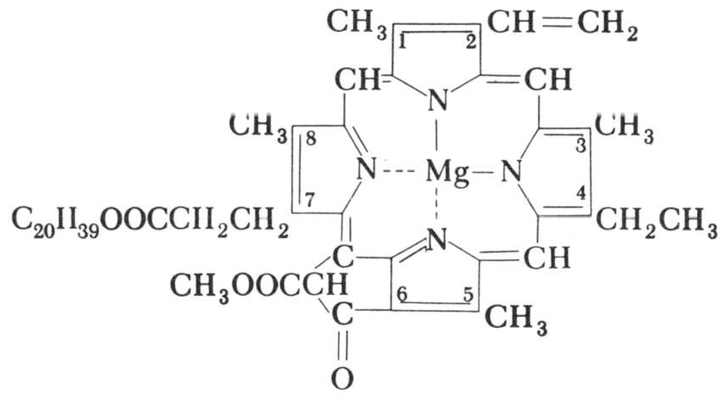

Chlorophyll a

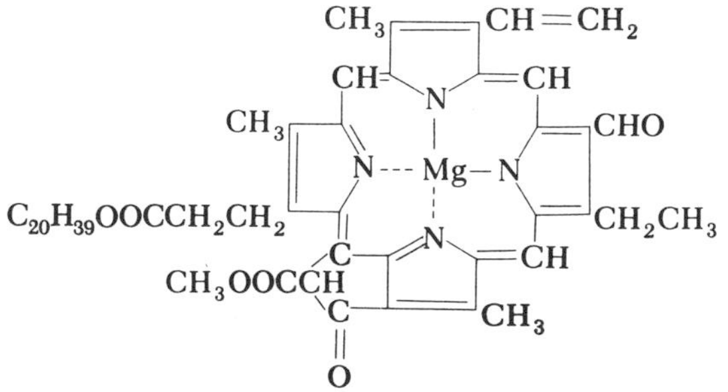

Chlorophyll b

You will notice that the chlorophyll a is an ester. The propionic residue on position 7 is esterified with phytol, while the acid joined to position 6 is esterified with methanol. Phytol, or phytyl alcohol, is an interesting high molecular weight alcohol because it occurs as an ester in a number of plant molecules. It has a chain which, although almost completely saturated, has an arrangement of carbons very similar to that in the carotenoids.

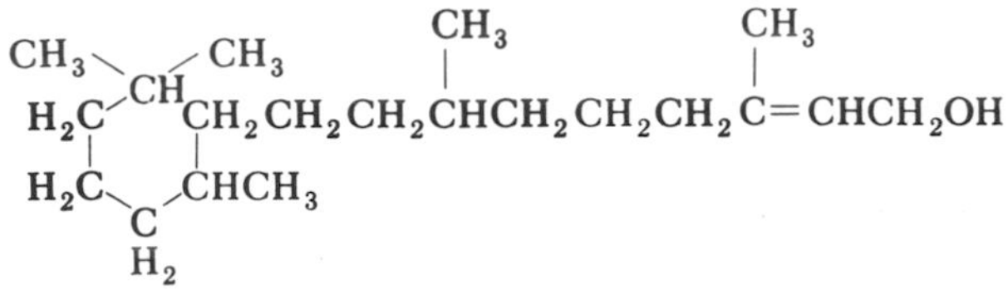

Phytol

Chlorophyll b differs in the occurrence of a —CHO (formyl) group in position 3 in place of the methyl group (—CH₃) in chlorophyll a. Neither chlorophyll has been synthesized as yet.

The chlorophylls are of great importance in the plant because of their role in photosynthesis and the formation of carbohydrates from carbon dioxide and water. This role has been recognized for almost a hundred years and a tremendous amount of work has been done on the mechanisms of photosynthesis. Only now are the reactions and the energy relations being elucidated.

The chlorophylls and other pigments are not the only components present in the chloroplasts. One analysis of chloroplasts of leaves found 47.7 per cent lipid, 37.4 per cent protein, 7.8 per cent ash, leaving 7.1 per cent undetermined. A more detailed analysis of the chloroplasts of cabbage leaves reported 9.3 per cent chlorophyll, 0.5 per cent carotene, 0.8 per cent xanthophyll, 17.5 per cent glyceride fatty acids, 12.3 per cent wax, 4.5 per cent sterols, 13.3 per cent undetermined unsaponifiable, and 18.4 per cent calcium phosphatides. It is now believed that the chlorophylls exist in the chloroplasts as conjugated proteins.

The chlorophylls are very unstable molecules when the living cell is killed and when the chemical and physicochemical relations in a cell are changed. Consequently, the chlorophylls are difficult to retain during any food processing and special care must be taken to produce food that retains a bright, attractive green color. Some workers have suggested that the chloroplasts disintegrate when plant cells are heated and that the chlorophyll is released into the cell. In 1940, Mackinney and Weast[45] reported that in spinach and string beans which were carefully sectioned after cooking the chloroplasts were shrunken and often collapsed onto the protoplasm, but they were still intact and the chlorophyll was still discernible in the plastids. The chlorophylls exist as protein complexes in the living cell; and when the cell is killed by heating, the protein is probably denatured and the chlorophyll released. This may account for the rapidity with which chlorophyll reacts in cooking food such as green beans. There is also the likelihood that the permeability of the membrane surrounding the chloroplasts changes on the death of the cell.

Chlorophyll changes to an olive green color and then to brown. The most likely reaction is the substitution of hydrogen for the magnesium which has been complexed in the porphyrin to form pheophytin. It is unlikely that hydrolysis of either of the esters can be involved. Ester groups usually hydrolyze slowly and this reaction of chlorophyll occurs rapidly. Also the reaction is rapid in acid solutions and does not occur readily in cooking waters where the pH is 8 or more. Esters, on the other hand, are more readily hydrolyzed in alkaline solutions than in acid. The possibility of oxidation as the reaction is likewise small since the brown color develops in conditions where there is little free oxygen available. It is thought, therefore, that when a vegetable becomes olive green on cooking, the chlorophyll has formed pheophytin. The reaction can be written schematically as follows:

$$C_{32}H_{30}ON_4Mg \Bigg\langle \begin{matrix} COOCH_3 \\ \\ COOC_{20}H_{39} \end{matrix} \quad + 2H^+ \rightarrow \quad C_{32}H_{30}ON_4H_2 \Bigg\langle \begin{matrix} COOCH_3 \\ \\ COOC_{20}H_{39} \end{matrix} \quad + Mg^{++}$$

Mackinney and Weast studied this change in green beans and peas and attempted to measure the amount of pheophytin produced in the food at various lengths of time during the cooking process. Their results with Kentucky Wonder beans are presented in Table 7.3.

When green beans are first dropped into boiling water, they, like other green vegetables, show a change in color. The green brightens, the velvety appearance disappears, and the beans become more translucent. These changes are probably caused first by the wetting of the fine hairs on the coat of the bean. Washing the bean and rubbing it produce the same results to a slight degree. Then as the bean is warmed, air is expelled and the intercellular spaces collapse or partially collapse. As cooking continues,

TABLE 7.3. KENTUCKY WONDER STRING BEANS HEATED IN WATER AT 100°C

Time (Minutes)	Pheophytin (Per Cent)	Color
0	0	Green
5	7	Green
10	37.5	Green
20	72.5	Brownish green
30	86.5	Yellowish green
60	100	Yellowish

MacKinney, G. and Weast, C. A., "Color Changes in Green Vegetables," *Ind. Eng. Chem.*, **32**, 392–395 (1940).

plant acids are liberated; and because the chlorophyll is released from the protein complex or because the membrane around the chloroplasts becomes more permeable or both, the acids react with the chlorophylls and form pheophytin.

Blair and Ayres[10] conducted an extensive study of methods for preserving the color in commercially canned peas. They found that on processing the pH of the peas fell from 6.6 to 6.1. By a preliminary treatment which raised the pH and by maintaining it about 8, the color was preserved and little of the chlorophyll was destroyed. We also know that when green vegetables such as spinach or cabbage, which produce considerable volatile acid during the early part of cooking, are cooked in a pot with a cover, the color very quickly changes to olive green and then to a dull brown. But if the lid is left off so that acids escape during the early part of cooking, color retention is much better.

Since it is possible to maintain the green color in the presence of alkali, you might wonder why it is always recommended that baking soda, sodium bicarbonate, be omitted in cooking vegetables. The answer lies, of course, in the fact that in food preparation the maintenance of color is only one of the important effects. With a high pH in the cooking or canning water, particularly if the cation is sodium or potassium, cellulose hydrolyzes rapidly and the texture of the vegetable becomes very soft and mushy. Some of the vitamins, particularly ascorbic acid and thiamine, are very sensitive to heating at high pH's; and in cooking water to which sodium bicarbonate has been added, the rate of destruction of these vitamins is accelerated.

Blair and Ayres,[10] in an attempt to preserve the green color in canned peas and still maintain a firm, but not tough, texture of the skins and cotyledon, used a 30 to 60 minute immersion of the peas in 2 per cent (0.19 M) sodium carbonate, a blanch in 0.005 M calcium hydroxide, and processing in a sugar-salt brine which contained a suspension of 0.02 to 0.025 M

magnesium hydroxide. The pH of the peas by this process is maintained around 8, but the cationic balance produces the desirable balance between the softening effect of the sodium ions and the toughening effect of the calcium and magnesium ions. The calcium and magnesium ions tend to react with pectic substances to form salts that toughen the cell walls.

Some metal ions react with the chlorophylls to form compounds with bright green colors. Ferric, zinc, and cupric ions will replace the magnesium in chlorophyll. They will also react with the pheophytins which have formed from the reactions of the chlorophylls with plant acids and produce the corresponding complex.

Studies by Strain[76] on chlorophyll reactions have significance for changes on cookery. He studied changes in green leaves boiled for 1 or 2 minutes, or frozen and then thawed. After fresh mallow leaves are boiled, there is a reduction in the amounts of chlorophyll a and b and the appearance of chlorophyll a' and b', isomers of chlorophyll a and b. Leaves frozen and then thawed yielded no other green pigments, but there was evidence that neither the oxidative nor the hydrolytic enzymes are inactivated. On drying the leaves in air for 24 hr at 20°C chlorophyll a and b formed not only chlorophyll a' and b', but pheophytins as well. The enzymes are inactivated on drying.

THE FLAVONOIDS[6]

The flavonoids are a group of compounds widely distributed in the plant kingdom. Every tissue studied so far has at least one of these compounds or a closely related one while most have many of them. They are water soluble and are often present in the juices of plants. Chemically the

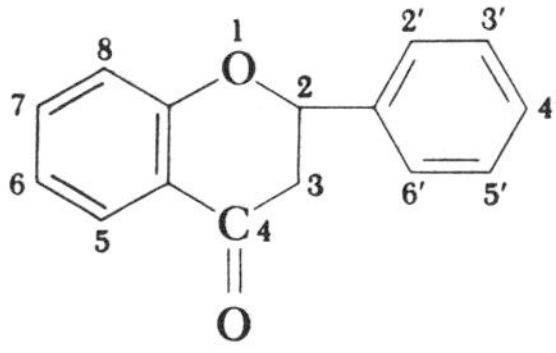

Flavone

flavonoids contain two benzene rings with a three carbon bridge. In most the three carbon bridge is condensed through an oxygen into an intermediate ring. The benzene rings hold hydroxyl groups.

The true flavonoids consist of the anthocyanins which are the red-blue-purple pigments of plants; the anthoxanthins which are yellow; the cate-chins; and the leucoanthocyanins. The last two groups of compounds are colorless but readily change to brownish pigments. They are probably the so-called "food tannins."

Compounds related to the flavonoids are numerous and are also widely distributed in nature. They are as follows: (1) cinnamic acid and the more complex caffeic and chlorogenic acids. Quinic acid, which forms part of the structure of chlorogenic acid, is a common fruit acid.

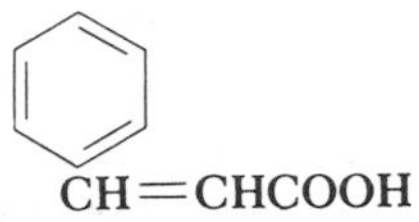

CH=CHCOOH

Cinnamic Acid

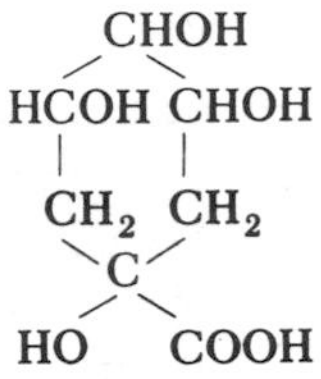

CHOH
HCOH CHOH
CH₂ CH₂
C
HO COOH

Quinic Acid

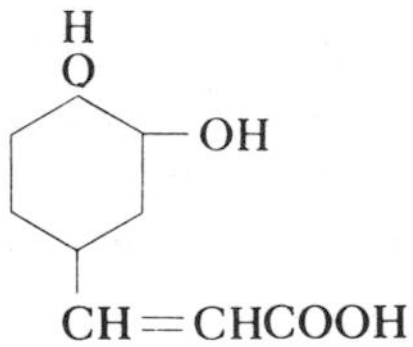

CH=CHCOOH

Caffeic Acid

CH=CHCOOCH

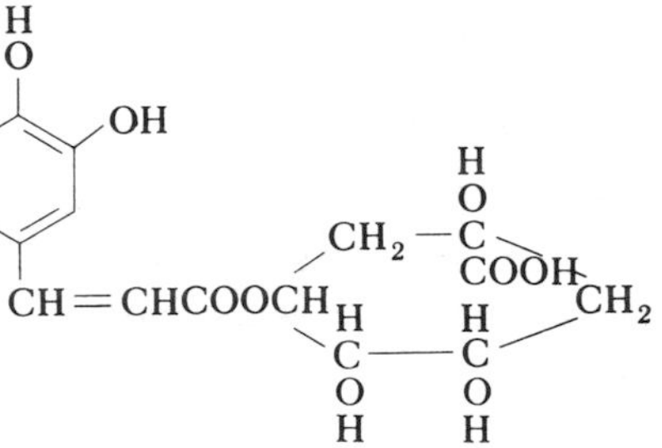

Chlorogenic Acid

(2) Coumarins, which contain only one benzene ring and a condensed side ring,

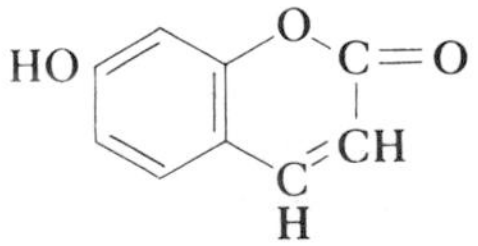

Umbelliferone, a coumarin

(3) Hydroxy acids, such as gallic acid, tannic acid, and others.

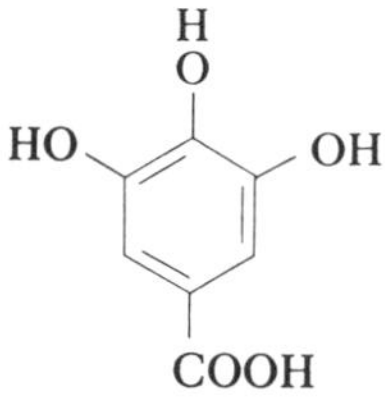

Gallic acid 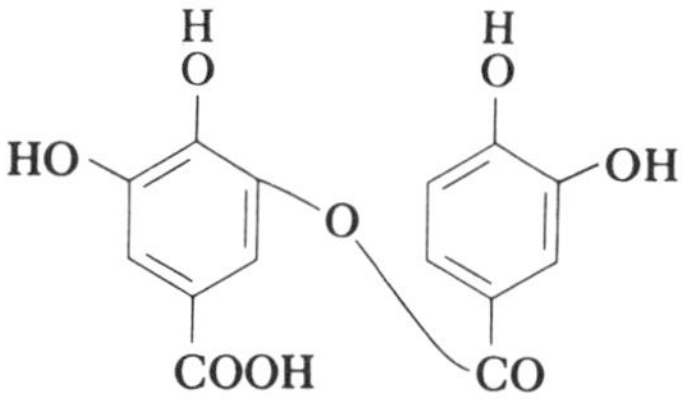   Tannic acid

The true flavonoids differ in the state of oxidation of the three carbon bridge which separates the two benzene rings. The most common anthocyanidin, cyanidin, and the most common flavone, quercetin, differ only in the oxidation of the three carbon bridge and have the same hydroxyl groups in the same positions on the benzene rings.

The Anthocyanins

Most of the red, blue, and violet pigments that occur in flowers, fruits, and other parts of plants belong to the group of pigments known as anthocyanins. These occur in plant cells as glycosides which are ethers of monosaccharides sometimes with one monosaccharide moiety and sometimes with two. The color results from the structure of the anthocyanidin which is combined with the monosaccharides. The carbohydrates commonly bonded to the anthocyanidins are glucose, galactose, rhamnose, and occasionally a pentose. Most of the anthocyanins are soluble in water, and it is only on boiling with fairly concentrated mineral acid, a condition which is not encountered in food preparation, that the pigments are hydrolyzed to form the anthocyanidin and the carbohydrate.

$$\text{Anthocyanin} \longrightarrow \text{Anthocyanidin} + \text{Monose}$$

Only three types of anthocyanidins have been identified in plant tissues, although a number of methyl derivatives of these three have been isolated. Pelargonidin, cyanidin, and delphinidin are wide spread in nature, with cyanidin the most common. Often a plant tissue will have a number of pigments composed of the same anthocyanidin but differing in the carbohydrate moiety. Formulas for the chlorides are presented at the top of p. 242. These compounds are formed when the corresponding anthocyanin is hydrolyzed with hydrochloric acid. The methyl ethers which have been identified, peonidin, syringidin, petunidin, malvidin, and hirsutidin, are all derivatives of the hydroxyl groups on the benzene ring. In anthocyanins

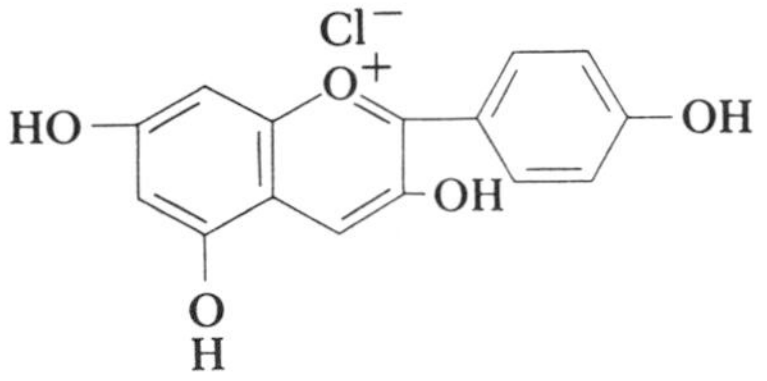

Pelargonidin Chloride

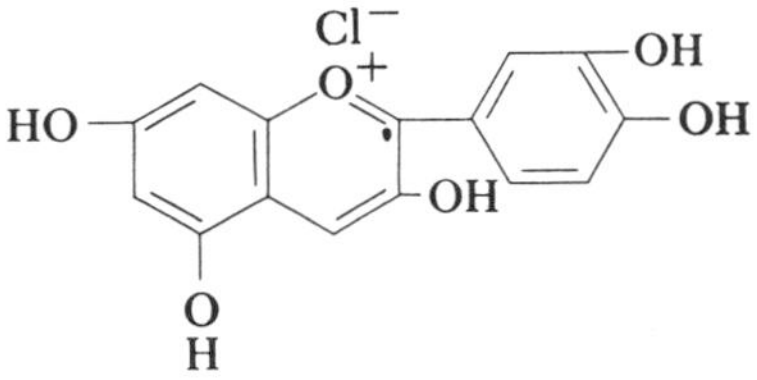

Cyanidin Chloride

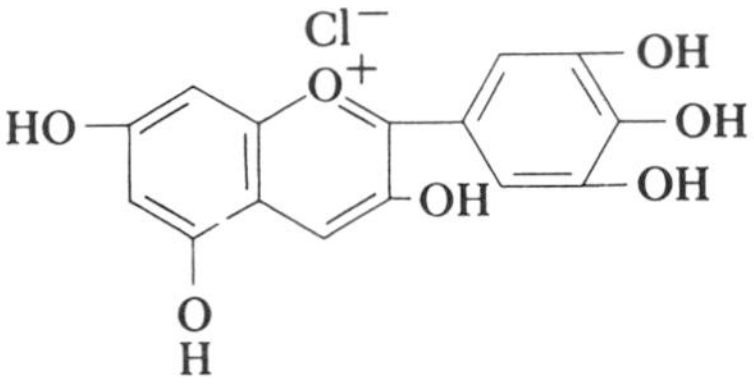

Delphinidin Chloride

carbohydrate is always attached to the hydroxyl on carbon 3 and occasionally one or more additional hydroxyls are etherified.

Some of the natural pigments occur as esters of organic acids; p-hydroxy benzoic acid, malonic acid, p-hydroxy cinnamic acid, and p-cumaric acid have been identified. These esterify either one of the hydroxyl groups on the anthocyanidin or one on the monosaccharide.

The Robinsons[43,64,65] published a remarkable survey of the distribution of anthocyanins in plant tissues in 1938 and 1939. Much of the research in this field has been carried out in their laboratory at Oxford. Their summary contains a list of the pigments present in many garden flowers, in autumn leaves, and in edible and ornamental fruits. The data for edible fruits are included in Table 7.4.

The great variety of colors, hues, and tints that occur in nature and the subtle shadings on the cheek of a fruit or in a blossom are the result of a number of factors.

(1) *At low pH these pigments are red; the hues may be different; but they are all reddish.* Thus pelargonidin (it occurs in the plant as pelargonin or some other derivative) is orange-red in acid solution while delphinidin is a bluish red. At high pH's the anthocyanins pass through a violet and then blue color. Some turn green and then yellow at very high pH's (not encountered in plants). Karrer and Jucker[34] point out that the pigment of the red rose and the blue cornflower are the same cyanin. However, in the red rose the pigment occurs as a salt of an acid while in the blue cornflower it is present as metal salts. Cornflowers also contain a flavone,

apigenin, which is colorless but causes blueing of cyanin. When an alkali such as potassium hydroxide is added to an anthocyanidin, it is believed that the following reaction occurs:

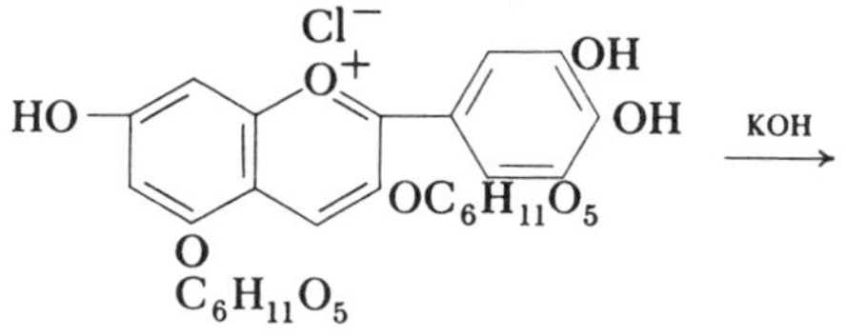

Cyanin
red at pH 3.0 or less

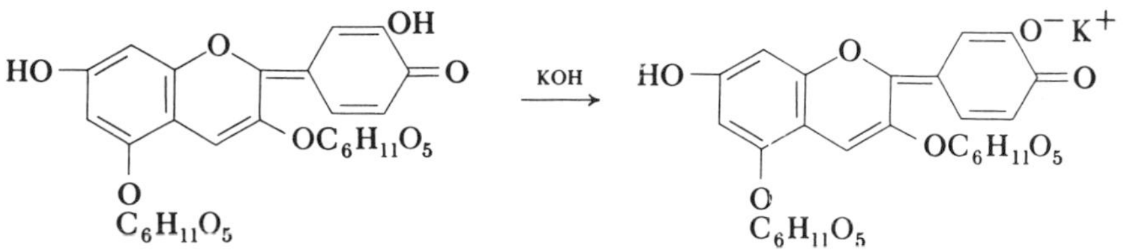

Color base
violet at pH 8.5

Salt
blue at pH 11

(2) *The concentration of the pigment alters the hue.* Robinson reports that when an acid solution of synthetic delphinidin is poured on filter paper, a dilute solution gives a blue color while a concentrated solution gives red. An intermediate concentration gives a purple color. When tannins are present this is more pronounced.

(3) *In the cell sap the anthocyanins are frequently adsorbed on colloidal particles, probably polysaccharides.* The pH is stabilized and the color influenced by this association. In blue corn flowers cyanin is adsorbed and the pH stabilized at 4.9.

(4) *Frequently the anthocyanins occur as mixtures.* As the composition of the mixture is altered, the hue changes. Thus blue grapes contain not only glucosides of delphinidin but also of syringidin, the dimethyl ether of delphinidin.

(5) *Sometimes plant cells will contain not only the anthocyanins as pigment but also some of the anthoxanthins which may be yellow and often yellow carotenoids.*

(6) *Tannins are often associated with anthocyanins and alter the color.*

The red color of beets results from a mixture of pigments. In 1948 Aronoff and Aronoff[1] reported the fractionation of the pigments of beets

TABLE 7.4. ANTHOCYANINS PRESENT IN EDIBLE FRUITS*

Botanical Name	Common Name	Anthocyanins†
Cornus mas Linn.	cornelian cherry	pel. 3-mono
Ficus caricus Linn.	fig	cyan. 3-mono
Fragaria vesca Linn.	wild strawberry	pel. 3-mono
F. virginiana Duch.	wild strawberry	pel. 3-mono
Morus nigra Linn.	black mulberry	cyan. 3-mono
Musa coccinea Andr.		pel. 3-mono
Oxycoccus macrocarpus Pers.	wild cranberry	peon. 3-mono
Pisum sativum Linn. var. (pods)	tall pea	cyan. 3-5 di, delph. di
Prunus avium Linn. var.	sweet cherry	cyan. 3-mono
P. communis Fritsh. var.	almond	cyan. 3-mono
Punica granatum Linn.	pomegranate	delph. di-
Sambucus nigra Linn.	European elderberry	cyan. 3-pent. gluc and mono-
S. racemosa Linn.	red berried elder	cyan. 3-, 5-di
Solanum melongena var. erculentum	eggplant	delph. 3-pent. gluc delph. 3-bio
Vaccinium Myrtillus Linn.		malv. 3-mono
V. Vitis-Idaea Linn.	whortle berry	cyan. 3-mono
Vitis hederacea	grape	cyan.? 3 sugar A.
V. quinquefolia	grape	malv. 3-mono
V. vinifera Linn.	vinifera grape	malv. 3-mono
V. labrusca	fox grape	malv. 3-mono
V. riparia		malv. 3-mono
V. aestivalis	summer or pigeon grape	malv. 3-mono
V. heterophylla Thunb.		malv. 3-5 di

*From Lawrence, W. J. C., Price, J. R., Robinson, G. M., and Robinson, R.
†Structure of anthocyanin shows anthocyanidin and carbohydrate.
Pel. = Pelargonidin (3,5,7,4′ tetra hydroxy flavylium chloride)
Cyan. = Cyanidin (3,5,7,3′,4′ pentahydroxy flavylium chloride)
Delph. = Delphinidin (3,5,7,3′,4′,5′ hexahydroxy flavylium chloride)
Malv. = Malvidin (3,5,7,4′ tetrahydroxy 3,5 dimethoxy flavylium chloride)
Peon. = Peonidin (3,5,7,4′ tetrahydroxy 3′ methoxy flavylium chloride)

into eleven different fractions on a chromatographic column. One pigment, betanin, has been extensively studied and appears to be the glucoside of an amino anthocyanidin. It also shows a change from red to blue as the pH is raised.

In Cookery and Processing. Fruits and vegetables which contain the anthocyanins, the red and violet foods, present a problem in cookery and processing because of the great solubility in water of the pigments. There is always a tendency for the pigment to leach out in the cooking or canning water or to run out in the juice. However, if the cell walls remain intact the pigment is not lost. Thus frozen red raspberries show excellent retention of the pigment, but in canned berries the color gradually passes out into the canning solution until the berries are practically colorless. When beets are cooked without removing the skins or even cutting off the root, color retention is much better than when they are peeled and cubed. When grapes are hot pressed, most of the color is removed in the juice; and if they are fermented in wine making for a short time before the pressing, the extraction is almost complete.

The effect of changes of pH on the color of the anthocyanins is often noticed in food preparation, and occasionally presents a problem. Most fruits contain sufficient acid so that the pigment remains red or bluish red through the cookery or processing. But if a small amount of the juice is added to dish water containing soap or a detergent and with a pH, therefore, of 8 or 9, a blue or a greenish blue color forms. When vinegar is added to beets in pickling, the color often reddens. When red cabbage is cooked in soft water with a pH near 7, there is a bluing of the color.

If vinegar is added to blue cabbage, the color will change back to red. When red cabbage is chopped, blue coloration often occurs along the edges cut. This is apparently not a reaction with iron ions, although a possibility, since red cabbage macerated with a Waring blender shows the same change on the surface exposed to air.

If red cabbage is cooked in hard water or, more important, if the water has been softened and has a pH of 8 or slightly higher, the cabbage becomes a grayish violet. If the cooking water is distinctly alkaline, usually from the addition of soda, $NaHCO_3$, the color becomes green. Most of the anthocyanins will form a green color at high pH's and sometimes even yellow at higher pH. These are irreversible changes, but we seldom see such a color change in food preparation since we do not encounter such high pH.

The anthocyanins form salts with metal ions, and have colors that depend not only on the particular anthocyanin but also on the metal ion. Most of the colors are grayish purple. The reaction is particularly important in canning and often in cookery. When tin cans (actually tin-coated iron) are used for canning those fruits or vegetables containing anthocyanins, it is necessary to lacquer the inside of the can. Tin cans without this lacquer will cause a discoloration of the fruit touching the side of the can. If a tin pie plate is used for preparing blueberry pie, the bottom of the crust becomes a grayish blue color. This is particularly noticeable if the tin has been scratched so that some of the iron is exposed. Small deposits of rust form, the fruit acids rapidly dissolve the rust, and the amount of reaction with the anthocyanins is increased. If a cherry pie is cut with a steel knife and the knife is allowed to remain in contact with the juice, the change to purple is observed. When aluminum salts are added to fruit

TABLE 7.5. SOME OF THE IMPORTANT PROPERTIES OF THE COMMON ANTHOCYANIDINS

	Pelargonidin	Cyanidin	Delphinidin	Peonidin 3' Methyl Cyanidin	Malvidin or Syringidin 3'-5' Dimethyl Delphinidin	Hirsutidin 7-3'-5' Trimethyl Delphinidin
Color of Aqueous Solution	Red	Violet red	Bluish red	Violet red	Violet red	Violet red
Solubility of Chloride in Water	Readily soluble	Only slightly soluble	Very soluble	Readily soluble	Slightly soluble	Slightly soluble
Ferric Chloride Reaction	Not definite	Intense blue	Intense blue	Faint— not definite	No reaction	No reaction
Behavior Toward Fehling's Solution	Reduces when warmed	Reduces in the cold	Reduces in the cold	Reduces in the cold	Reduces when boiled	Reduces when boiled
Color Change in Soda Solution	Violet then blue	Violet then blue	Violet then blue	Violet then greenish blue	Violet then blue	Violet then greenish blue
Behavior in Aqueous Solution	Color fades on standing	Color disappears on heating	Slow fading in the cold; when heated, rapid fading	Color disappears on heating	In very dilute solution color disappears when heated	In very dilute solution color disappears when heated

juices which contain the anthocyanins, a change in color occurs. But it is usually not as marked as with iron. In cooking fruits for jams and jellies in aluminum vessels some of the change in color may be the result of the reaction with aluminum ions. Iron vessels cause a marked alteration in the color of the fruit and must be avoided. Reactions of a few anthocyanidins are summarized in Table 7.5.

Prolonged storage of fruits with red or red-violet pigments is accompanied by bleaching of some pigment and the development of a red-brown and finally a brown color. Storage temperature is most important in the rate at which the change in color occurs. It has been found that strawberry spread stored at 1.1° C (34° F) showed little change in color, although it was fairly rapid at 18.3° C and 21.1° C(65° F and 70° F) and much faster at 37.8° C (100° F). Much the same results have been obtained with currants, raspberries and strawberries. The deterioration in the color of glass-packed pears, peaches, plums, and grape juice is much more dependent on high temperatures and oxygen content than light; ascorbic acid protects the pears, plums, and peaches against discoloration and the development of off-flavors, but it does not protect grape juice.

The Anthoxanthins and Flavones

One of the most important groups of pigments in plants are the anthoxanthins and the flavones. They are yellow pigments usually dissolved in the cell sap. The anthoxanthins are glycosides which on boiling with dilute acid yield one or two molecules of monosaccharides and a flavone or a flavone derivative such as a flavonal, flavanonal, or isoflavone. The basic ring structure of the flavones is:

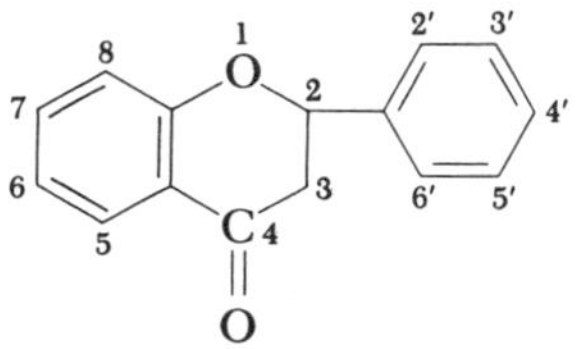

Flavanols have a hydroxyl group in position 3; flavanonols do not have a double bond between carbons 2 and 3; and isoflavones have the phenyl group in position 3 instead of 2. Most flavones contain a number of hydroxyl groups and some contain methoxyls. In the anthoxanthins one or more carbohydrates, usually monoses or dioses, etherify one or more hydroxyls.

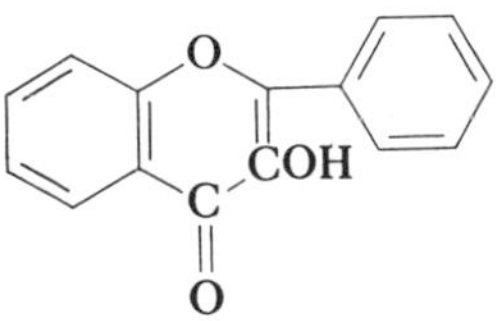

| Flavonol | Flavanonol | Isoflavone |

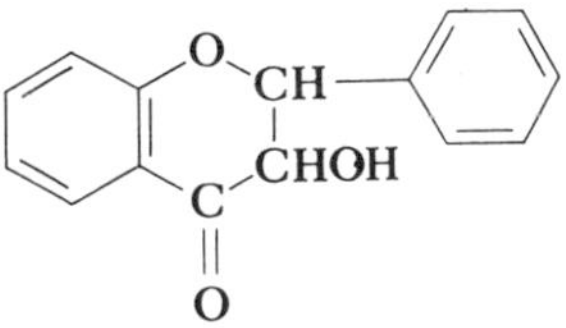

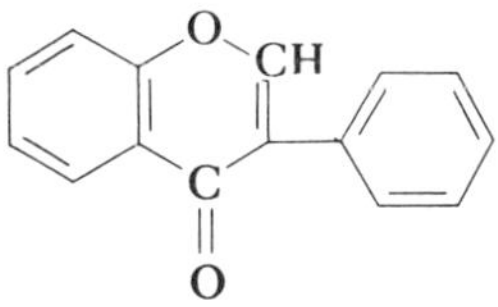

These pigments occur dissolved in the cell sap and are usually pale yellow or colorless but occasionally bright orange. Most bright yellow or orange fruits and vegetables are colored by carotenoids rather than anthoxanthins and flavones. Nevertheless the two latter are widely distributed and are probably present in all white fruit and vegetables as well as those colored green with chlorophyll or red, blue or purple with anthocyanins. During the past twenty years there has been considerable work on the organic chemistry of these pigments. Recently interest has centered on the role of these pigments in plant metabolism, in the relation of the genes and inheritance to flavone synthesis, and in their possible medicinal value to man.

The number of flavones and anthoxanthins isolated from plants is large. Seshadri[68] in a review in 1951 listed 18 flavones, 27 flavonols, 5 flavanonols, and 14 isoflavones. Many of these are from plants or parts of plants not used as food. The number isolated from foods are relatively small and many common foods have not been studied.

Examples of some of the flavones which occur in this group are:

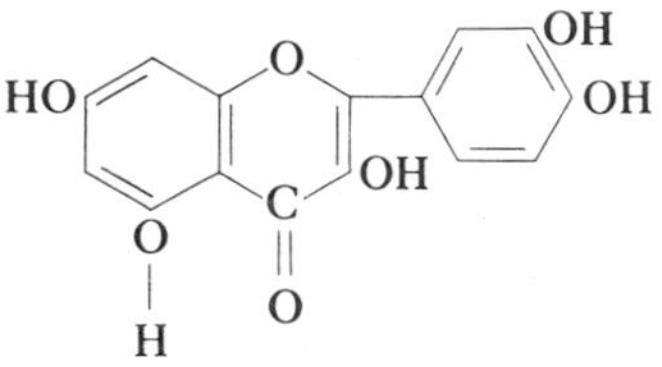

Quercetin

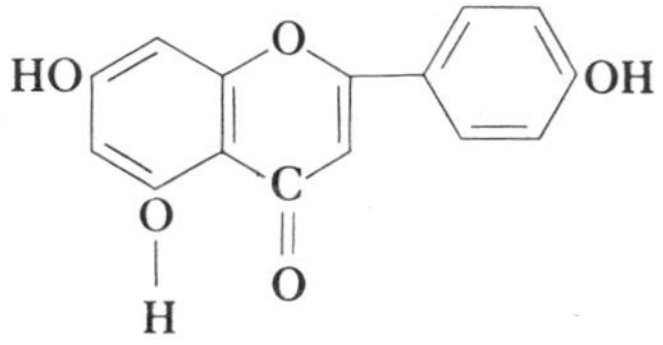

Apigenin

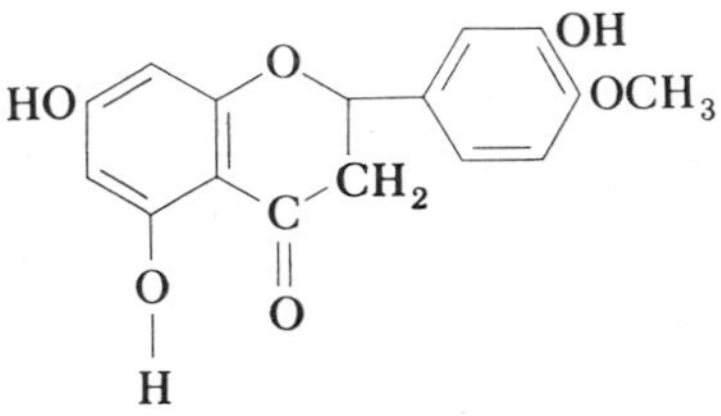

Hesperitin

Quercetin occurs in onion skins, tea, hops, horse chestnut, sumach, red rose, and the bark of the American oak and many other tissues. Its glycosides are likewise widely distributed. A few in foods are the 3-galactoside in Grimes Golden and Jonathan apples, and 3-glucoside in corn. Rutin is 3-rutinose (disaccharide of rhamnose and glucose) quercetin and occurs in many grains, in tomato stalks, in rue, elderberry blossoms, California poppies, etc. One of the flavones of lemons and tangerines is also a derivative of quercetin. Apigenin is present as a glucoside in parsley and is also one of the pigments of the yellow dahlia. Glycosides of apigenin also occur in field daisies, cosmos, and zinnias. Hesperitin occurs in oranges and lemons as a 7-rhamnoside.

The anthoxanthins are yellow to orange in color and are dissolved in the cell sap of flowers, stems, leaves, roots, and even wood. They are water soluble pigments and differ in this from the carotenoids, the other group of yellow pigments, which are lipid soluble. The anthoxanthins occur in small amounts in many fruits and vegetables which we ordinarily consider white or colorless. For many years a simple test has been used to demonstrate the occurrence of flavones and anthoxanthins in foods, and in fact, in all plant tissues. A drop of sodium hydroxide solution or ammonia vapor is applied to a slice or portion of the plant with the formation of a deep yellow, orange, or brown color. This color was supposed to show the presence of flavones. However, many other common plant components give this color change and the results are not specific to flavones and anthoxanthins. Some hydroxy acids are very widely distributed and many of them as well as some proteins give the color change. However, whether the change in color is caused by the occurrence of anthoxanthins and flavones or tannins, or both, it is of some interest in cookery.

Changes in Cooking or Processing. When a food is cooked in alkaline waters we often see the development of a yellow or cream color. Hard water will often have a pH as high as 8 and softened water which contains $NaHCO_3$ in place of $Ca(HCO_3)_2$ will have a pH even higher. Potatoes cooked in softened water often have a creamy color. The occurrence of this color can be prevented or the color removed by adding a little cream of tartar. If the potatoes are cooked in chunks, bands of yellow are sometimes seen where the pigment is more concentrated in some cells. Rice also shows this yellowing when cooked in softened water, and it too can be kept a clear, bright white by adding cream of tartar to the cooking water. This effect is most noticeable in onions, particularly yellow skinned, for the flesh turns pale yellow and the cooking water is bright yellow when alkaline water is used. Cauliflower and cabbage likewise sometimes show a yellowing. Occasionally cauliflower turns a brown or pinkish brown color on cooking.[69] Part of the color may be caused by flavones, but most is the result of reactions of ions with the tannins. Tea contains both the flavone, quercetin, and tannins. The change in color has probably most frequently been noticed when the acid of lemon juice causes a fading of the color.

Many of the flavones have definite physiological effects on man. Indeed, the active principle of some old drugs, used long ago, has been shown to be either one or a mixture of flavones or anthoxanthins. One of the most interesting is the mixture which has been called vitamin P or citrin.

In 1936, Szent-Györgyi and his collaborators showed that a factor other than ascorbic acid was important for the maintenance of normal capillary permeability. They named the factor vitamin P for permeability. In its absence guinea pigs and even rats develop fragile capillaries through which protein passes and which rupture readily to produce hemorrhages. They claimed that human patients with capillary fragility that does not respond to ascorbic acid are likely to be deficient in vitamin P. The factor is often more abundant in the rind of lemon and oranges than in the juice which is a rich source of ascorbic acid. A concentrate isolated from lemon peel was named citrin, a term still used occasionally. Citrin proved to be a mixture of flavones and anthoxanthins. It is composed of hesperidin, eryodictin, their glucosides, and rutin which is the rhamnoglucoside of quercetin. The effect is difficult to estimate and there has been conflicting opinion of the role of the flavones and of ascorbic acid in the maintenance of normal blood vessels. A number of derivatives of flavones have been tested and some seem to be effective. Other flavones have been shown to be diuretics, vasoconstrictors, heart stimulants, and cathartics.

Tannins

In prehistoric times it was discovered that some plant substances are able to react with components in the skins of animals and "tan" them. The leather produced is much more durable than the dried skin. In ancient times tanning materials isolated from plants became objects of commerce. Knowledge about extracting these substances, handling them, and applying them to skins developed through trial and error. In more recent times as science has been applied to an understanding of ancient skills, there have been attempts to develop methods for detecting these substances. The tannins react with a number of ions and form dark colors which have been used for inks; they are readily oxidized with permanganate and can be titrated. Both the ion tests and the permanganate test are nonspecific.

As interest developed in the chemistry of foods, dark colors and astringent tastes were ascribed to tannins. Since it was difficult to separate the compounds responsible for the darkening or astringency at that time, both the nonspecific permanganate and ferric chloride tests were used. Since certain compounds in foods react, the term "tannin" was used for them even when it was recognized that these compounds probably had little or no reaction in tanning leather.

Today the tannins of foods appear to be comprised of the catechins, the leucoanthocyanins, and some hydroxy acids. All of them give colors with metal ions. Those which are ortho and para dihydroxy benzene derivatives are readily oxidized by permanganate although the mono hydroxy or meta dihydroxy derivatives are not. The substances that react with the proteins in skins and bring about tanning are probably polymers of catechins with intermediate molecular weights. The low molecular weight compounds in many fruits and vegetables are related to them.

Catechin and epicatechin are reduced derivatives of flavones. They are isomers in which the ring and hydroxyl are probable *trans* in catechin

and *cis* in epicatechin. The structure of the leucoanthocyanins is still uncertain but they are probably closely related to the catechins.

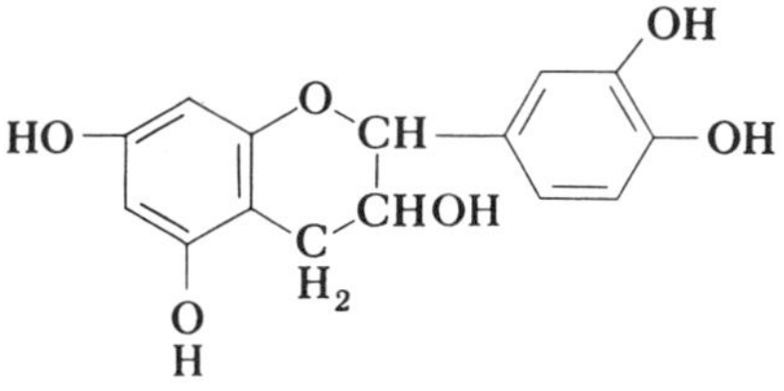

Catechin

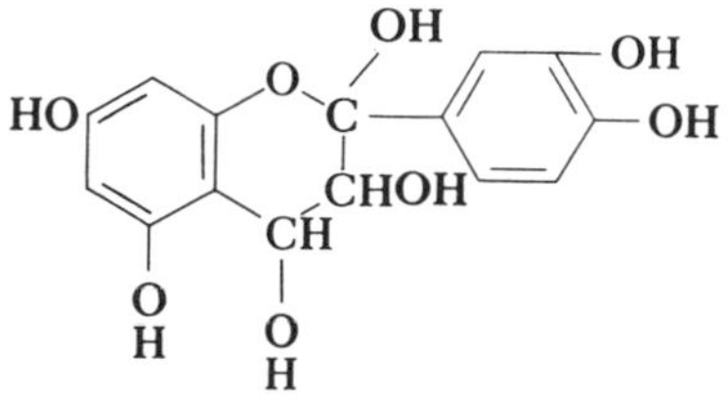

Leucoanthocyanin

Catechins and leucoanthocyanins are present in the tissues of those woody
plants studied such as apples, peaches, grapes, almonds, and some pears,
while they are absent in herbaceous plants. They are present in cereals
although the amounts vary.

Tea and cacao have been extensively studied with the newer analytical
methods such as chromatography. Tea contains a number of compounds of
catechin and epicatechin esterified with gallic acid. The most abundant are
3-galloyl epicatechin and 3-galloyl epigallocatechin. Epicatechin has been
identified in both types of cacao beans, Forastero and Criollo beans.

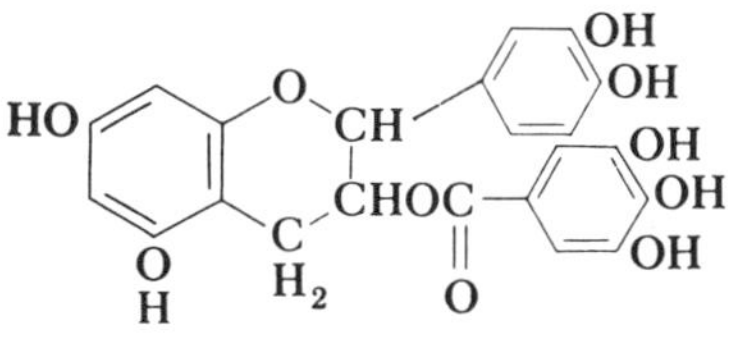

3-Galloyl Epicatechin (cis)

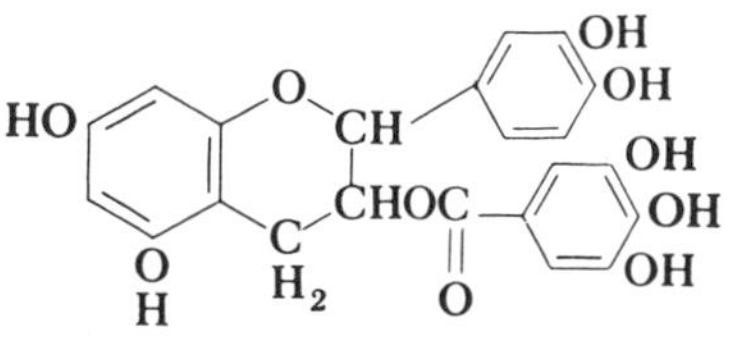

3-Galloyl Catechin (trans)

Some of the hydroxy acids are found widely distributed in plants. Caffeic
acid is the most common phenolic compound in leaves and one of its esters,

chlorogenic acid, is present in many tissues. Phenyl caffeate is another common ester of caffeic acid.

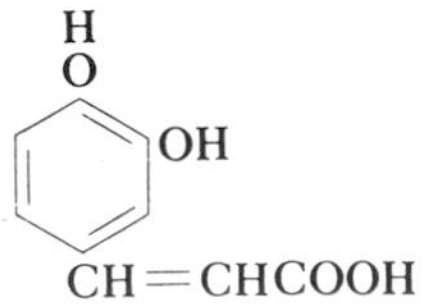

Caffeic Acid 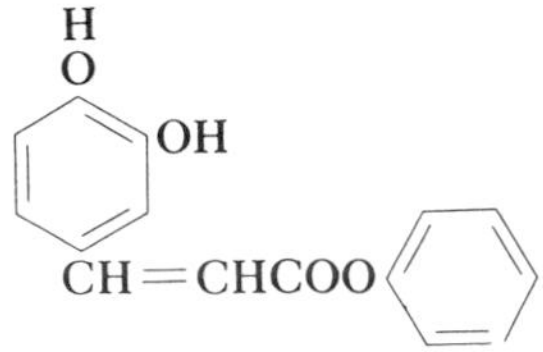   Phenyl Caffeate

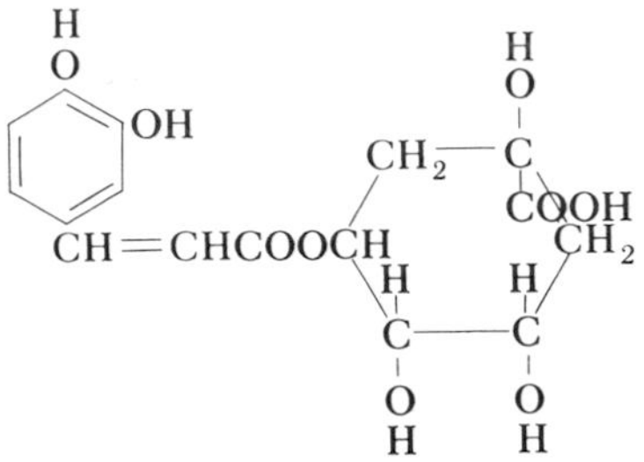

Chlorogenic Acid

If catechin is heated with dilute mineral acids, it forms an amorphous red precipitate which is highly insoluble. This is called "tannin red" or "phlobaphene" and is believed to be a polymer of the tannin. It is possible that the reaction is a general one for tannins or for one class of tannins. Phlobaphene is reported by Stansbury, Field, and Guthrie[74] to occur in small amounts in the red skins (testa) of peanuts. They found a large (7 per cent) amount of catechol tannins as well as some other pigments.

Colors in Foods from Tannins. The tannins are readily dispersed in hot water, some in cold, to form colloidal systems. When a fruit is pressed, as apples in the preparation of cider or grapes in making juice or wine, the tannins flow out in the juice. In extraction such as the brewing of tea or coffee some of the tannins are extracted. The exceedingly astringent brew produced by boiling tea in water for some minutes is probably the result of a thorough extraction of the tannins.

When the tea and coffee are brewed with hard water, a brown or red-brown precipitate forms on the surface of the liquid; and as the beverage cools, it appears throughout the liquid. Instead of a clear, sparkling infusion, the tea or coffee is distinctly muddy. In iced tea this is particularly noticeable, and with some waters the brew may be so full of precipitate that it is opaque. The precipitate clings to the side of the cup or glass, and the color is very noticeable. These transformations are believed to be caused by the reactions of the tannins in the tea and coffee with the calcium and magnesium ions of the water. If iron is present, very dark complexes are formed. Whether the precipitate is a simple calcium or magnesium salt of a tannin or whether it is a complex is not known. We usually speak of the formation of tannates. The change in color which occurs when lemon juice is stirred into black tea is also believed to be the result of a change in the tannins present.

Occasionally Seven Minute frosting flavored with coffee turns green. Knowlton[39] investigated the development of the color and attributed it to the formation of iron tannate. She found that the color occurs only when frostings containing egg are used. No color appears if the frosting is made with cream of tartar, lemon juice, vinegar, or brown sugar. She suspected that iron is introduced on the egg beater.

A greenish gray discoloration occasionally occurs in chocolate ice cream and has been accredited to iron tannates formed from specks of rust from the ice cream cans and tannins in the cocoa.

When evaporated milk is allowed to stand in an opened can for some days, it imparts a grayish green color to coffee. Gould[18] studied this reaction and found that it depended on the formation of rust on the can—the color is proportional to the amount of iron in the milk. This is probably a reaction of the iron ions with the tannins of the coffee.

Butland[11] investigated the darkening which sometimes occurs in the production of maraschino and glace cherries. He believed that the discoloration is the result of a reaction between tannins and iron or copper ions, although he presented no direct evidence that tannins in the cherries are responsible for the reaction. He also suggested that the tannin in the barrels in which the cherries are stored might be important. He found that only five parts per million of cupric ion causes darkening of the cherries while 25 parts per million turns them black.

Dark spots on canned sweet potatoes is attributed to reaction of tannins with the iron ions formed from the walls of tin cans. A gray color in sugar can occur when iron ions from the crushers react with tannins.

Astringency from Tannins. The presence of tannins in foods sometimes gives body and fullness of flavor to the food. The greatest difference in composition between cider apples and culinary apples is in the tannin content. The cider produced from those varieties with a relatively high tannin content has body and a mellow astringency while that from culinary apples is insipid.

The presence of tannins in hops has been a subject of considerable study and discussion among chemists associated with brewing. Some chemists find no evidence for tannins in hops although others get reactions attributed to them. It has been claimed that the tannins in wort and beer are combined with protein and therefore do not give the ordinary tests for tannins. In one report where the tannins and the place of their introduction into beer was detected by the color reaction with ferric chloride in the presence of gum arabic and ammonium hydroxide, it was found that both hops and malts contributed tannins. In 14 different samples of beer the quantity of tannins varied from 25 to 55 parts per million.

The chemical composition of tea and the organoleptic quality has been investigated. There is no consistent parallelism between the two but the tannins are the most reliable guide to quality. In tea the tannins are also believed to give body and a pleasing astringency to the beverage. Too much tannin can, of course, be very disagreeable and tea which is boiled or allowed to steep too long can be puckery and excessively astringent. Epicatechol, catechol gallate, and 5-hydroxy catechol have been isolated from Java tea.

In wine many compounds contribute to the quality, and tannins are one group. Valaer[82] says "A naturally high tannin content is desirable for slight astringency, in the rougher red types (for example Chianti) the tannin being derived principally from the seeds and skins." The grapes must be removed from the stems before fermentation begins so that excessive astringency is avoided. Also the wine cannot be allowed to remain in contact with the seeds for too long during fermentation, or too much tannin is extracted. A number of papers have appeared on methods of measuring tannins in wines and brandies.

THE BROWNING REACTION

When fruit and vegetable tissues are injured in any way or cut and peeled during processing, a darkening of the tissues called the *browning reaction* sometimes occurs. (See previous discussion p. 108.) This reaction has been extensively studied for a few fruits and vegetables but for most it has had little research. Some browning reactions are enzymatic and only occur in fresh living tissue or at least in tissues that still contain active enzymes. Thus when enzymes are denatured by heat or any other agent, this reaction no longer occurs. Consequently, although a fresh peach will turn brown after peeling, a canned peach will not. However, other darkenings may not be enzymatic. Thus when orange juice is concentrated, it often darkens with a deleterious effect not only on the appearance but also on the flavor. This browning reaction is nonenzymatic; it occurs at temperatures above those which denature most enzymes.

Enzymatic Browning

Enzymatic browning occurs in many tissues whenever they are injured. The injury can be the result of bruising, cutting, freezing, or disease. That part of the injured fruit or vegetable which is exposed to air undergoes a rapid darkening. Joslyn and Ponting[32] published an excellent review of the status of our knowledge about enzymatic browning in fruit in 1951. They point out that as early as 1895 Lindet recognized that the change in color occurring in freshly pressed cider is enzymatic. Since 1895 numerous studies have been made of the enzymatic browning reaction in fruits and vegetables, but little definitive information has as yet been gained. To begin with, the problem is extremely difficult and much of the work has been only qualitative or suggestive. However, the techniques of enzyme chemistry have developed in recent years. During the past ten years great strides have been made in elucidating many enzyme reactions, and doubtless before long the browning reaction will be understood.

Several theories of enzymatic browning have been suggested, but as yet none has been established to the exclusion of all others. During the 1920's Onslow[53] carried out extensive investigations of enzymatic browning in fruit. She did not isolate the compounds which undergo the changes but did carry out qualitative tests for their presence, concluding that browning occurs through the effect of an oxidase on a catechol compound to form either a peroxide or hydrogen peroxide. The hydrogen peroxide then oxidizes some other compound, the chromogen, to form brown pigment. Onslow studied the enzymes in numerous fruits and found some evidence for the presence of an oxidase and a catechol compound in the fruits that darken readily (apple, apricot, banana, cherry, grape, peach, pear, and strawberry), but no evidence of these enzymes in the fruits that do not darken readily (lemon, orange, lime, grapefruit, red currants, melon, pineapple, and tomato).

Other early work seemed to indicate that the enzymes are peroxidases and that hydrogen peroxide reacts in the presence of peroxidase with some compound to form the brown pigment. However, the possibility that peroxidases are the important enzyme systems in browning is now doubted. Ponting and Joslyn[32] believe that in apples peroxidase activity counts for little or none of the browning. Other attempts to find hydrogen peroxide or demonstrate its activity have not been successful.

Today most evidence suggests that oxidation of phenols or polyphenols by enzymes is the principal reaction in enzymatic browning. The nomenclature for the enzymes which cause oxidation of phenols or of polyphenols is not standardized. They are called "phenol oxidase," "polyphenol oxidase," "phenolase," and "polyphenolase." Most have been studied as crude extracts. They are very sensitive to decomposition at ordinary temperatures and are consequently difficult to isolate. Undoubtedly there are many different enzymes that can catalyze the oxidation of phenols and their derivatives by atmospheric oxygen. Not all catalyze the oxidation of the same phenols. Thus extracts of both the Royal apricot and Sphinx avocado are capable of catalyzing the oxidation of catechol and pyrogallol. However, although the apricot extract did not have any effect on phloroglucinol, the avocado extract caused slow oxidation. An extract of olives will catalyze the oxidation only of o-dihydroxy phenols and apple extract catalyzes the oxidation of catechol and pyrogallol but not hydroquinone, resorcinol, or phenol. This and other data suggest that the phenolase present in fruit tissues is not always identical and that the substrates they affect may well be quite different in different plants.

The substrates, the compounds affected by the enzymes, have not been isolated. Those tested which act as substrates for the phenolase complexes include catechol, tyrosine, 3,4 dihydroxyphenylalanine, caffeic, chlorogenic, gallic, and protocatechuic acids, urushiol, phloroglucinol, hydroquinone and a number of anthocyanins and flavonoids. Many of these compounds are widely distributed in the plant kingdom and now that their presence or absence in many plants is under study, precursors of the brown pigments in specific foods will soon be known.

The course of the reaction or reactions is not fully known. Oxygen is absorbed, carbon dioxide often evolved, a quinone formed and the final

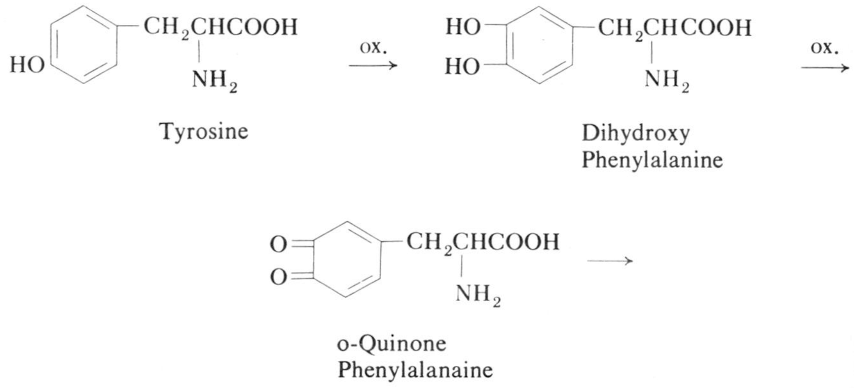

Tyrosine

Dihydroxy
Phenylalanine

o-Quinone
Phenylalanaine

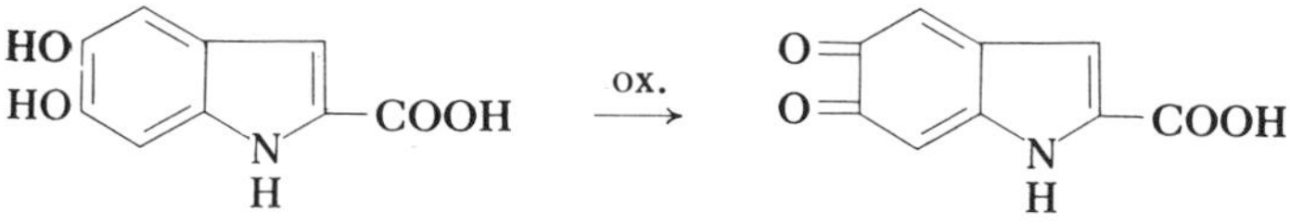

5,6 Di-hydroxy
indole-2-carboxylic
acid

5,6 Quinone
indole-2-carboxylic
acid (Red)

pigment is a polymer. It is evident with a number of substrates that quinone formation is not the only oxidative reaction. Catechol, for example, reacts with 2.4 atoms of oxygen during the sequence while its oxidation to the quinone requires only 1 atom of oxygen.

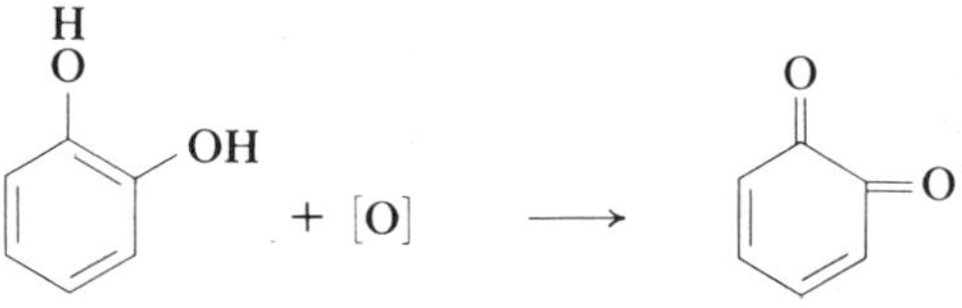

Catechol

Some fruitful work in the control of enzymatic browning has been done. The use of an antioxidant during processing has been used with some success. Thus in preparing peaches for freezing, the addition of small amounts of ascorbic acid not only prevents browning but also prevent loss of flavor. The effect of the ascorbic acid may be that of an antioxidant.

All fruits susceptible to browning should be processed as quickly as possible. Heating destroys the enzymes responsible for the reaction; thus when fruit is canned or made into jams or jellies, the browning reaction stops as soon as the fruit is heated sufficiently high to denature the enzyme. The exact temperature necessary varies with enzyme, rate of heat, pH, and other factors. Deoxygenation and vacuum closing have also been used to diminish oxidation.

In the preparation of fruit for freezing, sugars and sugar solution have been successfully used to prevent browning. The sugar solution coats the fruit and prevents direct contact with atmospheric oxygen. If sugar is added to the fruit, it dissolves in fruit juices and forms a concentrated solution on the fruit. Concentrated sugar solutions inhibit or depress the activity of plant oxidases, among them the phenolases, although at lower concentrations the activity is sometimes enhanced. Some sugars other than sucrose have been studied.

The effect of many salts and compounds such as sulfur dioxide, hydrogen sulfide, hydrocyanic acid, and thiourea has been tested and their effect on oxidases studied. Halides and sulfites inhibit darkening of fruit. Joslyn and Ponting[32] give the following list of compounds commonly used to prevent browning and another list of patented inhibitors:

(1) *Commonly Used or Unpatented Chemicals*: cystein, cystine, glutathione, sulfonamides, sulfurous acid and its salts, sodium sulfide, sodium chloride, citric acid, hydrochloric acid, ascorbic acid.

(2) *Patented Chemical Inhibitors*: (a) Sodium thiosulfate: Elion (1942); (b) thioamides such as thiocarbamide: Denny (1943); (c) sodium chloride, ascorbic acid, and sodium bisulfite: Johnson and Guagagni (1949).

Hydroquinone, toluhydroquine, hydroquinone + lecithin, hydroquinone + triethanolamine, diphenylamine, pyrocatechol, p-aminophenol, or resorcylaldehyde: Johnston, *et al.* (1943).

o-hydroxy aromatic oximes free of strongly acidic groups: Downing, *et al.* (1943).

Product of condensation of one mole of an alkali primary amino aliphatic carboxylate with at least one mole of an o-hydroxy substituted aromatic aldehyde. Preferred deactivators are salicyl derivatives of sodium glycinate, disodium glutamate, sodium tyrosinate and sodium cysteinate: Downing and Pedersen (1944).

Thiosemicarbazide and its derivatives: Clarkson (1946).

In the home preparation of fruits, pineapple juice and lemon juice have long been used to prevent browning. Pineapple juice has a relatively high percentage of sulfhydryl compounds which are active antioxidants while lemon juice contains both citric acid and ascorbic acid.

The pH has an important effect on the rapidity with which browning occurs. Acid dips are sometimes used to lower the pH and by this method delay or retard browning.

Many studies have been made on the darkening of potatoes when they are exposed to air by grinding, grating, or peeling as well as on the blackening that sometimes occurs when they are cooked. The reaction upon exposure to air is doubtless enzymatic. Wallerstein, *et al.*[85] showed that the tendency of white potato juice to pinking or graying increased on standing unless it was blanched at boiling-water temperatures for 50–60 sec. Some evidence has accumulated to indicate that the pigment formed is a melanin and that the enzyme is tyrosinase. Some investigators have suggested that the reaction which occurs on cooking likewise is enzymatic and is the result of the rupture of the cells with the release of tyrosinase. It seems improbable, however, that a reaction occurring at such high temperatures and even continuing after potatoes are cooked could be enzymatic. Other workers have presented evidence that the reaction in cooked potatoes is nonenzymatic. This will be discussed on p. 261.

Nonenzymatic Browning

Nonenzymatic browning has been the subject of numerous investigations but, nevertheless, this reaction or these reactions are also only slightly understood. There have been many attempts to control the browning that occurs on storage of citrus fruit juices; in grape, strawberry, and raspberry juices; and in dried apricots, peaches, pears, and apples. There have also been some studies of the possible course of the browning reactions in fruits. But the number of factors which must be controlled is so large, and the mixture of compounds in a fruit is so complex, that again knowledge is fragmentary. The subject is well reviewed by Stadtman.[73]

Three hypotheses have been suggested to explain nonenzymatic browning. There has not been sufficient work as yet to completely rule out any one hypothesis or to declare any other correct. (1) The browning reaction that occurs between carbohydrates and amino acids results in the formation of brown pigments. It is known as the Maillard reaction and is believed by many to explain the browning found in processed fruit. (2) Ascorbic acid undergoes oxidation with the formation of a compound which produces brown pigment. (3) Carbohydrates or carbohydrate acids (ascorbic acid is a carbohydrate acid) decompose to furfuraldehyde or related compounds which then polymerize or react with nitrogen compounds to form brown pigment.

The temperature of storage, the amount of moisture, and the exposure of the fruit or fruit juice to oxygen either during processing or storage are influential in the development of browning. For example, dried apricots which had been treated with sulfur dioxide were canned and stored at different temperatures. The samples stored at 46.1°C (115°F) darkened in 3 weeks, but those stored at room temperature (approximately 21.1°C or 70°F) did not darken for three months and those stored at 0°C (32°F) showed no darkening after six months. Similar data have been accumulated for canned orange juice, ground dehydrated apples, strawberry, currant, and raspberry juices. The data indicate that, in general, increase in temperature speeds up browning.

Studies on the influence of moisture on browning indicate that there is some effect in preventing discoloration, but the data are somewhat conflicting. Since Stadtman, *et al.* showed that in dried apricots it is not moisture alone, but oxygen uptake as well, that determines how rapidly darkening will take place, it is possible that some of the conflicting results are caused by lack of control of available oxygen. Nevertheless, it appears that lack of moisture favors the development of darkening.

Oxygen uptake either during processing or during storage has generally been shown to be a factor in browning. Some oxidation is apparently essential for the development of the reaction. For example, when the head space in canned orange juice is increased, the rate of browning is proportional to the increase in the space. Dried fruit is always exposed to oxygen during processing.

There has been some attempt to discover what compound originally present in the fruit is responsible for the browning reaction. The possibility that ascorbic acid is responsible has had considerable support from the work done on citrus juices, particularly orange juice. The ascorbic acid content of the juice falls off as browning occurs. Addition of ascorbic acid to orange juice causes a marked increase in the rate of browning. It has also been reported that a similar effect is shown in strawberry juice, but other work indicates that grape, apple, and cranberry juice show a decreased browning rate when ascorbic acid is added.

The possibility of a reducing sugar as the reactant has been studied by a number of workers. Slight decreases in these sugars appear to occur in orange juice concentrates as browning occurs. However, attempts to remove the sugars by fermentation from orange juice had little influence on the rate of darkening. Stadtman and his group were able to remove the reducing sugars in apricot syrups by fermentation but were only able to decrease the rate of the browning reaction by about half. There is at present the possibility that both ascorbic acid and reducing sugars may be important in the browning reaction.

The possibility that furfuraldehyde and its derivatives are involved in the browning reaction in apricots was tested by Stadtman and co-workers. They continuously extracted apricot syrup with ethyl acetate in which furfuraldehyde is soluble, and during the extraction period there was no browning. When extraction was discontinued, browning occurred. The extract contained a number of compounds among which were furfuraldehyde and hydroxyl methylfurfural.

$$\begin{array}{ccc} HC\!\!-\!\!CH & \quad & HC\!\!-\!\!CH \\ \| \quad \| & & \| \quad \| \\ HC \quad C\!\!-\!\!CHO & & HOCH_2C \quad C\!\!-\!\!CHO \\ {}^{\diagdown}O{}^{\diagup} & & {}^{\diagdown}O{}^{\diagup} \end{array}$$

Furfuraldehyde   Hydroxy Methylfurfural

When furfuraldehyde is added to apricot syrup, the rate of browning is accelerated.

Some work has been done on the isolation of the brown or black pigments formed during browning. By repeated precipitation, a product has been obtained from dried apricots which appears to be homogeneous. The exact nature of the compound is still unknown.

The black pigment formed when potatoes are cooked is most noticeable near the stem end. Much research has been devoted to this problem too. It has been found that some varieties of potatoes are much more prone to darkening than others. Northern grown potatoes are much more likely to give this browning reaction and careful evaluation seems to indicate that temperatures below 70°F at time of maturation favor the reaction. The occurrence of more extensive blackening at the stem end of the potato appears to be caused by a higher pH at this end of the potato. The pH of the cooking water is also of great importance, with more extensive darkening occurring at high pH's.

At present, then, it is known that a nonenzymatic browning reaction occurs when some fruits and vegetables are processed. We know that the reaction is influenced by atmospheric oxygen, moisture, and temperature. The compounds undergoing change may be ascorbic acid or some of the reducing sugars. It is quite possible that this is not a single reaction, but a group of reactions with one of them more important in certain fruits; however more needs to be learned about the reaction. It is of considerable economic importance to processors of these foods since not only is the appearance of the product seriously affected but the flavor and the ascorbic acid content as well.

PECTIC SUBSTANCES

Pectic substances (see discussion of structure, p. 87) are widely distributed in plant tissues and are present in rather large amounts in rapidly growing succulent tissues with a high water content. Some cells possess walls composed of a single layer, while others on cell division form walls composed of three layers: a primary wall, a secondary wall, and a middle lamella that is shared with adjacent cells. The middle lamella is believed to be composed chiefly of pectic substances that act as a cementing material and hold the cells together. By staining tissue sections one can readily see the cell walls and middle lamella under the microscope and the relative thickness or thinness can be easily determined. But when an attempt is made to demonstrate the composition of these layers, the results are not so clear cut. In many studies a stain, ruthenium red, has been used to reveal the pectic substances. It has been shown, however, that although this stain does color most of the pectic substances, it will also react with other compounds. Therefore, the work with staining techniques is still open to question since a stain that colors all the pectic substances yet does not color other substances has not been found.

It is believed that the middle lamella is composed of insoluble pectic substances, probably calcium pectates. The primary wall of many cells is rich in pectic substances believed to be protopectin. These substances undergo chemical changes as the cells mature and the fruit or vegetable ripens. Many parenchyma cells contain large vacuoles filled with cell sap. Some investigators believe that pectic substances are present in the cell sap and that this may even be the source of the pectic substances in the walls and middle lamella.

Apples have been extensively studied in all stages of growth and maturation of the fruit, both by staining and by isolation of pectic substances. Even when the fruit is extremely small, 1.3–1.5 cm in diameter, the middle lamella and cell walls are readily stained with ruthenium red. This observation has been interpreted to indicate the presence of pectic substances in the middle lamella and the cell walls. As the fruit grows, the total pectic substances increase; however after harvest as the apple softens, the amounts of soluble pectates and pectinates increase while the total pectic substances decrease. See p. 285. The crispness of the apple and its ability to resist puncture are the result of a number of factors. However, the presence of pectic substances in the middle lamella and in the cell walls is believed to be of great importance. When the pectic substances diminish in the middle lamella, the cells gradually loosen and can be torn apart more readily. When they diminish in the cell walls, the walls become thinner and more readily punctured. As the apple continues to soften and grow mealy, the pectic substances that can be stained finally disappear completely.

It has also been shown that in bananas a marked change in the pectic substances occurs during ripening. On storage the amount of soluble pectic substances increases, while the total pectic substances decrease. In one variety of bananas, it was found that there is only a trace of water soluble pectic substances in the green bananas; but the amount increases as the banana passes through the ripening stages.

Similar increases in the per cent of soluble pectic substances in mature peaches as compared to immature peaches has been measured for four varieties. There is a corresponding softening of the fruit as it ripens and the pectic substance becomes soluble.

Pears also show changes in the distribution and total amount of pectic substances as they ripen. The soluble pectic substances increase as the pear softens; as these compounds dissolve, the juice that can be pressed out increases in viscosity.

The viscosity of fruit juices is proportional to the amount of pectic substances dispersed colloidally in the juice. When tomato juice is pressed by the cold-break method, the cells are macerated and enzymes released. Two of the enzymes present in tomatoes are then free to catalyze the hydrolysis of pectic substances. The first, pectin methyl esterase, catalyzes the removal of methyl groups from the carboxyls on the galacturonic acid residues of pectinic acids with the formation of low-ester pectinic acids or pectic acids.

$$\text{Pectinic acids} + H_2O \xrightarrow{\text{pectin methyl esterase}} CH_3OH + \text{pectic acid}$$

The second enzyme, depolymerase, then attacks the low-ester pectinic acid or pectic acid and by hydrolysis forms products whose molecular weights are too low to contribute to the viscosity of the dispersion. The cold-break method forms a watery tomato juice. If the tomatoes are heated above 180°F (82.2°C), the enzymes are denatured, and the pectic substances extracted with the juice are pectinic acids with a sufficiently high molecular weight so that they form viscous colloidal dispersions.

The pectic substances in a juice not only influence its viscosity but also often contribute considerably to the ability of the juice to retain sediments in suspension. In tomato juice a desirable product has considerable suspended materials and a viscous juice can hold these without their settling out. Other factors such as the size and nature of the particles are important, but juices with a watery serum do not have the ability to hold even rather fine material in suspension.

Since a clear liquid is expected in some juices, the presence of a relatively high per cent of pectic substances may be objectionable. If it is necessary to filter the juice, a high viscosity slows down the filtration and may present an economic problem. The pectic substances are ordinarily removed by the use of pectinases, enzymes which depolymerize the pectic substances until the molecules are so small they no longer form colloidal dispersions or act as a protective colloid. "Pectinase" is an old term for a commercial enzyme preparation that contains a polygalacturonase and other enzymes. The polygalacturonase hydrolyzes pectic acids to low molecular weight polygalacturonides and galacturonic acid. In prepared clear apple juice, removal of pectic substances is almost always necessary because of the high percentage present. In wine the clarification and precipitation of tartrates and other sediments is sometimes difficult because the pectic substances hold them suspended. When the wine is treated with pectinase, precipitation occurs.

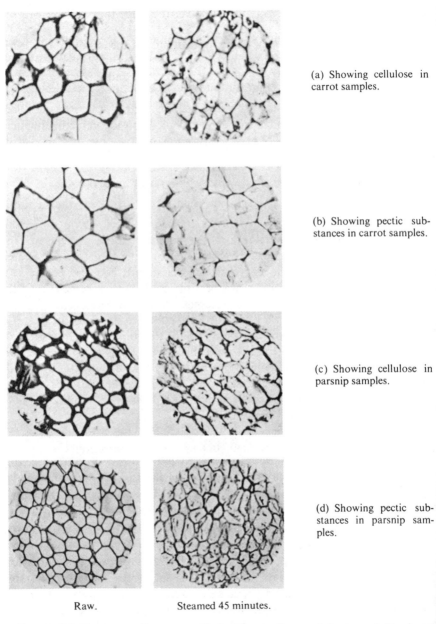

(a) Showing cellulose in carrot samples.

(b) Showing pectic substances in carrot samples.

(c) Showing cellulose in parsnip samples.

(d) Showing pectic substances in parsnip samples.

Raw.   Steamed 45 minutes.

FIGURE 7.5. TRANSVERSE SECTIONS OF XYLEM TISSUE. Raw on left, steamed 45 minutes on right. Magnification × 300. Courtesy of Simpson, J. I. and Halliday, E. G., *Food Research*, **6**, 189–206 (1941).

CHANGES ON COOKING AND PROCESSING

The chemical changes that occur when fruits and vegetables are boiled in water, steamed, canned, dried, or frozen are still for the most part unknown. We are aware that numerous reactions exist, since profound changes in the appearance, texture, flavor, and color of the fruit or vegetable occur. But exactly what these reactions are is still, for the most part, the work of the future. There have been a number of studies on potatoes and apples and a few on some other fruits and vegetables. The changes observed in a limited number of vegetables and fruits are suggestive of the reactions in others.

(1) As a result of changes in cellular structure, all fruits and vegetables undergo softening when cooked, no matter by what method. We can expect either that the cells are separating, rupturing, shrinking, or that a combination of these changes occurs. When cells separate during processing, the cementing substances between the cells, in the middle lamella, must alter. Thus, since the principal compounds in the middle lamella are believed to be the pectic substances, their change is the first one considered. (2) Since the cell walls are known to be composed of a matrix of cellulose sometimes encrusted with other materials, the possibility of changes in cellulose must be considered. (3) Starch granules can be seen in microscopic sections of most fruit and vegetable tissues and are known to swell rapidly in water. (4) The intercellular air changes in volume when subjected to higher temperatures because of expansion. (5) Alterations in pigments, formation of acids, and release of low molecular weight sulfur compounds are indicated from observations of gross changes. A large number of reactions are suggested by cursory observation, but careful and conclusive investigation of their role is difficult and has been applied to relatively few fruits and vegetables.

Changes in Pectic Substances

In 1941 Simpson and Halliday[70] carried out an interesting study on carrots and parsnips. See Figure 7.5. They determined the amount of pectic substances in the raw vegetable, vegetables steamed 20 minutes, and vegetables steamed 45 minutes and found that while there is a steady increase in the amounts of pectins and pectates as steaming progresses, there is a decrease in the protopectin as well as the total pectic substances. When sections of tissue were imbedded in paraffin and stained to show the middle lamella, vegetables steamed 45 minutes had a much thinner middle lamella than the fresh tissue. Simpson and Halliday concluded that at least one factor in the softening of carrots and parsnips that occurs on steaming is the changes in pectic substances—the hydrolysis of protopectin to pectins. See Table 7.6.

Many researchers have attempted to explain the difference in the culinary properties of some potatoes. Mealiness, waxiness, and sloughing are the result of cell separating. It has been found that the addition of calcium salts to potatoes during boiling greatly reduced splitting and sloughing. However, Freeman and Ritchie[17] did not find a correlation between the amount of pectic substances in raw and cooked potatoes and their mealiness.

The hardness of the water used in canning beans may have a profound effect on their texture. Addition of 1,000 ppm $CaCl_2$ to water produces beans so hard and tough as to be unacceptable. It is believed that the toughening action of calcium salts is through their effect on the pectic substances; calcium pectates are insoluble.

The pectic substances are also of considerable importance in the firming of canned tomatoes, apples, and other fruits by calcium salts. Canned tomatoes which are graded A and which therefore bring the highest price must be relatively firm, yet full-colored. However, during ripening, tomatoes pass through this period rapidly, soften, and on canning undergo considerable maceration and shredding. The addition of small amounts of calcium salts to the pack increases the firmness of the fruit. Calcium salts can be added to the dip or placed in the can with the salt. Calcium chloride is permitted in the United States at a level of 0.07 per cent; and salts such as calcium citrate, sulfate, or phosphate may be used at equivalent levels calculated on the basis of calcium ion. This level has no effect on the flavor of the fruit but produces a marked effect on the firmness. The method is also used widely for firming both canned and frozen sliced apples as well as baked apples. It has also been shown to be effective

TABLE 7.6. EFFECT OF STEAMING ON PECTIC SUBSTANCES
OF CARROTS AND PARSNIPS*

Fraction of Pectic Substance	Raw		Steamed 20 minutes		Steamed 45 minutes	
	Carrots (%)	Parsnips (%)	Carrots (%)	Parsnips (%)	Carrots (%)	Parsnips (%)
Pectin	3.7	4.7	6.0	6.1	8.8	7.9
Protopectin	14.1	10.2	9.0	7.7	3.6	5.7
Pectic Acid or Pectates	0.8	1.6	1.0	2.0	1.3	2.1
Total Pectic Substance	18.6	16.4	16.1	15.8	13.7	15.7

*Results represent the per cent of dry weight and are computed from the average of five weights of calcium pectate in each case. From Simpson, J. I. and Halliday, E. G., "Chemical and Histological Studies of the Disintegration of Cell Membrane Materials in Vegetables during Cooking," *Food Research*, 6 189–206 (1941).

in many other fruit products and will perhaps become commercially important in the future. Raspberries are firmer if they are treated with calcium salts before canning. The results are also good with canned potatoes, peaches, and olives. Kertesz, Tolman, Loconti, and Ruyle[37] studied this reaction extensively and concluded that the calcium ions react with pectic acids and low-ester pectinic acids which have numerous free carboxyl groups to form insoluble calcium pectate. It is interesting that mealy apples in which the amounts of pectic substances are low do not show an increase in firmness when they are treated with calcium salts.

Reeve and Leinbach[62] have studied the saucing properties of apples and their pectic substances but have not found a correlation between them. The insoluble protopectins were determined by the usual Carré-Haynes method. Gravensteins sauce rapidly and those from the 1948 crop disintegrated more rapidly than either the ripe or green 1947 Gravensteins. Newtown Pippins and Winesaps resist saucing, while Delicious and Jonathans are intermediate. Apples tending to sauce readily underwent complete cell separation when they were vacuum infiltered with water. When a $CaCl_2$ solution was substituted for the water, there was no increase in firmness. However, in Winesaps, Newtown Pippins, Jonathans, and some Delicious samples there was appreciable firming. These results indicate that, although the difference between saucing and nonsaucing apples is not entirely in the pectic substances, the firmness of some varieties is probably dependent to some extent on these compounds and their distribution. See Table 7.7.

You will remember that it has been shown that changes in pectic sub-

TABLE 7.7. SAUCING QUALITY AND PECTIC SUBSTANCES IN APPLES*

Variety and Storage Time at 33°F; Ripe, Except Where Noted	Cold Water Soluble Pectin (Per Cent)	Insoluble Pectin (Per Cent)	Saucing
Delicious, C Grade (1947) 1 Month	0.11	0.29	Intermediate
Delicious, Fancy (1947) 1 Month	0.09	0.34	Intermediate
Gravenstein (1947) 10 Days	0.09	0.16	Good
Gravenstein, Green (1947) 16 Days	0.20	0.30	Good
Gravenstein, Green (1948) 7 Weeks	0.06	0.35	Good
Jonathan, C Grade (1947) 1 Month	0.05	0.31	Intermediate
Jonathan, Fancy (1947) 1 Month	0.06	0.30	Intermediate
Newtown Pippin (1947) 1 Month	0.12	0.42	Poor
Winesap, Fancy (1947) 1 Month	0.05	0.34	Poor
Winesap, Fancy (1948) 1 Month	0.04	0.35	Poor

*From Reeve, R. M. and Leinbach, L. R., "Histological Investigation of Texture in Apples, I Composition and Influence of Heat on Structure," *Food Research,* **18**, 592-603 (1953).

stances occur as apples ripen and mellow. When apples are held in storage following harvest, there is a gradual increase in soluble pectic substances and a decrease in insoluble but the total remains fairly constant. As the apples become mealy and over-mature, there is a decrease in the total pectic substances and a change of the soluble to nonpectic compounds. (See p. 285). However, Reeve and Leinbach did find that the percentage of cold water extractable pectic substances from Cortlands (a saucing apple) is four times (0.31 per cent) what it is from Delicious (an intermediate—0.08 per cent).

Reeve[60] has also studied the seed coat of young peas. He finds that agents that react with the pectic substances have an effect on the young peas but not on the mature. Peas grown in soils with an abundance of calcium have tough coats. He found that as the pod becomes mature the pectic substances of the middle lamella in the cells that make up the seed coat become encrusted with pentosans or hemicelluloses. It is likely that in the mature peas, changes in pectic substances are not as readily available for reaction, since the pectic substances are partially covered with hemicelluloses.

It therefore appears likely that the softness that occurs on cooking or processing fruits and vegetables is partially the result of changes in the pectic substances. As our teeth sink into a piece of fruit or vegetable, few cells are ruptured; instead our teeth separate the cells. It is likely that the ease of biting and chewing results from the changes in the substances that cement the cells together, the pectic substances. The large molecules of insoluble protopectins are in some fashion hydrolyzed to smaller, so-called soluble pectic substances which are able to form colloidal dispersions in water. Reeve's findings on mature seed coats do not disagree with this conclusion, since it is a frequent observation that old woody vegetables cannot be softened by cooking.

There is the possibility, at present not established, that the pectic substances in other vegetables become encrusted with compounds such as the hemicelluloses incapable of partial hydrolysis under the mild conditions of cooking. Scott[67] found that leaves, stem, fruit, and flowers of a number of plants (not foods) contain a thin layer of material which she called "suberin," lining the entire system of intercellular air spaces as well as on the inner surface of cell walls. This substance or mixture is insoluble in 80 per cent sulfuric acid and also in chromic acid and increases in amount as the plant ages. Scott considered the material waxy in nature, but offered no evidence for this conclusion. Whatever its nature, it is likely that compounds resistant to 80 per cent sulfuric acid are not hydrolyzed during cooking and that this layer may account for the lack of softening that sometimes occurs. Until more evidence accumulates, we conclude that changes in pectic substances are among the important reactions that occur in processing vegetables and fruits.

Changes in Cellulose

Little attention has been directed to the possibility that the cellulose in cell walls undergoes some changes during cooking or other food-processing methods. The resistance of the cellulose from such well-known sources as cotton fibers, to the action of hot water, has thrown doubt on the possibility of hydrolysis. However, the work of Simpson and Halliday[70] with carrots and parsnips shows that on steaming, the cell wall thins out; they assume that this indicates a change in cellulose.

Changes in Starch Granules

Although starch granules occur in most plant tissues in storage leucoplasts surrounded by thin strands of cytoplasm, starch also occurs in the chloroplasts of leaf cells when light falls on the leaf. Chloroplast starch is thought to be only transient; i.e., rapidly hydrolyzed, carried in solution to the storage cells and then resynthesized into the storage starch. During processing the starch granules undergo the same changes that occur when starch and water are heated. See p. 105 for a discussion of these changes in simplified systems. During heating the starch granules swell if sufficient water is present. Tissue slices of steamed or boiled potatoes show swollen and often gelatinized starch granules. Occasionally when the cell is stained with iodine, the outline of a starch granule can be seen, but more often the whole cell is filled with the gelatinized starch. Sometimes the cells rupture and the gelatinized starch streams out.

A number of papers have been published on the changes in potatoes, particularly dehydrated potatoes. Weier and Stocking[89] studied (1) freshly mashed potatoes, (2) dehydrated mashed potatoes of poor quality, and (3) dehydrated potatoes of acceptable quality. They found that in all three cases (numerous samples were used) gelatinization of starch occurred although outlines of the swollen granules could be detected in most of the cells. In freshly mashed potatoes most of the cells remained intact with little gelatinized starch between the cells. However, in poor-quality potatoes many of the cells ruptured, resulting in much gelatinized starch between the cells. When water is added to these dehydrated potatoes, the starch granules swell, rupture the cells, and stream out between them making a thick, sticky starch paste. Weier and Stocking conclude that

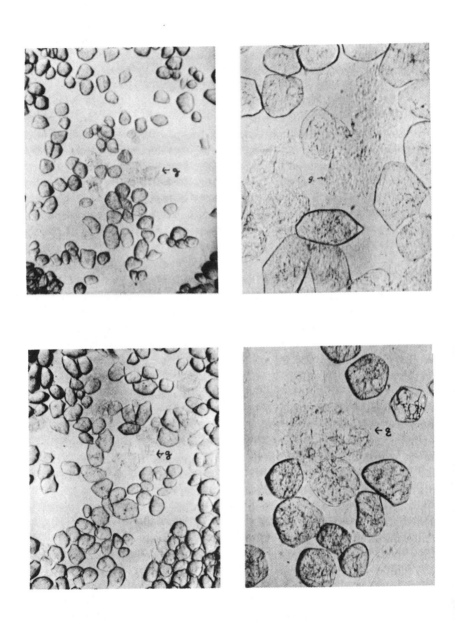

FIGURE 7.6. STARCH GRANULES IN POTATOES. Showing G-gelled starch escaped from ruptured cells. Magnification × 40 and × 100. Upper left and right: Unstained, separated cells of boiled, mature Russet Burbank potatoes. Lower left and right: Unstained separated cells of boiled, mature White Rose potatoes. Note more cell rupturing in White Rose. Swelled starch can be seen in all unruptured cells. *Courtesy of R. M. Reeve.*

fluffy mashed potatoes are composed of a mass of unbroken parenchyma cells with little starch in between.

Reeve[61] found that when White Rose (nonmealy) and Russet Burbank (mealy) potatoes are steamed or boiled, it is difficult to demonstrate the presence of individual starch granules because they are so swollen. See Figure 7.6. He found that in the nonmealy White Rose potato cell rupturing was more pronounced and there was much more gelatinized starch between the cells. He concluded that the escape of gelatinized starch from ruptured cells causes a sticky or gummy texture in cooked potatoes. However, he found that factors other than starch are also important. Young Russet Burbanks produced a texture in the steamed slices between mealy and waxy, but sometimes sticky, and yet showed no escaped starch. He had indications that the starch in the two types of potatoes had different hydration properties because of the difference in the colors obtained with iodine staining and the difference in the ability of some starch to escape without noticeable rupturing of the cells. In some specially treated samples of nonmealy White Rose potatoes, starch appeared between the cells although little or no cell rupturing occurred. This is probably low molecular weight amylose. See Tables 7.8 and 7.9.

The swelling of the starch granules in cells is often a factor in causing cells to break apart. However, the changes that occur in starch granules during processing is the result not only of gelatinization of starch but also activity of amylases and hydrolysis of starch to dextrins and maltose. Mann and Weier[46] reported that starch of carrots, although variable in amount and location, is generally found in the cells of the cambium. They observed that rate of heating during blanching has a marked effect: thus, if rapid, requiring less than 60 sec to reach 75° C, sections of carrot stain blue, with iodine showing the presence of starch; whereas if slow, sections stain purple, red, or pink, with iodine indicating presence of dextrins. The difference lies in the different temperatures at which carrot starch swells and the temperature at which the amylase is inactivated. Although carrot starch has a gelatinization temperature between 40° C and 50° C, Mann and Weier found that the inactivation temperature of carrot amylase is about 75° C. Thus, if the carrot is rapidly heated, the enzyme is inactivated before any extensive hydrolysis occurs. On the other hand, if the carrot is heated slowly, the starch granules swell and considerable amylase activity is possible before the temperature of inactivation is reached. Although the relevance of these phenomena to other fruits and vegetables is not really known, this work on carrots may be applicable; indeed in other fruits and vegetables blanching is usually carried to a point where enzyme inactivation occurs.

TABLE 7.8. COMPARISONS OF CELL SEPARATION, CELL RUPTURING, AND STAINING OF STARCH IN HEATED SLICES OF POTATO TUBERS AND THEIR RELATION TO FINAL TEXTURE QUALITIES*

Treatments and Raw Materials	Cell Separation	Cell Rupturing	Iodine Color†		Texture of Tissue after Final Heating
			Gelled Starch	External Solution or Filtrates (Amyloid Blue)	
Boiled					
Mature White Rose	slough in water	++++	reddish to blue-purple	++	slightly mealy soggy
Young White Rose	slough in water	++++	reddish to blue-purple	++++	soggy or sticky
Mature Russet Burbank	slough in water	++	blue-black	faint, if any	mealy when sloughing not severe
Young Russet Burbank	slough in water	+++	blue-black	+	slightly mealy to soggy
Steamed					
Mature White Rose	readily obtained on microslide	++	reddish purple	+++	slightly mealy
Young White Rose	readily obtained on microslide	++	reddish to blue-purple	++++	generally sticky or gummy, sometimes waxy
Mature Russet Burbank	slight sloughing	+	blue-black	generally none	mealy
Young Russet Burbank	readily obtained on microslide	+	blue-black or dark purple	faint	slightly mealy to waxy, sometimes sticky

*From Reeve, R. M., "Histological Survey of Conditions Influencing Texture in Potatoes," Food Research, 19, 334, (1954).
†With dilute aqueous solution of iodine and potassium iodide (approximately 0.1% total iodine).

TABLE ... CELL SEPARATION, CELL RUPTURING, AND STAINING OF STARCH IN SLICES OF SOAKED AND CHEMICALLY TREATED POTATO TUBERS*

Treatments and Raw Materials	Cell Separation	Cell Rupturing	Iodine Color†		Texture of Tissue after Final Heating
			Gelled Starch	External Solutions or Filtrates	
Boiled after Soaking at 75°C 1 hr					
Mature White Rose	readily obtained on microslide	slight	blue-purple to black	++++	rubbery, slightly sticky when slice is broken
Young White Rose	readily obtained on microslide	slight to none	reddish purple	++++	rubbery, slices sticky to gummy when broken
Mature Russet Burbank	readily obtained on microslide	none	blue-black	faint, if any	leathery, granular when slice is broken
Young Russet Burbank	readily obtained on microslide	none	blue-black	faint	rubbery, granular to waxy when slice is broken
Soaked 1 hr at 80°C in 0.5% Solutions of Pectic Solvents					
Mature White Rose	slight, by stirring	none	blue-purple	+	no final heating
Young White Rose	slight, by stirring	none	blue-purple	++	no final heating
Mature Russet Burbank	slight, by stirring	none	blue-black	none	no final heating
Young White Burbank	slight, by stirring	none	blue-black	none	no final heating

*From Reeve. R. M.. "Histological Survey of Conditions Influencing Texture in Potatoes," *Food Research*, **19**, 335 (1954).
†With dilute aqueous solution of iodine and potassium iodide (approximately 0.1% total iodine).

Changes in Intercellular Air

The parenchyma tissue of fruits and vegetables is composed of cells separated by tiny pockets and passages of air. In some fruits and vegetables the amount of air may be appreciable, in others quite small. Fresh peaches may have as much as 15 per cent air while plums have very little. When the fruit or vegetable undergoes processing, changes occur in the intercellular air. If the product is simply heated, the air will swell, force the cells apart, and often cause cracks in the food. Thus in a baked potato or baked apple, swelling occurs; however, on cooling, the product shrinks and cracks in the body are noticeable. Further changes may take place as the cell walls become more permeable. Cell sap, the solution held in the vacuoles, may escape into the intercellular spaces, causing a change in the appearance and juiciness of the food. Thus the chalky appearance of the product changes to one of translucence as the air is replaced with water.

This can be readily seen when green beans are placed in hot water. As soon as the bean is submerged, it begins to look greener because of the discharge of air from around the hairs on the surface and perhaps from between some of the surface cells. In many foods juiciness is markedly affected because the fluid that now fills the intercellular spaces readily escapes from the tissues. Weier and Stocking[87] point out that if the cellular sap is insufficient to fill the intercellular spaces the chalky appearance does not disappear; thus in steam blanching of cabbage it is retained. On boiling a fruit or vegetable a third change may occur: the discharge of intercellular air and the filling of the spaces with the cooking water. Reeve and Leinbach[62] were able to cause complete cell separation in saucing apples by vacuum infiltration of the apple tissue with cold water. They simply replaced the air in the intercellular spaces with cold water, and in some varieties of apples this caused the apples to disintegrate. Reeve[59] points out that the large intercellular spaces of the apple provide a tissue that is readily impregnated with syrups. He did not find that the amount of intercellular air (20–22 per cent for Delicious, Newtown Pippin, and Winesap; 23–24 per cent for Rome Beauty; slightly over 25 per cent for Gravenstein) explained the difference in texture of these varieties on cooking in water or steam.

Production of Volatile Acids

Many green vegetables contain volatile acids that are partially given off during cooking. These acids are of importance during cooking because they have a marked effect on the color and flavor of the cooked vegetable. If a lid is placed on the cooking vessel, these acids dissolve in the steam which condenses on the lid, drop back into the cooking water, and lower its pH. Chlorophyll, which is very sensitive to any pH below 7, will change to pheophytin and other olive green pigments. Green beans, spinach, or broccoli cooked in a covered kettle will be browner in color than another sample cooked without a lid for the same length of time and under the same conditions. When the lid is left off the vessel, these volatile acids are partially evaporated. They affect the flavor in two ways: (1) they have sourness and flavor themselves, (2) they speed up hydrolysis of sulfur-containing glycosides and produce distasteful sulfur compounds. The volume of water in which one of these green vegetables is cooked is likewise important since these volatile acids are readily soluble in water and can be diluted by using generous amounts of cooking water.

The chemical nature of the volatile acids evolved during cooking is at present unknown. Presumably they are low molecular weight organic acids such as formic, acetic, propionic acid, and perhaps lactic acid. The amount of acid evolved during the cooking of the vegetables has been measured by condensing the steam in a regular distillation outfit and titrating the acid evolved.

Exactly where these acids are located in the living tissue and their functions are for the most part unknown. They are probably dissolved in the cell sap since biting or squeezing a fruit or vegetable releases juice in which they are present. However, in some fruits at least part of the acids must occur in a different section of the cell from the anthocyanins that are dissolved in the cell sap. When plums, blackberries, or blueberries are cooked, the pigments redden. This indicates that the pH has dropped. Blackberries in a blackberry cobbler may become quite red. In some way acid and pigment which have been separated in the tissue have now reacted. When vegetables are cooked in water, the acids are quickly leached out.

Most investigation of organic acids in fruits and vegetables has been carried out on the whole fruit by macerating or grinding the tissue and extracting the acids. They have then been identified by typical reactions or occasionally isolated and identified. Malic and/citric acid are the most common and are present in small quantities even in fruits and vegetables not usually considered "acid." They appear to be present in all plant tissues. For example, beets contain 0.11 per cent citric acid, while cucumbers have 0.24 per cent malic acid and pumpkin 0.15 per cent. Oxalic acid occurs in a number of foods in small quantities and in rather large amounts in rhubarb and some leaves such as spinach, beet greens, lambsquarters, and purslane. Traces of tartaric acid have been reported in fruits and vegetables as different botanically as the avocado, artichoke, quince, and black raspberry, while Concord grapes contain as much as 1 per cent of this acid. The quantities of the acids vary with season, ripeness, and variety, of course. See Table 7.14, p. 287. Succinic, lactic, benzoic, and isocitric have also been found in small amounts in numerous fruits and vegetables. Cranberries contain quinic acid that also occurs in many plant tissues as part of the molecule of chlorogenic acid. See Table 7.10.

A few acids which are volatile have been identified along with numerous other types of organic compounds important in the flavoring of fruits and vegetables. Kirchner[38] lists both acetic and n-caproic acids in the ether ex-

TABLE 7.10. ORGANIC ACID CONSTITUENTS OF FOODS*

Food Items	Citric Acid (Per Cent)	Malic Acid (Per Cent)	Other Acids
Apples:			
Crab	0.03	1.02	—
Delicious	—	0.27	—
Grimes' Golden	—	0.72	—
Jonathan	—	0.75	—
McIntosh	—	0.72	—
Rome Beauty	—	0.78	—
Winesap	trace	0.50	—
Yellow Transparent	0.02	0.97	—
Apricots, Canned	1.06	0.33	—
Dried	0.35	0.81	Trace of oxalic
Artichokes	0.10	0.17	Trace of tartaric
Asparagus	0.11	0.10	—
Avocados	—	—	Trace of tartaric
Bananas	0.32	0.37	—
Beans, Lima	0.65	0.17	—
String, Green	0.03	0.13	—
Beets	0.11	—	—
Blackberries	trace	0.16	Trace of oxalic and succinic, 0.92 per cent isocitric
Blueberries	1.56	0.10	Trace of oxalic
Broccoli	0.21	0.12	—
Cabbage	0.14	0.10	—
Cantaloupe	—	—	—
Carrots	0.09	0.24	—
Cauliflower	0.21	0.39	—
Celery	0.01	0.17	—
Cherries	—	0.56–1.99	—
Montmorency, Canned	—	1.45	—
Corn, Sweet	—	—	—
Cranberries	1.10	0.26	Benzoic, 0.065 per cent; quinic, 1 per cent (Isham, 1935)
Cucumbers	0.01	0.24	—

TABLE 7.10. (*Continued*)

Food Items	Citric Acid (Per Cent)	Malic Acid (Per Cent)	Other Acids
Currants	2.30	0.05	Traces of oxalic and succinic
Figs	0.34	trace	—
Gooseberry	present	0.50–2.08	—
Grapes	—	0.65	0.43 per cent tartaric
Juice, Concord	0.02	0.31	1.07 per cent tartaric
Grapefruit	1.33	0.08	—
Kale	0.35	0.05	—
Lemons	3.84	trace	—
Juice	6.08	0.29	—
Lettuce, head	0.02	0.17	—
Mushrooms	—	0.14	—
Okra	0.02	0.12	—
Onions	0.02	0.17	—
Oranges	0.98	trace	—
Parsnips	0.13	0.35	—
Peaches	0.37	0.37	—
Canned	0.05	0.69	—
Pears	0.24	0.12	—
Bartlett, Canned	0.42	0.16	—
Peas, Fresh	0.11	0.08	—
Persimmons, Japanese	—	0.09	—
Pineapple	0.84	0.12	—
Plum, California	0.03	0.92	—
Damson	—	2.48	—
Potatoes, Idaho	0.51	—	—
Sweet, Cuban	0.07	—	—
Prunes, Italian Style	—	1.44	—
Pumpkin	—	0.15	—
Raspberries, Black	1.06	—	—
Red	1.30	0.04	—
Rhubarb	0.41	1.77	0.12 per cent oxalic
Spinach	0.08	0.09	—
Squash	0.04	0.32	—
Strawberries	0.91	0.10	—
Strawberries	1.08	0.16	—
Tomatoes	0.30	0.20	—
Tomatoes	0.47	0.05	—
Turnips, White	—	0.23	—
Watermelon	—	0.20	—
Whole Wheat Flour	0.05	—	—
Youngberries, Canned	0.62	0.24	—

*Prepared from Table 28 in Bridges, M. A. and Mattice, M. R., "Food and Beverage Analysis," Lea & Febiger, Philadelphia, Pennsylvania, 1942.

tract of strawberry juice; butyric acid in extract of carrot seed oil; acetic and propionic acids in molasses; and octanoic, valeric, and n-nonoic acids in cocoa.

Volatile Sulfur Compounds

Some vegetables have volatile sulfur compounds as an important part of the flavorful materials present; others produce volatile organic sulfur compounds when the cells of the vegetable are crushed, enzymes are released, and hydrolysis occurs. In still other vegetables volatile sulfur compounds are formed during cooking when acids are released from the cells and reaction with larger molecules takes place. Our knowledge of these compounds is fragmentary because few vegetables have been studied in this respect and even these few not intensively.

Garlic, onions, and related species owe their peculiar penetrating odor and flavor to sulfur compounds. In 1892, Semmler reported that garlic oil contains 60 per cent of diallyl disulfide, $C_6H_{10}S_2$; 6 per cent allyl propyl disulfide, $CH_2=CHCH_2SSCH_2CH_2CH_3$; and 20 per cent of diallyl trisulfide, $C_3H_5SSSC_3H_5$. In more recent years investigation of garlic indicates that "garlic oil" is not present in the whole garlic, but is formed on crushing. Crushing inaugurates an enzymatic reaction that releases the following:

$$\underset{\text{Alliine}}{CH_2=CHCH_2\overset{O}{S}CH_2\overset{NH_2}{C}HCOOH} \qquad \underset{\text{Allicin}}{CH_2=CHCH_2\overset{O}{S}SCH_2CH=CH_2}$$

Alliine   Allicin

S-allyl Cystein Sulfoxide   Diallyl Thiosulfinate

Stoll and Seebeck[77] have shown that the parent substance in garlic is alliine, which forms allyl thiosulfinate on crushing. Onion has been shown to contain allylisothiocyanate, $CH_2=CHCH_2CNS$, as well as allyl propyl disulfide.

The cruciferous plants are a large family (Brassica) noted for their peppery flavor and their content of organic sulfur compounds. They include the cresses, radishes, mustards and cole vegetables as well as many weeds and garden flowers. Synge and Wood[79] have identified S-methyl-cystein-S-oxide in cabbage and have detected it in turnip, cauliflower,

$$\begin{array}{c}CH_3 \\ SO \\ CH_2 \\ CHNH_2 \\ COOH\end{array}$$

S-methyl-cystein-S-oxide

kale, sheperds' purse, and white mustard but not in water cress, radish, or some other members of the family which have been tested.

The isothiocyanates are a group of organic compounds, commonly called "mustard oils." The name arises from the occurrence of allyl isothiocyanate in mustard oil. It has also been detected in many other members of this plant family.

$$CH_2=CHCH_2NCS$$

Allyl Isothiocyanate

"Mustard Oil"

Steam distillation of radishes produces an isothiocyanate which is either 4-(n-butylthio)butylisothiocyanate, $C_4H_9SC_4H_8NCS$, or 4-(n-butylthio)-crotonylisothiocyanate, $CH_4H_9SCH_2CH=CHCH_2NCS$. β-Phenylethylisothiocyanate has been isolated from ground water cress.

 CH_2CH_2NCS

β-Phenylethylisothiocyanate

Simpson and Halliday[71] measured the amount of hydrogen sulfide evolved during the cooking of cabbage and cauliflower by condensing the distillate and precipitating the sulfide ion as cadmium sulfide. The organic sulfur compounds in the filtrate were estimated by oxidizing to sulfate with bromine water and precipitating as barium sulfate. They obtained determinations of the rate of formation of hydrogen sulfide and organic sulfur compounds but did not identify the organic sulfides. They found that the amount of hydrogen sulfide increases between 5 and 20 min in boiling cabbage and the organic sulfur between 7 and 30 min. Cauliflower gives off more hydrogen sulfide and volatile organic sulfur during the same cookery periods. Since the most acceptable products are formed when cooking time is only long enough for the vegetable to become soft, they believe prolonged cooking produces disagreeable flavors and odors through the formation of hydrogen sulfide and organic sulfur compounds.

The evolution of hydrogen sulfide during the cooking of corn can be demonstrated. Since corn is not cooked for great lengths of time, hydrogen sulfide formation is no problem; but during canning, where it is important to use higher temperatures and longer periods of processing, the volatile sulfide formation may account in part for the "canned corn" flavor. So far no reports on sulfide formation during "flash" canning of corn are available.

In one study volatile sulfur compounds were measured in a number of vegetables. During processing a small amount splits off from peas, asparagus, and kohlrabi but more from spinach and beans. The products of open-kettle canning contain less volatile sulfur compounds than the fresh vegetable since the sulfides are rapidly vaporized from the kettle. During the cooking of carrots, chanterelles (a kind of mushroom sometimes called "brick tops"), and celery enough was volatilized so that none was detected in the cooked food.

The formation of volatile sulfur compounds is probably responsible for the development of the cooked flavor of many vegetables and the over-cooked flavor of some on prolonged boiling. Some vegetables such as onions become increasingly mild with cooking—probably the result of the solution and vaporization of the sulfur compounds, so prominent a part of their flavor. Others such as cabbage become increasingly strong flavored, which indicates that more and more volatile sulfur compounds are synthesized as cooking progresses. Mellowing or ripening of flavor occurs when stews and soups containing vegetables such as onions, garlic, or leek are allowed to stand; such mellowing is probably the result of the dispersion of these substances through the mix.

The origin of these volatile sulfur compounds is believed by some investigators to be in glycosides hydrolyzed by enzymes, or from the elevated temperatures and the release of acids during processing. Little isolation work has been done, and we are still ignorant of the precursors of most of these compounds. Much study is needed for a complete understanding of the role of sulfur compounds in the preparation of fruits and vegetables.

POST HARVEST CHANGES IN FRUIT

It has been recognized for many years that fruit continues to undergo chemical changes after harvest until finally spoilage occurs as it is attacked by fungi, yeasts, or bacteria. Since the eating quality changes with these reactions and the monetary value of the crop depends on it, a few fruits have been studied extensively. Both apples and bananas have been the subject of many investigations, and a considerable body of knowledge has been built up through the years. Pears and cherries have been studied less frequently, while some fruits have scarcely been investigated. Apples, pears, citrus fruits, and bananas are stored for variable lengths of time before they are consumed; and the changes which occur in them are significant economically. Cherries, plums, and berries of all kinds are more perishable and reach market soon after picking. If these fruits are kept for any length of time, they must be either canned or frozen.

The changes of fruit after harvest are numerous and include changes in (1) respiration, (2) water content, (3) carbohydrate, and (4) organic acids and pH.

(1) At the time of harvest the respiration rate of apples and pears has sunk to a low level. However, soon after picking, the uptake of oxygen and the production of carbon dioxide begin to speed up until finally a climax is reached, called the *climacteric*. It is followed by a steady decrease in respiratory rate, often called *senescence*. Pearson and Robertson[54] have studied the respiration of Granny Smith apples in Australia and have found that apples left on the trees for longer than 250 days past petal fall

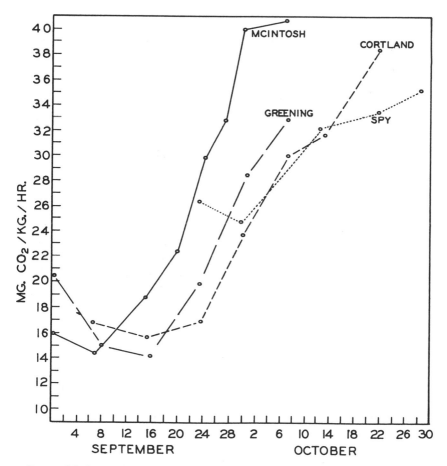

FIGURE 7.7. RESPIRATION RATE OF FOUR VARIETIES OF APPLES. Picked at different dates in 1944 at 74° F (23° C). Each point on each curve is the respiration rate made 24 hours after each harvest date. *Courtesy of R. M. Smock.*

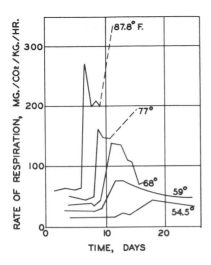

FIGURE 7.8. RATE OF PRODUCTION OF CAR-
BON DIOXIDE BY BANANAS (INITIALLY UNRIPE)
AT DIFFERENT TEMPERATURES. The curves at
77° F and 87.8° F are terminated by mold
growth indicated by the broken lines. Reproduced from Von Loesecke, H., "Bananas," Interscience Publishers, Inc., New York, 1950
from data of Gane, R., *New Phytologist*, **35**, 383
(1936).

(normal harvest is at 170 days) show a climacteric similar to picked fruit. Bananas are normally picked green, and they show a climacteric as they ripen. Tomatoes and possibly other fruit also show this surprising change in respiration.

(2) When fruit is picked and severed from the plant, water no longer flows into the fruit although the loss of water continues. Usually water cannot be taken in through the skin. In apple storage one problem is to prevent water loss so that the fruit does not wither and decrease in value. In dry atmospheres, and particularly at high temperatures, water loss is rapid. Apples rapidly cooled after delivery to the storage barn have a much smaller water loss than those cooled slowly. During the ripening period of bananas, the pulp increases in water content and the peel decreases. Water loss is checked in bananas (and probably other fruit) by the waxy layer of the skin. Golden Delicious apples, which are very susceptible to water loss, have a thin skin and, what is more important, this skin is covered with pits and fissures through which water evaporates rapidly.

(3) Many changes occur in the carbohydrate fraction of fruit during ripening, during the climacteric, and during senescence. See Figure 7.9. Green fruit usually contains an abundance of starch, but is short on the soluble sugars that give ripe fruit its sweetness. On ripening, however, starches decrease and sugars increase in concentration. Since these changes have often been observed, it has been assumed that the sugars are produced at the expense of the starch. However Hulme[26] has found that in Bramley's Seedling and Early Victoria apples the loss of starch does not parallel the increase in sucrose or reducing sugars. Also important in this

context is the work of Barnell[3] on ripening bananas: he records a steady loss of starch; whereas his data for sugars show an increase in some cases but a decrease in others. See Table 7.11. These data indicate that the relation between starch and sugars in fruit is complex. In apples and pears sucrose, glucose, and fructose are the important sugars, with fructose the most abundant. See Table 7.12. The source of these sugars and their role in metabolism during the climacteric have been the subject of some speculation, but as yet the course of the reactions has not been established.

One of the most obvious changes in fruit is the alteration in texture. Since pectic substances are present in the cell walls, they have been the subject of many investigations, so that we have a fairly clear picture of the changes in these substances after harvest. Apples in storage slowly soften, and the rate depends on the temperature of the storage barn as well as the variety of the apple. The graph (Figure 7.10) shows the course of softening and the close correlation with protopectin and soluble pectin content. Protopectin falls and soluble pectin rises until late in the storage period when a reversal of this trend occurs. A decrease in chain length as well as

FIGURE 7.9. DISAPPEARANCE OF STARCH ON RIPENING OF APPLES. Photograph taken on August 27, 1958 of apples stained with iodine. From left: green Jonathan, ripening McIntosh, ripe Wealthy. *Courtesy of Kalamazoo Gazette.*

TABLE 7.11. CARBOHYDRATE COMPOSITION OF BANANAS DURING "EATING RIPE" STAGE EXPRESSED AS PERCENTAGE OF FRESH PULP[a]

Days from Harvest	Starch	Total Sugars	Sucrose	Glycosidic Glucose
Heavy 3/4 Full. Ripened at Tropical Temperature				
7	1.56	13.450	3.660	0.344
9	0.654	12.770	1.700	0.227
11	0.357	12.600	0.990	0.209
Heavy 3/4 Full. 14 Days at 53°F, Ripened at 68°F				
19	7.52	6.586	1.888	0.241
21	3.67	10.210	2.241	0.630
24	1.62[b]	11.245[b]	2.356[b]	0.656[b]
Standard 3/4 Full. 14 Days at 53°F, Ripened at 68°F				
20	5.25	11.610	3.755	0.307
22	2.89	12.720	3.890	0.208
24	1.61[b]	12.895[b]	2.220[b]	0.235[b]

[a] From Barnell, H. R. "Studies in Tropical Fruit XIII Carbohydrate Metabolism of the Banana Fruit During Storage at 53°F and Ripening at 68°F," *Ann. Botany*, **5**, 608 (1941).
[b] Values thus marked are for fruit which would be judged by the eye to be slightly overripe.

loss of methyl groups probably occurs during the softening period and accounts for the rise in soluble pectin.

Pears are picked in the hard stage and held at low temperatures until required for ripening. On return to room temperature they rapidly ripen and soften. If pears are held too long at low temperatures, they develop "sleepiness" and instead of ripening satisfactorily they turn brown as they soften. In pears as in apples there is a change in the pectic substances during softening with a rapid drop in the amount of protopectin and a rise in soluble pectin. See Figure 7.11. During cold storage the 'amount of protopectin increases. Bartlett pears that develop "sleepiness" do not show the rise in soluble pectins on exposure to ripening temperatures during January or February. It is believed that during cold storage inactivation of pectic

TABLE 7.12. SUGARS IN APPLES AND PEARS*

	Apple Juice	Pear Juice (26 Varieties)
Sucrose	6.6– 56.6 g/L	1– 24 g/L
Glucose	12.3– 58.0	5– 35
Fructose	69.2–113.8	65–112

*Data from Tavernier and Jacquin, reported by A. C. Hulme, "Some Aspects of the Biochemistry of Apple and Pear Fruits," *Advances in Food Research* **8**, 297–413 (1958).

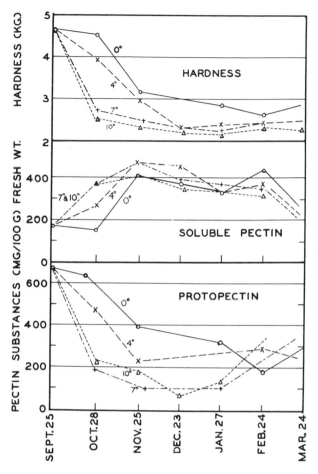

FIGURE 7.10. CHANGES IN HARDNESS, SOLUBLE PECTIN, AND PROTOPECTIN. As shown in fruit for Canada apples during storage at various temperatures. Adapted by A. C. Hulme from Ulrich.

enzymes slowly occurs and normal hydrolysis of protopectin does not occur.

Bananas undergo loss of protopectin and increase in the smaller molecules which make up soluble pectins during ripening. Von Loesecke[84] has presented data (Table 7.13) for various varieties of bananas.

Loss of protopectin and rise of soluble pectins have been found in peaches, plums, and tomatoes as they ripen. Raspberries do not show a correlation between the pectic substance fractions and the development of mushiness in the fruit as it passes its peak of ripeness.

Changes in celluloses, hemicelluloses, and lignins have been followed in a small number of fruits and with a few varieties. Jermyn and Isherwood[28] have found that during ripening Conference pears show a rapid rise in

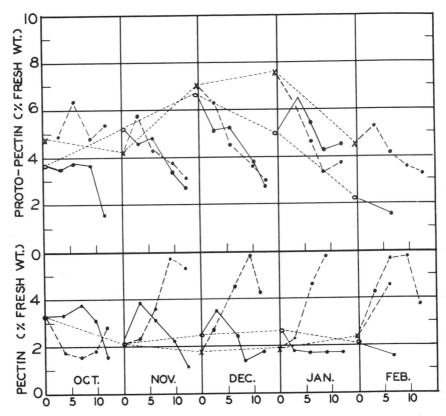

FIGURE 7.11. PECTIC CHANGES IN BARTLETT (o) AND ANJOU (x) PEARS. Stored at
-1.11° C to -.56° C and removed at monthly intervals for ripening at 20° C to 21.1° C
Dotted lines show changes during cold storage while interrupted lines (Anjou) and continuous
lines (Bartlett) show changes occurring in the ripening room. Adapted from Date and
Hansen, *Proc. Indian Acad. Sci.,* **B39,** 17 (1954) by A. C. Hulme, *loc. cit.*

xylans and arabans and a decrease in cellulose. Pears contain lignins in the
stone cells and are unlike other fruits in this respect. Hulme does not consider a few studies on apples definitive for hemicellulose. The alcohol insoluble fraction of apples is partially soluble in dilute acid and contains
along with pectic substance a so-called "hemicellulose fraction." This fraction slowly decreases during storage.

(4) Organic acids decrease in apple pulp and in pears during storage.
Both of these fruits contain a large number of organic acids in low con-
centration, a small amount of citric acid, and larger amounts of malic acid.
With the development of the techniques of chromatography, it is now pos-

TABLE 7.13. CHANGES IN PECTIN AND PROTOPECTIN IN PULP OF BANANAS*
(Per Cent of Fresh Pulp)

Variety		\multicolumn{6}{c}{Days in Ripening Room}					
		0	3	5	7	9	11
Gros Michel	Pectin	—	0.27	0.36	0.34	0.37	0.40
	Protopectin	0.53	0.56	0.31	0.34	0.21	0.22
Lady Finger	Pectin	0.21	0.48	0.52	0.57	0.58	0.68
	Protopectin	0.50	0.76	0.27	0.29	0.31	0.35

*Unpublished data from Manion, J. T., United Fruit Co., Research Department, 1933 in von Loesecke, H. W., "Bananas: Chemistry, Physiology, Technology," 2nd ed. Interscience Publ., Inc., New York, N. Y., 1950.

sible to separate and identify compounds present in low concentration which were extremely difficult to find by other classical methods. The technique has not been widely utilized to follow the changes in concentration of various acids, but some information has been obtained. Quinic acid and shikimic acid are evidently much more widely distributed in plant tissues than has previously been recognized. Hulme[26] reports that in storage at 15°C Bramley's Seedling apples show changes in total acids, with a different pattern for different acids in pulp or peel. (Table 7.14) The changes in organic acids during storage should be of considerable interest, since many organic acids are related to metabolic processes. The Krebs cycle for the metabolism of two carbon fragments has not been demonstrated completely in apples and pears as yet, but the possibility of its importance is

TABLE 7.14. CHANGES IN ORGANIC ACIDS IN BRAMLEY'S
SEEDLING APPLES STORED AT 75°C*

	15 Days mg/100 g	40 Days mg/100 g	100 Days mg/100 g
Pulp			
Citric	6–8.5 fresh tissue	—	10
Quinic	45	80	50
Shikimic	—	—	1–2
Peel			
Citric	1–2	—	1–2
Quinic	400	400	—
Shikimic	5	—	8
Atramalic	—	10 at 25 days	25

*From Hulme, A. C., "Some Aspects of the Biochemistry of Apple and Pear Fruits," *Advances in Food Research*, **8**, 329 (1958).

great since it is used by many forms of life as a metabolic pathway. The function of acids such as quinic and shikimic is as yet unknown, but must surely have some role in metabolism.

REFERENCES

1. ARONOFF, S. and ARONOFF, E. M., "Thermal Degradation of Dehydrated Beets," *Food Research*, **13**, 59–66 (1948).
2. ARTHUR, J. C. JR. and McLEMORE, T. A., "Properties of Polyphenolases Causing Discoloration During Processing of Sweet Potatoes," *J. Agr. Food Chem.*, **4**, 553–555 (1956).
3. BARNELL, H. R., "Studies in Tropical Fruit XIII Carbohydrate Metabolism of the Banana Fruit During Storage at 53° F and Ripening at 68° F," *Ann. Botany*, **5**, 608–646 (1941).
4. BATE-SMITH, E. C., "Common Phenolic Constituents of Plants and Their Systematic Distribution," *Proc. Roy. Dublin Soc.*, **27**, 165–176 (1956); *C. A.*, **51**, 4514e (1957).
5. BATE-SMITH, E. C., "Leucoanthocyanins," *Biochem. J.*, **58**, 122–125 (1954).
6. BATE-SMITH, E. C., "Flavonoid Compounds in Foods," *Advances in Food Research*, **5**, 261–295 (1954).
7. BATE-SMITH, E. C. and SWAIN, T., "'Leucoanthocyanins, the 'Tannins' in Foods," *Chem. and Ind.*, 1953, 377–378.
8. BENNETT, E., "On the Nature of Cellulose from Cranberry Pulp," *Food Research*, **21**, 207–208 (1956).
9. BENNETT-CLARK, T. A., "Organic Acids of Plants," *Ann. Rev. Biochem.*, **18**, 639–655 (1949).
10. BLAIR, J. S. and AYRES, T. B., "Protection of Natural Green Pigment in the Canning of Peas," *Ind. Eng. Chem.*, **35**, 85–95 (1943).
11. BLAKE, M. A. and DAVIDSON, O. W., "Tannins and Acidity in Peaches," *Proc. Am. Soc. Hort. Sci.*, **39**, 201–204 (1941).
12. BRAVERMAN, J. B. S. and LIFSHITZ, A., "Pectin Hydrolysis in Certain Fruits during Alcoholic Fermentation," *Food Technol.*, **11**, 356–358 (1957).
13. BUCH, M. L., "Bibliography of Organic Acids in Higher Plants," U. S. Dept. Agriculture, Agriculture Research Service, Oct. 1957. Reissue Mar. 1960.
14. BUTLAND, P., "The Darkening of Maraschino and Glacé Cherries," *Food Technol.*, **6**, 208–209 (1952).
15. CRAFTS, A. S., "Cellular Changes in Fruits and Vegetables," *Food Research*, **9**, 442–452 (1944).
16. DE EDS, F. and COUCH, J. F., "Rutin in Green Asparagus," *Food Research*, **13**, 378–380 (1948).
17. FREEMAN, M. E. and RITCHIE, W. S., "Pectins and the Texture of Cooked Potatoes," *Food Research*, **5**, 167–175 (1940).
18. GOULD, I. A., "The Color of Coffee—Evaporated Milk Mixtures," *Mich. Agr. Expt. Station Quart. Bull.*, **26**, No. 4 (1944).
19. GRIFFIN, J. H. and KERETSZ, Z. I., "Changes Which Occur in Apple Tissues upon Treatment with Various Agents and their Relation to the Natural Mechanism of Softening during Maturation," *Botan. Gaz.*, **108**, 279–85 (1946).

20. GUADAGNI, D. G., SORBER, D. G., and WILBUR, J. S., "Enzymatic Oxidation of Phenolic Compounds in Frozen Peaches," *Food Technol.*, **3**, 359–364 (1949).
21. HASSELSTROM, T., HEWITT, E. J., KONIGSBACHER, K. S., and RITTER, J. J., "The Composition of the Volatile Oil of Black Pepper, Piper nigrum," *J. Agr. Food Chem.*, **5**, 53–55 (1957).
22. HENZE, R. E., "Inhibition of Enzymatic Browning of Chlorogenic Acid Solutions with Cysteine and Glutathione," *Science*, **123**, 1174–1175 (1956).
23. HERMANN, K., "Tanninlike Substances I Enzymatic Oxidation of Various Phenols," *Pharmazie*, **9**, 212–220 (1954); *C. A.*, **50**, 15748f (1956).
24. HEULIN, F. E. and GALLOP, R. A., "Studies in the Natural Coating of Apples I Preparation and Properties of Fractions," *Australian J. Sci. Research* B4, 526–532 (1951); *C. A.*, **46**, 3177g (1952).
25. HOLMES, A. D., SPELMAN, A. F., and WETHERBEE, R. T., "Comparison of Light vs. Darkness for Storing Butternut Squashes," *Food Technol.*, **3**, 269–271 (1949).
26. HULME, A. C., "Some Aspect of the Biochemistry of Apple and Pear Fruits," *Advances in Food Research*, **8**, 297–413 (1958).
27. ISLER, O., OFNER, A., and SIEMERS, G. F., "Synthetic Carotenoids for Use as Food Colors," *Food Technol.*, **12**, 520–526 (1958).
28. JERMYN, M. A. and ISHERWOOD, F. A., "Changes in the Cell Wall of the Pear During Ripening," *Biochem. J.*, **64**, 123–132 (1956).
29. JOHNSON, G., MAYER, M. M., and JOHNSON, D. K., "Isolation and Characterization of Peach Tannins," *Food Research*, **16**, 169–180 (1951).
30. JOSLYN, M. A., "The Use of Liquid Sugars in Freezing Apricots, Peaches and Nectarines," *Food Technol.*, **3**, 8–14 (1949).
31. JOSLYN, M. A. and PETERSON, R. G., "Reddening of White Bulb Onions," *J. Agr. Food Chem.*, **6**, 754–765 (1958).
32. JOSLYN, M. A. and PONTING, J. D., "Enzyme-catalyzed Oxidative Browning of Fruit Products," *Advances in Food Research*, **3**, 1–37 (1951).
33. KARIYONE, T. and HASHIMOTO, Y., "Detection and Colorimetry of Flavanol Derivatives," *J. Pharm. Soc. Japan*, **71**, 433–436 (1951); *C. A.*, **45**, 9799a (1951).
34. KARRER, P. and JUCKER, E., "The Carotenoids," D. Van Nostrand Co., Inc., Princeton, N. J., 1950.
35. KELLY, L., BENNETT, F. B., RAFFERTY, J. P., and PAUL, P., "The Nutritive Value of Canned Foods V Changes in Carotene Content of Vegetables Prior to Canning," *Food Technol.*, **4**, 269–272 (1950).
36. KERTESZ, Z. I., "The Pectic Substances," Interscience Publishers, Inc., New York, N. Y., 1951.
37. KERTESZ, Z. I., TOLMAN, T. G., LOCONTI, J. D., and RUYLE, E. H., New York State Expt. Station Bull. 252, 1940.
38. KIRCHNER, J. G., "The Chemistry of Fruit and Vegetable Flavors," *Advances in Food Research*, **2**, 259–289 (1949).
39. KNOWLTON, H., "Green Color in Coffee Icings," *J. Home Econ.*, **30**, 26–27 (1938).
40. KOEHN, C. J., "Relative Biological Activity of Beta Carotene and Vitamin A," *Arch. Biochem.*, **17**, 337–343 (1948).

41. KURODA, C. and UMEDA, M., "The Pigments and Related Compounds in the Onion," *J. Sci. Research Inst.*, (*Tokyo*), **45**, 17–22 (1951); *C. A.*, **45**, 6698i (1951).
42. LAWRENCE, J. M. and GROVES, K., "Determination of a Soluble Pectin in Apples," *J. Agr. Food Chem.*, **2**, 882–885 (1954).
43. LAWRENCE, W. J. C., PRICE, J. R., ROBINSON, G. M., and ROBINSON, R., "A Survey of Anthocyanins," *Biochem. J.*, **32**, 1661–1667 (1938); "Distribution of the Anthocyanins in Flowers, Fruits and Leaves," *Trans. Roy. Soc.* (*London*), **B230**, 149–178 (1939).
44. MACGIBBON, C. H. and HALLIDAY, E. G., "Color Changes in Large Quantity Cooking and Service of Green Vegetables, *J. Home Econ.*, **29**, 40–44 (1937).
45. MACKINNEY, G. and WEAST, C. A., "Color Changes in Green Vegetables," *Ind. Eng. Chem.*, **32**, 392–395 (1940).
46. MANN, L. K. and WEIER, T. E., "Fruit Products," *J. Am. Food Manuf.*, **23**, 309–311 (1944).
47. MASON, H. S., "Comparative Biochemistry of Phenolase Complex," *Advances in Enzymol.*, **16**, 105–184 (1955).
48. MAYER, F., "The Chemistry of Natural Coloring Matters," *Ann. Rev. Biochem.*, **21**, 472–492 (1952).
49. MOSIMAN, G., "The Question of the Uniform Designation of the Coloring Matter of Cacao: 'Cacao-red' or 'Cacao-purple'?" *Intern. Chocolate Rev.*, **1**, 111–113 (1947); *C. A.*, **44**, 9585g (1950).
50. NEBESKY, E. A., ESSELEN, W. B. JR., MCCONNELL, J. E. W., and FELLERS, C. R., "Stability of Color in Fruit Juices," *Food Research*, **14**, 261–274 (1949).
51. NIEGISCH, W. D. and STAHL, W. H., "The Onion: Gaseous Emanation Products," *Food Research*, **21**, 657–665 (1956).
52. OKAJIMA, J., "Constituents of the Pigmented Outer Skin of Onion Bulb., I Ceryl Cerotate from the Waxy Matter of the Skin," *J. Sci. Research Inst.* (*Tokyo*), **48**, 281–283 (1954); *C. A.*, **50**, 5093f (1956).
53. ONSLOW, M. W., "The Browning of Fruit," *Biochem. J.*, **14**, 535–540, 541–547 (1920); **15**, 107–117 (1921).
54. PEARSON, J. A. and ROBERTSON, R. N., "The Physiology of Growth in Apples," *Australian J. Biol. Sci.*, **6**, 1–20 (1953).
55. POLITIS, J., "A New Method for the Microchemical Localization of Chlorogenic Acid and Tannins in Plants," *Compt. Rend.*, **225**, 954–956 (1947); *C. A.*, **42**, 2304a (1948).
56. PONTING, J. D. and JOSLYN, M. A., "Ascorbic Acid Oxidation and Browning in Apple Tissue Extracts," *Arch. Biochem.*, **19**, 47–63 (1948).
57. POSTLMAYR, H. L., LUH, B. S., and LEONARD, S. J., "Characterization of Pectin Changes in Freestone and Clingstone Peaches During Ripening and Processing," *Food Technol.*, **10**, 618–625 (1956).
58. PRICE, J. R. and STURGESS, V. C., "A Survey of Anthocyanins," *Biochem. J.*, **32**, 1658–1660 (1938).
59. REEVE, R. M., "Histological Investigation of Texture in Apples, II Structure and Intercellular Spaces," *Food Research*, **18**, 604–617 (1953).
60. REEVE, R. M., "Histological Observations on the Seed Coats of Succulent Peas," *Food Research*, **12**, 10–23 (1947); *ibid.*, **14**, 77–89 (1949).

61. REEVE, R. M., "Histological Survey of Conditions Influencing Texture in Potatoes, I Effects of Heat Treatments on Structure," *Food Research*, **19**, 323–332 (1954); "II Observations on Starch in Treated Cells," *ibid.*, **19**, 333–339 (1954).

62. REEVE, R. M. and LEINBACH, L. R., "Histological Investigation of Texture in Apples, I Composition and Influence of Heat on Structure," *Food Research*, **18**, 592–603 (1953).

63. RENIS, H. E. and HENZE, R. E., "Cysteine Derivatives in Mature Onion Bulbs," *Food Research*, **23**, 345–350 (1958).

64. ROBINSON, G. M., "Notes on Variable Color in Flower Petals," *J. Am. Chem. Soc.*, **61**, 1606–1607 (1939).

65. ROBINSON, G. M. and ROBINSON, R., "The Colloidal Chemistry of Leaf and Flower Pigments and the Precursors of the Anthocyanins, *J. Am. Chem. Soc.*, **61**, 1605–1606 (1939).

66. SANDO, C. E., "The Coloring Matter of Grimes Golden, Jonathan and Stayman Winesap Apples," *J. Biol. Chem.*, **117**, 45–56 (1937).

67. SCOTT, F. M., "Internal Suberization of Tissues," *Botan. Gaz.*, **111**, 378–394 (1950).

68. SESHARDI, T. R., "Biochemistry of Natural Pigments," *Ann. Rev. Biochem.*, **20**, 487–512 (1951).

69. SHIROYA, M. and HATTORI, "Browning and Blackening of Plant Tissues," *Physiol. Plantarum*, **8**, 358–369 (1955); *C. A.*, **50**, 7966a (1956).

70. SIMPSON, J. I. and HALLIDAY, E. G., "Chemical and Histological Studies of the Disintegration of Cell Membrane Materials in Vegetables During Cooking," *Food Research*, **6**, 189–206 (1941); *ibid.*, **7**, 300–305 (1942).

71. SIMPSON, J. I. and HALLIDAY, E. G., "The Behaviour of Sulfur Compounds in Cooking Vegetables," *J. Home Econ.*, **20**, 121–126 (1928).

72. SMOCK, R. M. and NEUBERT, A. M., "Apples and Apple Products," Interscience Publishers, Inc., New York, N. Y., 1950.

73. STADTMAN, E. R., "Nonenzymatic Browning in Fruit Products," *Advances in Food Research*, **1**, 325–372 (1948).

74. STANSBURY, M. F., FIELD, E. T., and GUTHRIE, J. D., "Tannins and Related Pigments in the Red Skins (Testa) of Peanuts," *J. Am. Oil Chemists' Soc.*, **27**, 317–321 (1950).

75. STRAIN, H. H., "Chlorophyll Reactions: Oxidation and Isomerization of Chlorophylls in Killed Leaves," *J. Agr. Food Chem.*, **2**, 1221–1226 (1954).

76. STRAIN, H. H. and MANNING, W. M., "Isomerization of Chlorophylls a and b," *J. Biol. Chem.*, **146**, 275–276 (1942); **148**, 655–656 (1943); **151**, 1ff. (1943).

77. STOLL, A. and SEEBECK, E., "Concerning Alliine, the Precursor of Garlic Oil," *Helv. Chim. Acta*, **31**, 189–210 (1948).

78. STRODTZ, N. H., BLUMER, T. E., and CLIFCORN, L. E., "Retention of Carotene During the Canning of Tomato Juice," *Food Technol.*, **6**, 299–301 (1952).

79. SYNGE, R. L. M. and WOOD, J. C., "(+)-(S-Methyl-L-Cysteine S-Oxide) in Cabbage," *Biochem. J.*, **64**, 252–259 (1956).

80. TING, S. V., "Enzymatic Hydrolysis of Naringen in Grapefruit," *J. Agr. Food Chem.*, **6**, 546–549 (1958).

81. TRESSLER, D. K., JOSLYN, M. A., and MARSH, G. L., "Fruit and Vegetable Juices," AVI Publishing Co., New York, N. Y., 1939.

82. VALAER, P., "Alcoholic Beverages" in Jacobs, M. B., ed., "Food and Food Products," 2nd ed., Interscience Publishers, Inc., New York, N. Y., 1951.

83. VAN BLARICOM, L. O. and MARTIN, J. O., "Retarding the Loss of Red Color in Cayenne Pepper with Oil Antioxidants," *Food Technol.*, **5**, 337–339 (1951).

84. VON LOESECKE, H. A., "Bananas: Chemistry, Physiology, Technology," 2nd ed., Interscience Publishers, Inc., New York, N. Y., 1950.

85. WALLERSTEIN, J. S., BERGMANN, L., BYER, A., ALBA, R. T., and SCHADE, A. L., "Determination of Enzymatic Discoloration in White Potatoes," *Food Research*, **12**, 44–54 (1947).

86. WEIER, T. E., "Carotene Degradation in Dehydrated Carrots," *Am. J. Botany*, **31**, 342–346 (1944).

87. WEIER, T. E. and STOCKING, C. R., "Histological Changes Induced in Fruits and Vegetables by Processing," *Advances in Food Research*, **2**, 297–342 (1949).

88. WEIER, T. E. and STOCKING, C. R., "The Stability of Carotene in Dehydrated Carrots Impregnated with Antioxidants," *Science*, **104**, 437–438 (1946).

89. WEURMAN, C. and SWAIN, T., "Chlorogenic Acid and Enzymatic Browning in Apples and Pears," *Nature*, **172**, 678 (1953).

90. WHISTLER, R. L. and THORNBERG, W. L., "The Development of Starch Granules in Corn Endosperm," *J. Agr. Food Chem.*, **5**, 203–207 (1957).

91. WHITTENBERGER, R. T. and NUTTING, G. C., "Observations on Sloughing of Potatoes," *Food Research*, **15**, 331–339 (1950).

92. WILSON, C. W., "Study of Boric Acid Color Reaction of Flavone Derivatives," *J. Am. Chem. Soc.*, **61**, 2303–2306 (1939).

93. YANG, H. Y. and STEELE, W. F., "Removal of Excess Anthocyanins by Enzymes," *Food Technol.*, **12**, 517–519 (1958).

CHAPTER EIGHT

Milk and Milk Products

MILK AND MILK PRODUCTS have formed an important part of the diet of Western man since the dim reaches of history. It is not known when animals were first domesticated but soon after man became a farmer, he began to keep and raise animals. All mammals produce milk after the birth of the young and man has used the milk of many animals for his own food. The cow is, of course, the most important of all these animals as a supplier of food for man, but yak, reindeer, buffalo, or goat milk is more important in some parts of the world.

For many generations numerous families kept their own cow and the milk was supplied fresh from the udder, morning and night. However, with the rise of cities and the growth of specialties, the dairy industry developed. Both governments and dairies attempted to develop checks on the purity of milk, both from the chemical and bacteriological standpoints. The United States has extensive federal regulations on the milk and milk products that pass in interstate commerce, while each state has many laws for the regulation of their production and sale within its boundaries.

The interest in obtaining and producing high grade milk and milk products has resulted in a large number of studies of the composition of milk. The studies for the most part have been slanted to answering the problems of control, but nevertheless much general information about milk and milk products has accumulated.

In the United States when the term "milk" is used, it always refers to the milk of cows. If the milk of another species is intended, the name of that species precedes "milk"; i.e., "human milk" (sometimes "woman's milk"), "goat milk," etc.

The Food and Drug Administration of the U. S. Department of Agriculture defined milk in August, 1926. "Milk is the whole fresh, clean, lacteal secretion obtained by the complete milking of one or more healthy cows, properly fed and kept, excluding that obtained 15 days before and 5 days after calving, or such longer period as may be necessary to render the milk practically colostrum free."

FORMATION OF MILK

The production of milk by the mammary gland of any female mammal depends on three phases; (a) the development of the mammary gland, (b) secretion of milk, and (c) emptying of the gland. After conception and the implantation of the embryo into the uterine wall, a series of changes begin in the mammary glands. These are under the influence of hormones elaborated by the ovary and the corpus luteum. The mammary glands be-, gin to proliferate; the epithelium of the lobes increases rapidly and the secretory cells develop. Toward the end of pregnancy these glands begin to form a secretion. When parturition occurs, the influence of the placental hormones is removed from the pituitary, permitting it to form a hormone, prolactin, which causes secretion in the prepared mammary gland.

The first secretion of the gland differs from milk and is called *colostrum,* the composition of which is compared below with that of milk. It is a relatively thick fluid during the first days postpartum, gradually changing in appearance and composition to those of milk. It differs from milk chiefly in its protein content as well as some of the vitamins and amino acids. Rather recent work with cow's colostrum gives the following composition compared with normal milk. [Quoted from Sutton's report in Macy, I. G., Kelly, H. J., and Sloan, R. E., "Compositions of Milk," *Natl. Acad. Sci. Natl. Research Council, Publ.* (1953)].

Per Cent of Total Nitrogen

	Casein	Albumin	Globulin	Nonprotein N
Colostrum	31	6	55	8
Mature Milk	71	8	18	3

In a summary of the work to 1953 on the vitamin composition of the two fluids, Sutton points out that vitamin A activity is much higher in colostrum than in milk. The concentration is 10 to 15 times higher during the first few days of secretion than the average for milk. Likewise, riboflavin is present at a 3 to 3.5 fold concentration. Colostrum is higher in thiamine and biotin, about the same in niacin, and lower in pantothenic acid. It also contains more of the amino acid, tryptophan, than normal milk.

In women it is common to consider the period of colostrum formation the first five days postpartum; of transitional milk formation, days 5 to 10; and mature milk formation after the tenth day. Actually, evidence indicates that the period of colostrum formation differs with individuals and varies from 1 to 5 days. Cows and goats have a shorter period of colostrum formation.

Once the secretion becomes characteristic of that of milk, its composition remains relatively constant. It continues to flow for some months in most species, if the gland is emptied regularly. If the milk is not removed, it disappears, and the breast tissue undergoes involution and returns to the pre-pregnancy state.

COMPOSITION OF MILK

Milk is a complex mixture of lipids, carbohydrates, proteins, and many other organic compounds and inorganic salts dissolved or dispersed in water. Macy, Kelly, and Sloan[7] list over 100 compounds in it. Some of these compounds such as the carbohydrate, lactose, and most of the salts and vitamins are soluble in water. Others such as the lipids, proteins, and dicalcium phosphate are dispersed through the water in the colloidal or near colloidal state.

The lipid is composed primarily of fat although there are also small amounts of phospholipids, sterols, the fat soluble vitamins A and D, carotenes and xanthophyll. The proteins (see p. 140) are numerous and are traditionally classed in the following fractions which do not consist of pure proteins: (1) casein precipitated by rennin or acid, (2) lactalbumin, and (3) lactoglobulin precipitated by heating. The carbohydrate of milk, whether from cow or other species, is lactose and it is the only carbohydrate present. The ash of milk contains all of the minerals which occur there, but not necessarily in the same form. The salts are not only those of inorganic acids but of some organic ones such as citric acid as well. Part of the phosphate occurs as phosphoprotein and phospholipid, part combined with calcium. It is the ash of milk which has been most extensively studied rather than the minerals in the milk itself. Calcium phosphate is known to be colloidally dispersed in the water (it has a very low solubility in water), and it is known that milk contains potassium, sodium, magnesium, and chloride as well as small amounts of copper, iron, zinc, manganese, aluminum, and iodide. Sulfur is present in some of the amino acids of the proteins.

The vitamins are plentiful in milk. Only ascorbic acid is present in limited amounts, and it may even be absent from pasteurized milk. But the other water and fat soluble vitamins are there in goodly quantity. The B complex vitamins are dissolved in the water and vitamins A and D in the fat. Winter milk is not a rich source of vitamin D unless the diet of the cow is enriched with it.

Milk contains a number of enzymes, some of them apparently secreted in the milk and some of them formed by the microorganisms which inhabit

TABLE 8.1. COMPOSITION OF MILK

	Range (%)	Average (%)
Water	82.0–90.0	87.3
Fat	2.3– 7.8	3.67
Total Protein	2.0– 4.5	3.42
Lactose	3.5– 6.0	4.78
Ash	0.6– 0.9	0.73
Solids	10.0–18.0	12.69
Solids-not-fat	7.5–10.6	8.77

From Jacobs, M. B., "Milk, Cream, and Dairy Products," in Jacobs, M.' B., ed., "The Chemistry and Technology of Food and Food Products," 2nd ed., Interscience Publishers, New York, N. Y., 1951.

the milk. Amylase, lipase, tryptase, peroxidase, catalase, and reductase have all been demonstrated. Galactase, lactase, and aldehydase have been reported. Some of the enzymes are destroyed or reduced on pasteurization, and tests for their presence or level may be used as a test for effective pasteurization.

The milk sold on the consumer market is pooled milk from many herds and consequently is relatively constant in composition. The ranges in composition are given in Table 8.1. Variations in the composition of milk occur with breed of cow, time of year, time of day, portion of the milking, time in the lacteal cycle, and the nutritional status of the cow.

Some breeds of cows give milk which is always higher in butterfat than that of others. See Table 8.2. Thus Jerseys and Guernseys are noted for the high percentage of fat. The protein of their milk is likewise a little higher than most other breeds, and the percentage of water is consequently lower.

The time of year has an effect on the composition of the milk although this may be indirect, since the diet of the herd and the amount of green pasture consumed varies with the time of year. Jacobs[4] presents tables

TABLE 8.2. VARIATION IN COMPOSITION DUE TO BREED

Breed	No. of Cows	Fat %	Protein %	Lactose %	Ash %	Total solids %	Water %	Solids-Not-Fat %
Jersey	29,495	5.37	3.92	4.93	0.71	14.91	85.09	9.54
Guernsey	32.562	4.95	3.91	4.93	0.74	14.61	85.39	9.66
Brown Swiss	721	4.01	3.61	5.04	0.73	13.41	86.59	9.40
Ayrshire	6,999	4.00	3.58	4.67	0.68	12.90	87.10	8.90
Shorthorn	6,155	3.94	3.32	4.99	0.70	12.81	87.19	8.87
Holstein	37,598	3.40	3.32	4.87	0.68	12.26	87.74	8.86

From Jacobs, M. B., *loc. sit.*, p. 848.

from Gamble, J. A., Ellis, N. R. and Besley, A. K., *U. S. Dept. Agr. Tech. Bull.*, **671**, (1939), for both Holstein and Jersey cows which show variations in the amount of fat, protein, lactose, ash, iron, copper, calcium, and phosphorus. The variations are not too regular. Protein in Holstein milk for January through December shows the following monthly values: 3.19, 3.09, 3.08, 3.16, 3.14, 3.10, 3.06, 3.19, 3.18, 3.30, 3.34, 3.17. Woodman[8] says that the poorest milk is produced in the spring and early summer when green pasture is bountiful and the richest in autumn and winter. See Table 8.3. However, the table from Gamble, *et al.* that was previously mentioned does not bear out this trend. The fat for Jersey milk from February through December showed the following monthly values: 5.1, 5.4, 5.5, 5.4, 5.5, 5.4, 5.2, 5.0, 5.5, 5.2, 5.5 while for Holstein: 3.8, 3.7, 3.7, 3.8, 3.2, 3.4, 3.5, 3.2, 3.5, 3.6, 3.5. September is the low month for both breeds, but there is no regular pattern through the year.

The composition of milk varies with the time of day and portion of the milking. The evening milk tends to contain more butterfat and slightly less water than morning milk. Likewise the milk which is first removed from the udder contains a smaller amount of fat than the "strippings"—that removed in the final phase of milking.

The time in the lactational cycle, i.e., the time after calving also has an influence on the composition of the milk. For example Duncan, Watson, Dunn and Ely[3] found that the threonine level in milk proteins is 15 per cent higher at the end of the lactational period than 60 days after birth of the calf.

The diet of cows has a very marked influence on the quantity of milk produced, but only a limited effect on the composition. Diets have been widely studied, but complete knowledge of all of the factors that influence milk production is still in the future. Any dietary deficiencies of protein, total energy, etc. are immediately reflected in a decrease in the total volume of milk formed. It is true that the vitamin D and carotene content of the milk reflect the level in the cow's diet, but for other components the quan-

TABLE 8.3. VARIATION IN MILK WITH SEASON

Time of Year	Total Solids %	Fat %	Solids-Not-Fat %
November–January	13.04	4.11	8.93
February–April	12.72	3.88	8.84
May–August	12.66	3.89	8.77
October–November	13.03	4.25	8.78

From Woodman, A. G.[8] (Data of Richmond).

tity present in the milk is relatively constant, and it is the total secretion of all components which is influenced by a lowered intake.

Diet does have an effect on the constituents of the milk responsible for the development of "oxidized flavor." This is a flavor which is variously described as "cardboardy" or "oily" and it may be so marked as to make the milk completely unpalatable. It appears to develop when milk is in contact with oxygen and particularly rapidly if the milk contains small amounts of copper. Milks vary considerably in their tendency to develop this flavor and a fair amount of work has gone into attempts to understand and control it. Various compounds have been implicated as the active agent— phospholipids, ascorbic acid, and the tocopherols—but so far it is not unequivocably known. Krukovsky, et al.[6] have shown that diet has a role. Cows fed roughage from late cut hay, timothy in full bloom, and alfalfa, whether fed as silage or as field cured hay, did not produce a single sample of milk which developed an oxidized flavor. The milk was tested by adding 0.1, 0.5 or 10 mg copper per 1 of milk and holding the samples at $0°-5°C$ for 10 days. However 53 per cent of the samples from cows fed early-cut hay, either silage or dried hay, developed oxidized flavor—some of them very strong.

Milk possesses a fine delicate flavor that results from the blending of many compounds which affect taste and olfactory endings. A flavor different from that of normal milk is objectionable and milk with an unusual flavor is not accepted by the consumer. Josephson[5] discussed flavors in milk in his American Chemical Society Borden Award address and attributes the "sunlight" flavor which is produced when milk is exposed to sunlight to the photolysis of the amino acid, methionine. A "cowy" flavor occurs when acetone bodies appear in the milk. Cows are prone to develop ketosis, a condition in which fatty acids are not completely oxidized to carbon dioxide and water and in which some acetone bodies are formed. The acetone bodies are acetone, β-hydroxybutyric acid, and acetoacetic

CH_3COCH_3	$CH_3CHOHCH_2COOH$	CH_3COCH_2COOH
Acetone	β-Hydroxybutyric Acid	Acetoacetic Acid

acids. When they are produced in an animal, they are carried by the blood and excreted in urine and breath. In a lactating animal they also appear in small amounts in the milk.

The flavor of milk changes when it is heated for any length of time and a "cooked" flavor develops. This is noticeable in evaporated milk even when great precautions are used to avoid overheating. Josephson attributes the "cooked" flavor to heat denaturation of lactalbumin. He attributes the "carmelized" flavor which sometimes develops to changes in lactose with the formation of furans.

Variations in composition are balanced by pooling the milk from a large number of cows. Many dairies produce "standardized milk" which contains approximately the same amount of butterfat throughout the entire year. Usually the standard is that required in the particular area served by the dairy, and it can be met by adding cream or skimmed milk to the pool if this is allowable. The regulations governing the sale of milk are quite stringent. Even when a dairy serves a relatively small area, its product is often subject to local laws. These laws usually have some regulations regarding microorganisms and filth, as well as some regarding butterfat content. Many state laws allow standardization by the addition of either cream or skimmed milk.

The bacteriology of milk is very important. In the udder, milk is relatively free of microorganisms, bacteria, molds and yeasts, but it is readily contaminated by organisms on the outside of the udder, on the hands of the milker or the milking machine, and on the vessels in which it is transported and stored. Not only is milk an excellent food for the nourishment of man, but it is also fine for the nourishment of microorganisms. When they are introduced into milk, they flourish and multiply rapidly. A food with a large population of microorganisms is not only worthless as food but highly dangerous.

CHEMICAL ANALYSIS OF MILK

Butterfat is the most common constituent of milk that is determined. In the past milk has often been bought on the basis of its butterfat content. The practice of paying for the milk solely on the basis of fat, if it comes up to standards as far as bacterial count, is not as common as formerly, but the amount present is still determined. Every little dairy in the country has the simple equipment necessary for the determination of butterfat. The Babcock method is used widely for this determination and although it is not too simple to carry out with accuracy, it is used by many people who have had no training in chemistry. Some states have legal restrictions on the equipment used in the determination, in order to insure accuracy in able hands.

The Babcock method is one which releases the fat from the milk emulsion and measures the percentage directly in bottles calibrated for this. A standard volume of milk, 17.6 ml is treated with 17.5 ml of concentrated sulfuric acid. The acid denatures and partially hydrolyzes the protein, so that it no longer acts as a protective colloid. The fat rises to the top. It is kept liquid by heat from the reaction and by warming the milk-acid mixture, as the determination proceeds. The bottles are whirled in a centrifuge which causes the fat to rise. Hot water is added and the process is repeated. The fat which accumulates in the neck of the bottle and which is read as percentage butterfat is "crude fat" and contains all the lipid materials in the milk.

The Röse-Gottlieb (because of the umlaut, sometimes spelled Roese) method requires extraction of the milk with a mixture of ethyl ether and petroleum ether, distillation of the solvent and weighing the fat residue. The fat is then dissolved in petroleum ether and any insoluble residue weighed. The difference gives the butterfat content of the sample. This is also a "crude fat" determination since it includes all compounds which are soluble in petroleum ether. This comprises all lipids, the fat soluble vitamins, and pigments.

Milk solids include all of the compounds present in milk except water, and are readily determined by evaporation of a sample in a weighed flat bottom dish at approximately 100°C to constant weight. It is necessary to avoid heating the sample above this temperature since browning of the residue will occur. Milk solids can be determined approximately by measuring the specific gravity with a Quévenne lactometer and applying the formula (%Solids = $0.25L + 1.2F + 0.14$) where L = lactometer reading and F = percentage of fat in the milk. The lactometer reading must be corrected to 60°F before the formula is applied. The total solids of milk are fairly constant and this is a useful determination in indicating some types of adulteration. Often the solid-not-fat is calculated by subtracting the amount of crude fat from the total solids. Since fat shows wider fluctuations than the other components, this value has slightly more constancy than total solids.

Lactose may be determined by a number of methods. Those which are commonly used are not specific for lactose but since it is the only carbohydrate present in milk, they serve well to give fairly accurate and reproducible values. Either the determination of the optical activity or the reducing capacity with a copper reagent are used. In the determination of the optical rotation, the protein is precipitated and the filtrate read in a polarimeter. The assumption that lactose is the only remaining compound that possesses optical activity is not quite valid since some of the other components are optically active but are present in very low concentration.

The copper reducing method is also applied to the protein-free filtrate. The proteins are commonly removed by adding copper sulfate to the diluted milk and precipitating most of the cupric ion with either sodium or potassium hydroxide. The cupric hydroxide which forms carries down the protein as well as the fat. An aliquot of the filtrate is treated with Fehling's Solution and the cuprous oxide formed either filtered off, dried, and weighed or determined by one of the other standard methods.

Proteins of milk are ordinarily determined by the Kjeldahl method (see p. 131) and the percentage of nitrogen multiplied by 6.38. Use of this method for total proteins in milk suffers from the same limitations as does the use of the method for the protein content of any food. It will be noted that the constant used for milk differs slightly from the general constant, 6.25. Casein and albumin fractions are sometimes determined but these are crude fractions, not pure proteins. Casein is precipitated by warming the diluted milk and adding acetic acid to pH 4.3. The precipitate is analyzed for nitrogen by the Kjeldahl method and the percentage of nitrogen determined is multiplied by 6.38. The albumin is precipitated from the filtrate obtained after removal of casein, by neutralization, addition of the correct amount of acetic acid and heating on a steam bath. This precipitate is likewise analyzed for nitrogen by the Kjeldahl method and the percentage of nitrogen multiplied by 6.38.

Pasteurization is an important process in insuring that some of the microorganisms are destroyed. A number of chemical tests have been devised to determine whether or not a given sample of milk has been pasteurized or if the process has been carried out satisfactorily. Enzymes are sensitive to heat and some of them are denatured by the pasteurization. Pasteurization is carried out by heating the milk to 142 to 150°F for 30 minutes or by heating it (flash pasteurization) to 160°F and holding for 15 seconds. Although this is a relatively low temperature, most of the enzymes of milk are destroyed. Tests for the enzymes can therefore be used as a chemical means of differentiating raw from pasteurized milk or of detecting incompletely pasteurized milk.

The Schardinger test detects the presence of peroxidase in milk. An alcoholic solution of methylene blue, formaldehyde and water is added to the milk samples. The tubes are placed in a water bath at 45°C. In less than 20 minutes the raw milk will decolorize the methylene blue while the pasteurized milk will take much longer.

The Kay and Graham test for phosphotase is more delicate and can readily detect faulty pasteurization procedures. A sample of milk is incubated with phenyl phosphate in a diethyl barbiturate buffer for 18 to 24 hours at 34–37°C. In the presence of phosphatase the phenyl phosphate is hydrolyzed and phenol is formed.

$$C_6H_5OPO_3H_2 + H_2O \rightarrow C_6H_5OH + H_3PO_4$$

If the milk has been pasteurized sufficiently, most of the phosphotase will have been destroyed and little hydrolysis will occur. The phenol is determined colorimetrically and any values over 0.047 mg phenol/0.5 ml milk indicate progressively inadequate pasteurization.

CHECKS FOR PURITY

The constant vigilance of the governmental agencies as well as the development of dairying into a big business anxious to maintain quality, have resulted in pure milk for the consumer in the United States. Milk sold is unadulterated chemically and fairly pure bacteriologically. (Absolute sterility except in canned products is impossible.) In generations past and in some parts of the world today this is not true. In some instance there has been an addition of water and skimmed milk to increase the volume when milk has been sold on a fluid basis, of thickeners such as gelatin and calcium sucrate to make cream appear richer, and of preservatives and coloring matter. During the past 50 years a large literature has developed on these adulterants and methods of detecting them.

Since the composition of milk is variable, it is difficult to detect watering, skimming or the addition of skimmed milk. Watering is illegal in all states of the Union, but standardization is permissible in many. The addition of water decreases the percentage of all components, but if the amount added is small, it may simply reduce them to the lower limits of the common range. Fat shows the greatest variation in milk and consequently the determination of the solids-not-fat is frequently used as an index of watering. In Table 8.1 the range for solids-not-fat is given as 7.5–10.6 per cent, but if the value for a sample of milk is close to the bottom of this range, it would surely be suspicious.

Milk serum or whey, that portion of the milk from which both fat and casein have been removed, is quite constant in composition and analysis of it is often used in attempting to detect watering. The refractive index, the specific gravity, or the total solids are determined. If the milk has been watered, the solids are present in lower concentration and the refractive index and specific gravity will consequently be closer to those of water than for pure milk.

One of the most reliable physical constants of milk is its freezing point. The presence of dissolved substances, the salts, the lactose, and other molecules, depresses the freezing point of milk below zero while colloidal substances have a slight effect on the freezing point. Pure milk freezes between −0.530 and −0.566° C with an average value of −0.545° C. Milk which has been watered will have a freezing point closer to zero. Here, too, the range in value does not allow a sharp differentiation. A small amount of water might be added to milk which contained an unusually large amount of soluble substances without raising the freezing point above the normal range.

Skimming of milk or the addition of skimmed milk to ordinary milk does not change the amount of any of the components except fat. It can be detected by calculating the ratio of the fat to protein, lactose, solids, or solids-not-fat. But here again, since the ratio of these constituents is not constant in different samples of milk, this is difficult to prove. If the fat falls below the standard set for that area of the country, but the protein, or solids do not, then skimming has doubtless been used.

Preservatives in milk are prohibited by law. The milk delivered must be kept in fresh condition by the use of sanitary procedures, equipment, packages and adequate refrigeration. In the early days, preservatives were commonly used to extend the time during which the milk was saleable. Formaldehyde in very small quantities has the ability to preserve the milk; boric acid or borax, salicylic acid, benzoic acid, hydrogen peroxide, and fluorides have all been used. Tests for all of these preservatives and methods for detecting them even when present in small amounts, have been widely used in the past.

Thickening agents and coloring matter have more frequently been added to cream than to milk. Cream is expected to be slightly viscous and yellowish, if it is chilled and has a relatively high fat content. The addition of gelatin, calcium sucrate, or agar-agar has sometimes been used to create the illusion of a higher fat content than is actually present. The use of these thickeners, as well as coloring matter, is prohibited in milk and cream. In chocolate drinks, egg nog and prepared milk drinks they are often permissible and are sometimes used.

SPECIAL MILKS

Most American dairies sell more than one kind of milk today and some have quite a list of products. The names of some of the more common milks and a very brief description of their special composition or characteristics will be given.

Certified milk is milk which reaches the bacteriological standards of the American Association of Medical Milk Commissions. It is produced under strict conditions from cows known to be free of tuberculosis and brucellosis and has a low count of bacteria. Most certified milk was formerly sold raw but much of it is now pasteurized. Sometimes it is designated as certified-raw and certified-pasteurized.

Homogenized milk is milk in which the size of the cream globules has been reduced sufficiently so that the cream does not separate out in 48 hours. The milk is forced through very small orifices or between disks which mechanically reduce the size of the globules.

Vitamin D milk is widely sold and is the most common fortified milk product. Milk has a relatively low content of vitamin D, particularly in the winter. Summer milk may have approximately 30 International units in each quart but in the winter, the content falls very much below this level. Since the need for vitamins D during infancy and early childhood is great, vitamin D milk was introduced in 1932 and is widely sold now in the United States. The vitamin D content of milk can be improved by (1) irradiation with ultraviolet light, (2) addition of a vitamin D concentrate, or (3) feeding a diet high in vitamin D to the cow. When ultraviolet light comes in contact with some of the sterols (but not cholesterol), a chemical reaction occurs with a change in the ring structure of the sterol. The new product possesses vitamin D activity in young animals. The sterol which is present in milk is 7-dehydrocholesterol, the same sterol present in human skin, and it is changed by ultraviolet light to the vitamin which is designated Vitamin D_3. Irradiated milk usually contains 135 to 200 International or U.S.P. units per quart. When a concentrate is added to milk, it is usually added so that the level of vitamin D is 400 International or U.S.P. units per quart. Milk produced from cows which are fed a special source of vitamin D, usually irradiated yeast, produces milk which is often called "metabolized milk" since the added vitamin D passes into the milk during the metabolic processes of the cow. Where this procedure is used, the cow is given sufficient vitamin D so that the level in the milk produced is 400 International units per quart.

Today almost all of the Vitamin D milk produced in the United States is fortified by the addition of a concentrate. If milk is irradiated sufficiently to raise the vitamin D content to 400 U.S.P. units, off flavors develop. The milk with lower levels has not been able to compete with fortified milk. Also since the idea of adding a concentrate has been accepted by the American Medical Association, this far easier method which does not require special equipment or special care to avoid contamination has been widely adopted. Metabolized milk has likewise largely disappeared from the market, since the amount of vitamin D consumed by the cow is far greater than the amount which appears in the milk.

Fortification of milk with other vitamins is now practiced in many areas. The milk is sold under various names and is usually several cents a quart more expensive than ordinary milk. The vitamins used are often the B complex, or some of the B complex, as well as ascorbic acid and vitamin A. Some nutritionists are enthusiastic about this type of fortification, but many are very much opposed. Their opposition rests on the widespread distribution of these vitamins and the ease with which they are provided in a well-rounded diet.

Filled milk is skimmed milk to which some fat such as coconut oil has been added before evaporation and canning. In some states the sale of filled milk is prohibited.

Reconstituted milk is milk produced by the addition of water to milk powder, skim milk powder, evaporated or frozen milk or some combinations of them. During World War II and since, reconstituted milk has been used in many parts of the world where the supply of fresh milk is either insufficient or unsafe. It is made up so that the fat and the solids-not-fat equal the concentrations in fresh milk. In some states its sale is illegal.

Many fermented milks are available in the United States, made by the addition of cultures of microorganisms to the milk. These organisms grow and feed on milk components and form acid. Buttermilk is produced either from whole milk or from skimmed through the action of the bacterium *streptococcus lactis*. The raw milk is heated to kill most of the micro-organisms present, is cooled and inoculated with a culture. Incubation for 24 hours allows the organism to grow and form acid. Buttermilk is sometimes stabilized with a small amount of gelatin so that separation of the casein does not occur.

Acidophilus milk is prepared in a similar manner by using *Lactobacillus acidophilus*. The milk is sometimes more thoroughly sterilized than with ordinary buttermilk. Twenty or thirty years ago, there was quite a fad for this milk since it was believed that the flora of the intestinal tract has a profound influence on the health of the individual and that the establishment of *L. acidophilus* would lead to improved health. It has been shown that the identity of the bacteria present in the large intestine can be changed by continued ingestion of a large number of organisms, such as *L. acidophilus*. But it has not been proved that the health of the individual depends on this.

More recently yogurt has been one of the foods extolled by the faddists. It is sold in many areas of the United States. It is an excellent food but not superior to many other milk products. It is also called "bulgarlac" or "matzoon" and is likewise a fermented milk. In this case the organism added to the partially sterilized milk is *Lactobacillus bulgaricus*. Both acidophilus and bulgaricus milk usually have a smaller amount of acid than buttermilk and a creamier texture.

BUTTER

Butter is a milk product composed principally of fat. It must contain at least 80 per cent crude fat and is sold with or without the addition of salt and with or without the addition of coloring matter. The 20 per cent not-fat is composed of approximately 16 per cent water, 2.5 per cent salt, and 1.5 per cent milk solids. While neutral fat is the principal ingredient of the "fat," phospholipids, sterols, pigments, and fat soluble vitamins are also present. A small amount of lactic acid is also present in butter and contributes to its flavor. The amount is controlled by washing the butter. Too high an acid content is undesirable but if the amount is below 0.02 per cent the butter is more liable to putrefaction. The manufacture of butter is discussed briefly on p. 51.

Butter must contain at least 80 per cent crude fat and be free of filth. It is scored on a grading system which allows the following points: flavor, 45; body, 25; color, 15; salt, 10; and package, 5 to give a total of 100 possible points. Butter which scores 93 or 94 must be very mild, sweet and fresh in flavor. Only the most expensive butter on the consumer market scores this high; most is 92 score and the cheaper grades are below this. Any butter scoring below 75 is called grease and is considered unfit for food. On the letter scale AA is U. S. 93 score; A, 92; B, 90; C, 89; and below 89, CG or cooking grade.

The addition of salt to butter inhibits the growth of some microorganisms and kills others, so that keeping quality of salted butter is better than for unsalted. However the relatively high moisture content of butter and the presence in the water, not only of salts and vitamins but also of protein, make it a fine nutrient medium for some microorganisms. Butter has a relatively short life unless it is stored at low temperatures. Frozen butter can be kept for some months without serious deterioration in flavor. Butter grease which is prepared by melting butter and separating the fat from the whey, has a much longer storage life, since the moisture content is very low.

CHEESE

For hundreds of years cheese has been made by man from the curd precipitated from milk by acid or rennin. The curd is given various treatments—heating, pressing, the addition of seasonings—and many cheeses are allowed to ripen before they are used. The ripening process is one in which microorganisms—sometimes bacteria, sometimes molds—grow and use the curd products for food. These microorganisms form gases, acids, and other compounds that produce flavors characteristic of the particular cheese. They also modify the protein so that the texture is altered and becomes characteristic of the cheese. In centuries past when nothing was known of microbiology, techniques developed so that cheeses were produced in various regions of the world that, from year to year and from kitchen to kitchen, were approximately similar. These techniques developed through happenstance and through trial and error. The cheeses of one type are not always exactly identical but they are so similar that expectations for texture and flavor of say, Gorgonzola, or Cheddar are possible. Not only is the strain of microorganism responsible for the flavor developed, but also the type of milk and its cream content. Some cheeses are made from skimmed, others from whole milk and some have added cream. Cow's milk is the chief milk used but sheep and goat milk are used for some types. In order to achieve fairly reproducible results, the techniques all along the line must be standardized. The curd must be treated the same way, the ripening must be at the same temperature and for the same length of time, etc. Considerable skill and attention to details are required to produce repeatedly cheeses with similar textures and flavors.

Composition

Since the curd used for the preparation of all cheeses is a casein precipitate, they are rich in protein which varies from approximately 20 to 30 per cent. The exceptions are the cheeses which are very moist and which consequently have a slightly lower percentage of protein. Other components vary considerably in their levels depending on the ingredients used. Fat varies from 1 per cent for skimmed-milk cottage cheese to 38.3 for American red, 39.9 for French demi-sel Cream, and 51.3 for Ricotta salata.[2] Other components do not show this very wide range of values, but there are nevertheless differences and there are also some similarities. Most cheeses except cottage and English cream cheese are excellent sources of calcium.

Unripened Cheese

Unripened cheeses are cream and cottage cheeses. Cream cheeses are prepared by coagulating the casein in milk with rennet after it has soured slightly. An inoculation of a culture of bacteria is added to the milk and a short incubation period allowed before the rennet is added. The precipitate has a fine creamy texture, hence the name. It may or may not be made from cream. Cottage cheese (also called, Cup, Dutch, Clabber, Pot and Smeerkaas) is prepared from either whole or skimmed milk in which the incubation has been carried on long enough to develop more acid (0.65 to 0.7 per cent expressed as lactic acid). Sometimes rennet is used to complete the precipitation and sometimes the acid alone is the agent. The curd formed is heated to a temperature of about 120° F, the whey drained and, after chilling and salting, the curd is "creamed" by beating. Sometimes cream is added at this point.

Rennet is an enzyme preparation widely used for the precipitation of

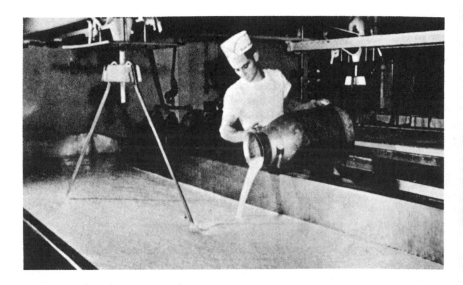

FIGURE 8.1. STARTER CULTURE. One of the agents that causes the milk to coagulate is added to the vat as paddles rotate to stir it thoroughly into the milk. Stainless steel vats of this type hold 18,000 lbs of milk—enough to make 1800 lbs of cheese.

Courtesy of Kraft Foods.

FIGURE 8.2. AFTER A CURD HAS FORMED. It is cut with stainless steel wire knives into small cubes, to help expel some of the whey. Following this, automatic paddles will stir the contents of the vat while it is heated, to shrink and firm the curd and expel more whey.

Courtesy of Kraft Foods.

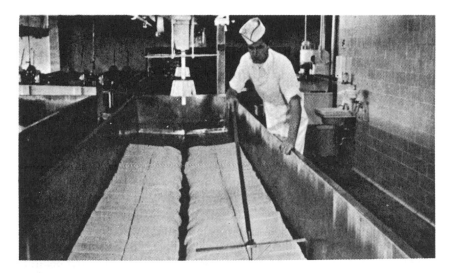

FIGURE 8.3. THE MATTED OR CHEDDARED CURD. It is cut with stainless steel knives into slabs. These will be piled one on top of the other two deep, turned, and piled three deep, using the curd's own weight to matt it out into thinner slabs.

Courtesy of Kraft Foods.

FIGURE 8.4. THE SALTED CURD. It is being packed into stainless steel cheese hoops, lined with heavy parchment that will form the wrapping for the finished block of cheese. Until recent years, round hoops were used and each was lined with cheesecloth. From here the cheeses go to the press, where they are held overnight.

Courtesy of Kraft Foods.

casein. It is a mixture of the enzymes pepsin and rennin and is used as an extract or occasionally as a powder. It is prepared by extracting the lining of the fourth stomach of calves with sodium chloride. In the presence of calcium ions, rennin forms a thick gelatinous curd of casein. See p. 141 for a short discussion of the reaction. The casein curd formed either through the action of rennet or acid is the basis for all cheeses.

Ripened Cheese

A good example of a ripened or cured cheese is Cheddar. This yellow cheese was first made in Cheddar, England and now accounts for more than half of the cheese manufactured in the United States. It is sold under many names—Cheddar, American, "store cheese," the name of the community where it is made, or even "rat trap" cheese. Although the final characteristics of other ripened cheese may be quite different from Cheddar, the methods of preparation are very similar. Differences in hardness and consistency of the cheese are the result of slight difference in the temperature at which the curd is formed, the amount of heating it receives and the amount of evaporation allowed during curing. Differences in flavor and also texture and consistency result from the enzymes and microorganisms introduced into the cheese and developed during ripening. Some of these are present in the milk used, others are introduced in the culture, and many come from the air of the cheese factory.

Cheddar cheeses are produced by adding a "starter" to milk in order to produce acid. The culture commonly used for butter manufacture containing *Streptococcus lactis, S. citrovorus,* and *S. paracitrovorus* is often used. When the acidity calculated as lactic acid is approximately 0.2 per cent (lactic acid is not the only acid present but is the most abundant), rennet and sometimes color is added and the curd forms. Great skill is required in handling the curd so that the maximum yield is obtained and so that the characteristics necessary to a desirable cheese are developed. The curd is held until the proper degree of firmness results, a matter of 6 to 8 hours, and during this period a rapid increase in acid occurs. The curd is cut to form small cubes and then gently stirred to separate the whey. The mixture is then heated to 100° F. The curd is filtered and the final amount of whey is removed by cheddaring or matting. This consists of rotating and turning the slabs of curd until they are sufficiently dry. During this process, acid continues to develop and the proteins probably undergo some changes in chemical constitution since their physical characteristics change. Salt is now added which hastens the removal of the last of the whey, depresses the action of the acid-forming organisms as well as the putrefactive ones. The curd is gently pressed, wrapped in cheese cloth and pressed again. It must now be ripened.

During ripening the enzymes present in the milk as well as those formed by the microorganisms catalyze the transformation of many of the compounds in the curd into other products. Sometimes an enzyme preparation is added at the time of salting. Ripening may be carried out at a temperature of 40° to 65° F, usually about 55° F; the lower the temperature the slower the ripening. A short time after ripening begins, the green cheeses are dipped in paraffin so that the loss of moisture can for the most part be prevented.

Cheddar-type cheese and all hard cheeses have relatively long keeping times. The cheese ripens uniformly through the mass and is not subject to the growth of organisms on the surface until after it is cut. Often these cheeses are made in large 70 lb molds.

Soft cheeses are made in small sizes, so that the microorganisms which develop on the surface can penetrate and the flavors developed will pass through the cheese.

Ripening

The changes which occur on ripening are numerous. Many investigations have attempted to trace the changes in chemical composition and the microorganisms and enzymes responsible. Numerous reports continue to appear in the literature as the attempt to understand and control the process, so important to the economic success of cheese making, persists. Only in very general terms can we describe the changes that occur on ripening. Since the organisms in some green cheese are quite different from those in others, the particular reactions will vary. In general, there are three types of reactions in all cheese plus some others: (1) hydrolysis of protein under the influence of proteases, (2) hydrolysis of lipids under the influence of lipases, and (3) changes in amino acids and fatty acids under the influence of specific enzymes to form flavorful compounds.

(1) The hydrolysis of proteins results not only in change in texture but also change in flavor. The longer the ripening, the more extensive the changes which occur. Large protein molecules are fragmented into smaller insoluble proteoses, into soluble peptones, into low molecular weight peptides and amino acids. The enzymes which catalyze these changes include rennin and pepsin introduced in the rennet extract, as well as those synthesized by the microorganisms in the starter by which the milk was soured or from the air of the cheese factory. Sometimes enzymes are introduced at the time of salting. As the molecular weights of the proteins decrease, the

physical properties of the cheese changes. If most of the proteins are hydrolyzed to relatively low molecular weight soluble compounds, the cheese will get very soft and creamy. This occurs as Limburger ripens. Even when the fragmentation of the protein is not so extensive, the physical properties of the cheese change. The ability of ripened Cheddar cheese to melt readily and blend with liquids reflects a change in the protein. The low molecular weight products of protein hydrolysis are flavorful. Some peptones have a bitter taste but others are meaty. The amino acids for the most part have relatively mild flavors. Some are sweet, some brothlike, and others slightly bitter. All contribute to the blended flavor of the cheese.

(2) The action of lipases on lipids, particularly fats, has a marked effect on flavor. Butterfat contains a much higher percentage of low molecular weight fatty acids which are soluble in water and volatile and consequently flavorful, than any other food fat. Hydrolysis of the fat liberates these acids and adds sharpness and their particularly distinctive flavor to the cheese. Butyric acid has a very strong, rancid odor which is objectionable but small amounts blended with other substances account for the odor of some cheeses.

(3) Amino acids and fatty acids also undergo change which results in the formation of some of the compounds most important for flavor. Numerous enzymes catalyze the deamination of amino acids with the formation of ammonia and either hydroxy acids or simple acids. Both the ammonia and the acids are flavorful. Decarboxylation of amino acids results in the formation of amines, some of which have very penetrating odors.

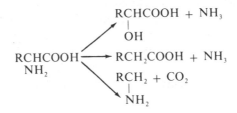

Changes in fatty acids can result in the formation of shorter acids either through beta oxidation of the fatty acid or through oxidation of unsaturated acids at the double bond. Oxidations along the chain can result in the formation of many molecules of low molecular weight acids or of ketones and aldehydes. The high molecular weight acids, aldehydes, and ketones are not water soluble nor volatile and consequently do not possess flavor, but the low molecular weight ones are capable of stimulating either the taste buds or the olfactory endings or both. For example, lauric acid can undergo many changes depending on the enzymes present,

$$C_{11}H_{23}COOH \begin{cases} C_9H_{19}COOH + CH_3COOH \\ 6CH_3COOH \\ COOHC_8H_{16}COOH + CH_3COOH \end{cases}$$

while oleic can be oxidized at the double bond as well:

$$CH_3(CH_2)_7CH{=}CH(CH_2)_7COOH \begin{cases} CH_3(CH_2)_7CHO + CHO(CH_2)_7COOH \\ CH_3(CH_2)_7COOH \end{cases}$$

In Roquefort-type cheese the flavor depends to a great extent, although not entirely, on the presence of methyl ketones and butyric acid.

Numerous other changes are possible and occur in some cheese. It has been reported, for example, that the lactic acid formed during curd precipitation disappears rapidly on ripening of some cheese while other acids appear. Any lactose which may still be present in the green cheese also disappears.

The organisms introduced from the air of the cheese factory have been mentioned several times and since the character of the final ripened cheese depends so much on their activity, a little more should be said about them. When a large number of a certain strain of microorganisms develops in one room or building, the air and dust of that place contain these living organisms sometimes for years. If they are provided with a nutrient medium in which to grow, the descendants flourish and in turn populate the air and dust with living members. Thus a wine factory where yeast fermentations are carried out year after year contains many strains of yeast, but particularly those established in that place. A kitchen in which milk has never soured will be practically free of acid-forming bacteria and if milk is left open to the air, it will putrify before it sours. But in a kitchen where milk has frequently soured, the air is so laden with these bacteria that on exposure to the air milk sours in a few hours.

As the art of cheese making developed through the centuries, cheese from one area tended to be different from that of another even when ingredients and methods were identical because the microorganisms established in the different factories were not identical. Roquefort cheese was first prepared at Roquefort, France by ageing the cheese in caves. Until microbiology became a science, this cheese was only prepared in this town. Now it is possible to culture the mold which traditionally has been used for the ripening and add it to the curd.

In the United States many of the cheeses developed in the Old World, particularly in Europe, are successfully copied. By duplicating the ingredients and the methods, and by adding the correct microorganism cheeses with very similar, if not identical, characteristics are produced.

Process Cheese

Processed cheese has become very popular in the United States because of its mild, standardized flavor and soft texture. It is produced by blending aged cheese with green, young cheese in the presence of an emulsifier and some water. The food sold as "processed" or "process" cheese has a moisture content not more than 1 per cent greater than the average of the cheese used or in no case more than 43 per cent. That sold as "cheese food" can have a moisture content up to 44 per cent while that sold as "cheese spread" can have a moisture content up to 60 per cent.

The expectation of the American consumer for uniformity in flavor is doubtless one of the factors which has led to the popularity of these cheese mixtures. Aged cheese from a single factory is usually *similar* in flavor and texture but not completely uniform, while that from two factories which use approximately the same techniques may differ considerably. A cheese company catering to a nationwide clientele, is able to sell a product with the same flavor and texture in all parts of the country and at all times of the year only by blending. Mild flavors in cheese products are likewise demanded by a large section of the American public. These mild flavors can be produced by blending a sharp aged cheese with green, young cheese or with the curd.

The emulsifying agents used in the blending are salts: frequently sodium or potassium phosphates such as NaH_2PO_4, Na_2HPO_4, Na_3PO_4, $NaPO_3$, $Na_4P_2O_7$, $Na_2H_2P_2O_7$; potassium, calcium, or sodium citrate, $Na_3C_6H_5O_7$; sodium tartrate, $Na_2C_4H_4O_4$; or sodium potassium tartrate.

Processed cheese is produced by grinding aged cheese, warming, and mixing thoroughly with young cheese, water, and emulsifier. Other foods such as pimientos, olives, etc. may be added. Sometimes blends are manufactured which contain two or more aged cheeses. An example is a cheese blend which contains, Cheddar, Swiss, and Limburger.

Cheese *foods* are prepared by blending aged cheese, young cheese or curd with milk or milk products to form a smooth soft product. Cheese *spreads* often have gums or gelatin added to promote smoothness and bind the spread together. Some cheese spreads contain no aged cheese at all, but only the curd. They have sugar, vinegar, and other flavoring substances added, but the "cheese" flavor is very mild. Cheese spreads are so soft that they are packaged in jars.

REFERENCES

1. BISHOV, S. J., HENICK, A. S., and MITCHELL, J. H., JR., "Rate of Mineral Removal from Milk by Ion Exchange," *Food Research*, **24**, 428–433 (1959).
2. BRIDGES, M. A. and MATTICE, M. R., "Food and Beverage Analysis," Lea and Febiger, Philadelphia, Pa., 1942.
3. DUNCAN, C. W., WATSON, G. I., DUNN, K. M., and ELY, R. E., "Nutritive Value of Crops and Cow Milk as Affected by Soil Fertility II The Essential Amino Acids in Colostrum and Milk Proteins," *J. Dairy Sci.*, **35**, 128–139 (1952).
4. JACOBS, M. B., "Milk, Cream, and Dairy Products," in Jacobs, M. B., ed., "The Chemistry and Technology of Food and Food Products," 2nd Ed., Interscience Publishers, New York, N. Y., 1951.
5. JOSEPHSON, Borden Award Address, *J. Am. Diet. Assoc.*, **30**, 855 (1954).
6. KRUKOVSKY, O. V., TRIMBERGER, G. W., TURK, K. L., LOOSLI, J. K., and HENDERSON, C. R., "Influence of Roughages on Certain Biochemical Properties of Milk," *J. Dairy Sci.*, **37**, 1–9 (1954).
7. MACY, I. G., KELLY, H. J., and SLOAN, R. E., "Composition of Milk," *Natl. Acad. Sci. Natl. Res. Coun. Publ. 254*, Washington, D. C., (1953).
8. WOODMAN, A. G., "Food Analysis," McGraw-Hill Book Co., Inc., New York, N. Y., 1941.

CHAPTER NINE

Cereals and Their Use

LONG BEFORE THE BEGINNING of the historic period, man learned to use cereals. Undoubtedly when primitive man was a food gatherer, he discovered that the seeds of many grasses are valuable because they can, under proper conditions of dryness, be stored for long periods. Later, he learned to grow grains in a little patch of cleared ground and assured himself of food through the winter months. In the Western hemisphere maize, or corn to Americans, was the cereal domesticated by the early Indians. When Columbus discovered America, the inhabitants were growing it in most regions of the continent. Soon afterward maize was introduced into Europe, and the grains of the Old World were brought to America by the early colonists. In Asia Minor and Asia wheat and rice were the most important early cereals developed. Since the first wheat was grown in prehistoric times and since before the time of recorded history many changes occurred in the grain either through purposeful or chance breeding, geneticists as yet do not know what wild seeds were the parents of today's strains.

THE CEREALS

The cereals comprise a group of plants from the grass family, Gramineae, whose seeds are valuable for food either for man or domestic animals. Wheat, barley, corn or maize, oats, rice, rye, and sorghum belong to this group; and although buckwheat is not a member of the family Gramineae, it is usually classed with the true cereals. In some parts of the world one cereal is more important in man's diet than another. Usually the difference depends on the requirements of the particular cereal for soil, moisture, and length of time to maturity. Diet patterns are the result of centuries of custom where the availability of a particular food has established it firmly. Thus rice, which requires warm, humid conditions and a long growing period (between 120 to 185 days from seeding to maturity), is eaten widely by the inhabitants of almost all the tropical regions of the world. Rye, which flourishes in poor soil and in cool climates, can be grown in northern countries and is relished by the inhabitants of northern Europe.

All cereals give relatively high yields with a small amount of labor. Undoubtedly this also was a factor in their early establishment as a cultivated crop.

Structure of the Grain

Although the seed (grain) of each of the cereals differs from the others and there are likewise differences between subspecies like popcorn, sweet corn, flint, and dent corn, all of these seeds are closely related botanically. Since detailed descriptions of each grain can be found in many textbooks of botany, a generalized description will be given here.

The Primary Unit: The Cell. Like all living things, the primary unit in grain is the cell. Some descriptions of grains and other plant products by chemists, home economists, or food technologists seemingly ignore the fact that cells are present and that any cell is a complex mixture which always has a number of proteins, salts, many organic compounds in small amounts such as the vitamins, molecules formed during synthesis or degradation, and usually at least one form of carbohydrate, as well as lipids. Descriptions which ignore this fact are so oversimplified that they are highly misleading. The cells in cereals are living cells until milling occurs. Respiration continues during storage, although if the moisture content is low or if the temperature is very low the rate may be very slow.

Parts of the Seed. The grain or seed of a cereal is in general composed of three main parts: (1) the embryo or germ from which the root and leaf of the new plant are formed when it sprouts; (2) the endosperm, the storage portion of the seed which supplies the sprouting embryo with food in the period before the root and leaf begin to function; and (3) the bran, which forms the covering or protecting layers. The three main portions of the seed can readily be seen with the naked eye. The embryo is small and is attached to the base of the seed, while the endosperm makes up the major portion of the seed. The branny coats are composed of a variable number of layers of cells. Since the functions of each of these three parts are quite different in the sprouting seed, their chemical composition is very different.

These structures are readily differentiated in a wheat kernel. The bran is shown to possess six layers although botanically the aleurone layer, the innermost one, is classed as the outer layer of the endosperm. However, in milling the aleurone layer separates with the bran. The peculiar property of toughness of the branny layers, particularly after the wheat is tempered by the controlled addition of water several hours or more before milling, allows the ready separation of bran from other parts of the wheat seed. The bran layers also have a very low density which assists in their separation from pieces of endosperm during milling.

Composition of Seed Parts. Chemically the bran is very different from the rest of the seed. The bran has an unusually high per cent of crude fiber and ash and a fair amount of crude fat. The crude fiber, it will be remembered, is composed of those organic molecules which are not soluble in dilute acid or dilute alkali (see p. 107) and includes for the most part the cellulose, hemicelluloses, and some lignins. These substances make up the cell walls, encrust the cell walls, or exist in the middle lamella between the cells.

The embryo (germ), like the bran, has distinct properties which make it possible to remove most of this part during milling and which depend on its chemical composition. The germ is high in lipids and rather high in total nitrogen and ash. During milling, the crushing and shearing action of the rollers squeezes out some of the lipids and causes a flat adherent flake of the germ cells to form. The high lipid content causes the density of the flake to be low, and it is therefore readily separated from the endosperm particles and the fine flour by shaking and bolting.

The endosperm is made up of cells containing large quantities of starch granules and cytoplasm, whose protein content varies with the location of the cell in the endosperm. Those cells close to the bran possess more protein and less starch than the cells in the center of the endosperm. These proteins are most important in determining baking quality of the flour.

Storage of Grain

Easy *storage* of grains for relatively long periods makes them particularly valuable to man during winter and has no doubt been an important factor in the prominent role that cereal grains have played in his history. Under proper conditions of dryness and temperature grains can be stored for long periods with little or no change in either their fertility or milling and baking qualities. However, there are some problems. High humidity and high temperatures lead to mold growth, and infestation with insects is difficult to avoid so that long periods of storage are particularly hazardous.

Seeds are composed of living cells and during storage respiration continues with the utilization of oxygen and the formation of carbon dioxide and heat. Oxygen is readily available in the air held between the grains. Wheat in bulk is one-third air. At low moisture levels, the rate of respiration is slow; but as the amount of moisture in grains increases, so does the rate of respiration. All grains have a critical level of moisture above which respiration increases rapidly, causing heating of the grain and consequent damage. It has been demonstrated that the critical moisture level, about 14 or 15 per cent in wheat, is the level at which molds present on the grain and in the bran begin to grow. Their respiration is added to that of the seeds and gives an accelerated production of carbon dioxide and heat. Although the amount of heat produced is sufficient to raise the temperature of the grain, it is small and difficult to measure. However, carbon dioxide is relatively easy to measure, and since it parallels heat production, it serves as a convenient laboratory method for estimating heating.

Heating of grain may start in one localized area where the humidity is particularly high. This can occur in grain where the average moisture content is quite low when the grain is stored in large bulk and when irregular cooling of it occurs. Moisture diffuses from warm areas to cool and the moisture content in a cool area may rise to a point where mold growth begins. Grain is a poor conductor of heat, but as the mold flourishes, the area becomes warmer and more favorable for this fungus. However, the temperatures may become so high that the enzymes of the mold will be inactivated and the mold killed. This biological heating may be followed by a second period of nonbiological heating in which the temperature may reach the ignition point. The whole process is termed "bin burning" and whether or not fire results, there is marked damage to the grain in that location.

Insects also contribute heat to grain since they have relatively high respiratory rates. The presence of insects in stored grain is always undesirable because of the damage which occurs to the grain. Insect infestation is very difficult to prevent and is one of the great problems of grain storage.

When molds grow on grains, they invade the seed and by their enzymes produce compounds useful to their own development and life. These enzymes affect lipids, proteins, and carbohydrates, and serious changes in the composition of the seed occur. The germ may die and the milling and baking qualities are markedly affected.

Since the moisture content of the grain is of such critical importance for its keeping quality, grain is often dried before storage at the elevator. Drying must be carefully controlled so that sufficient time is allowed for diffusion of water to the surface of the seed and so that damage to the grain does not occur.

WHEAT AND WHEAT FLOUR

To prepare a wheat flour, the miller faces the problem of separating the parts of the wheat so that they can be blended into a flour with a chemical composition offering desirable baking characteristics.

Milling

Milling is a complicated process in which grain is ground in successive steps that gradually separate portions of it. Thus the bran is broken off and flattened, the germ is pressed into a flake, and the endosperm is powdered. Clean wheat is "tempered" before grinding by treating it with water so that the bran will be tough and readily separated from the endosperm. The tempered wheat is crushed between corrugated rollers called break rolls. The first break rolls are set relatively far apart and grind the wheat lightly, while successive breaks yield finer and finer products. The first break is separated by sieving or bolting into very fine particles (flour), intermediate particles (middlings), and coarse particles (chop, or stock). The chop or stock is then sent to the second break rolls. This process may continue through 5 or 6 breaks. The chop contains pieces of endosperm and bran, and the chop from the last break is principally bran. The middlings contain endosperm, bran, and the germ. The middlings are classified and some of the bran removed and sent to the reduction rollers. These are smooth rollers, but like the break rolls they are graduated so that successive reduction becomes finer and finer. After each reduction, sifters separate the flour, middlings, and chop. This process is continued until most of the endosperm has been removed as flour and most of the bran has been separated in the sifters. The remnant consists of fine middlings, bran, and a little germ. It is used for animal feed.

Types of Flours Produced. The flour streams from the various rollers are named for the roller—"first break stream," etc. They vary in chemical composition because they vary in the amount of bran and germ which they contain as well as in the portion of the endosperm which has been released. Many different flour streams are produced, especially in large mills. These streams are combined and blended to form flours with particular baking characteristics. *Straight flour* is a combination of all the flour streams. It is seldom produced. *Patent flours* are flours from the more refined streams and vary considerably in the percentage of the total flour represented. Geddes[8] gives the following approximate percentages:

> Family patent: 70–75 per cent total flour
> Short patent: 75–80 per cent total flour
> Long or standard patent: 90–95 per cent total flour

The flour remaining after the patent is removed is called *clear flour* and it too may be separated into different grades. Clear flours are used in combination with rye for dog biscuits and in other products where high quality is not essential. The flour from the last reduction is called *red dog* because it is dark in color and contains relatively large amounts of bran and germ. It is usually sold as animal feed.

Relation to Dough. Milling does not involve any chemical procedures. It is only a physical separation of parts of the wheat seed, but it does affect the performance of the flour in the dough stage. Starch is held in granules in the cells of the wheat seed, and on milling the cells are ruptured and the starch granules escape. The grinding also damages some of the granules so that starch flows out of the granule case when the flour is made into dough. This starch is available for hydrolysis through the action of α-amylase. In bread flours the amount of damaged granules is around 4 per cent. These damaged granules are called "ghosts" and the level at which they occur is important in determining the amount of α-amylase activity at room temperature in the dough.

The analyses of the various mill products must be considered as only representative since wheat, like all plant products, shows variation in composition with variety, climatic conditions, and soil fertility. Milling does not completely separate bran, germ, and endosperm. Thus, red dog contains considerable amounts of bran and germ particles, and the analyses reflect this in the relatively high crude fat and crude fiber concentrations. See Table 9.1.

Flour Proteins

The flour proteins are of most importance as far as baking quality of flours is concerned. The large molecules of proteins are readily modified by mild treatment, both physical and chemical, and consequently their

TABLE 9.1. CHEMICAL COMPOSITION OF CERTAIN MILL STREAMS AND BY-PRODUCTS OBTAINED IN WHEAT MILLING*

Product	Moisture (%)	Total Nitrogen (%)	Fat (%)	Fiber (%)	Pentosans (%)	Ash (%)	Total Sugars (%)	P_2O_5 (%)
Wheat	10.3	2.05	2.1	—	5.1	1.73	2.6	0.85
First Patent Flour	11.5	1.82	1.0	0.2	2.9	0.40	1.3	0.22
First Clear Flour	11.0	2.13	1.7	0.2	3.6	0.81	1.8	0.44
Second Clear Flour	10.4	2.33	2.0	0.3	3.4	1.34	2.1	0.70
Red Dog	9.2	2.87	5.4	2.4	8.4	3.15	6.4	1.63
Bran	8.8	2.33	4.1	10.8	25.1	6.38	5.4	3.15
Shorts	8.9	2.47	5.2	8.4	16.3	4.10	6.0	2.23
Germ	8.5	4.84	11.9	1.8	6.2	4.80	15.1	2.73

*Condensation of data of C. H. Bailey by Geddes, W. F., "XXIV, Cereal Grains" in Jacobs, M. B., ed., "The Chemistry and Technology of Food and Food Products," Interscience Publ., New York, N. Y., 1951.

physical properties are changed by these treatments. Some of these modifications occur in doughs and batters.

At the beginning of the century Osborne studied the proteins of a number of cereals and separated them on the basis of solubility (i.e., their ability to form colloidal dispersions in certain water solutions) into five categories: (1) the proteins soluble in 70 per cent alcohol, called prolamines, (2) the heat-coagulable proteins soluble in water, called albumins, (3) the proteins soluble in neutral salt solutions, called globulins, (4) the proteins soluble in dilute acid and alkali, called glutelins, and (5) an ill-defined fraction called proteose. In the many years since Osborne's pioneer work many proteins have been fractionated on the basis of their solubility. However, now that the ultracentrifuge, electrophoresis, and diffusion make it possible to demonstrate whether molecules in a given fraction have the same size or range of sizes, it has been found that fractions based on solubilities do not give single proteins but a mixture of them. Wheat flour globulin contains at least three proteins while the albumin fraction has six. Keeping in mind the limitations of the methods of separation based on solubility, we can nevertheless learn a little about the proteins of cereals by considering the results of this method. In all cereals the chief proteins are prolamines and glutelins. In corn or maize the prolamine (zein) fraction is the most abundant protein, while the amounts of glutelins are low. In oats and rice it is the glutelins (called avenin in oats and oryzenin in rice) which are most abundant. In wheat, rye, and barley the proportions of prolamines and glutelins are intermediate. In the wheat endosperm a prolamine (called gliadin) and a glutelin (called glutenin) are present in approximately the same concentrations; in the bran a prolamin is most abundant with fair amounts of an albumin and globulin. Although all cereals are more or less similar in their protein content, the unique presence of glutenin and gliadin in the wheat endosperm is important to the baking operation. In the presence of water and with mechanical agitation these protein fractions form a tough, elastic complex termed gluten which is capable of retaining gases and by so doing makes a leavened product possible. Cereals other than wheat cannot form a large light loaf since gluten is not developed.

Gluten. It must be remembered that the terms gliadin and glutenin do not indicate homogeneous proteins but rather protein fractions. Likewise the term gluten does not apply to a group of identical molecules. Gluten can be readily prepared by adding 60 to 65 per cent water to a hard wheat flour, allowing the dough to stand approximately 30 minutes, and then washing out the starch granules and soluble compounds under a stream of water. A tough, elastic, gummy product is obtained which consists of approximately two-thirds water and one-third protein. There are also small amounts of lipids, starch, and ash. Sullivan[21] gives a typical analysis of dried gluten: protein, 85 per cent; lipid, 8.3 per cent; starch, 6.0 per cent; and ash, 0.7 per cent. The analyses will vary somewhat with the source of the flour and the method of handling. The lipid is held strongly to the protein and cannot be extracted with ethyl or petroleum ether.

Sullivan also reviews briefly the evidence which Hess has presented to show that gluten does not occur as such in the endosperm but is formed through mechanical treatment of the flour in the presence of water. Protein separated from flour and freed of the starch granules by methods which do not use water shows a swell in contact with water of 25 per cent. Gluten swells in contact with water 200 per cent. The swelled particles show different diffraction patterns with X rays. These observations indicate they must be different proteins.

The nature of the chemical reaction by which gluten is formed has concerned cereal chemists for many years. Although the product is always sticky, gummy, elastic, and tough, the degree to which these properties are developed varies from flour to flour. Both the "machining qualities," the way in which a dough can be handled by a mixing machine, and the baking properties depend on a nice balance between stickiness, extensibility, toughness, and tenderness. As yet there is no chemical explanation for the differences in gluten produced from different flours and the only way in which the quality of the flour for machining and baking can be estimated is by carrying through the operation. It was once believed that differences in the relative proportions of glutenin and gliadin might explain the differences in the gluten. But flours that differ markedly in their baking properties show little or no difference in the classical fractions of glutenin and gliadin. Other investigators once hoped that when methods for the determination of the amino acid content of proteins had been perfected, this might account for the difference in quality. Today those methods are well developed; and although the proteins of flour have an unusual distribution of amino acids (high in the amides of aspartic and glutamic acid, particularly glutamic, and relatively high in leucine and proline), there are no significant variations in the amino acid content of glutens of widely different properties.

It has been traditional to consider the development of gluten in a flour and water mixture as the reaction of two protein components, gliadin and glutenin. Many types of studies have shown these to be mixtures rather than single proteins. Attempts to disperse gluten, gliadin, and glutenin in various solutions have resulted in fractions which are heterogeneous. Electrophoresis, diffusion, and sedimentation by the ultracentrifuge indicate numerous components present in each of these proteins. The problem of the structural composition of gluten is far from solved; and until the time when it is solved, it is convenient to use the old names—gluten, gliadin, and glutenin.

Strength of Gluten. The properties of gluten formed from different flours or with variations in procedure, differ. The stickiness, toughness, elasticity, brittleness, and coherence are all part of the "quality" of the gluten formed. A gluten which is elastic but still fairly tough is called "strong" while one which is quite sticky and not very elastic, which spreads when a ball is placed on a plate, is called "weak." Many factors influence the strength of a gluten formed from a single sample of flour, and account for the difference in glutens formed from different samples. The effect of all factors is not completely understood.

In general, hard wheat gives gluten of good strength while soft wheat forms gluten of low strength. In the past it has been suspected that the total protein content of a flour is a measure of gluten strength. This is not true since flours with the same percentage of protein may have gluten strengths which are widely different. It is true, however, that hard wheat flours which form the strongest gluten have the highest percentage of protein. Protein content has not explained the difference in gluten strength and cannot be used as a means of evaluating the gluten strength of flours. Unfortunately, the knowledge of the chemistry of gluten has not advanced sufficiently to allow measurement of differences in gluten strength by chemical analysis.

The problem of gluten development is further complicated by the fact that hydration of the protein occurs as the gluten develops. Many proteins become hydrated and hold water tenaciously. The protein swells and the water is not readily removed by ordinary drying methods. Gluten shows this same phenomenon. Hydration capacity is influenced by many factors such as pH, total ion concentration, and presence of soluble molecules such as sucrose. The properties of a dough or batter also vary with pH, ion concentration, and sugar concentration; and doubtless the effect of these conditions on gluten hydration is one of the factors which produce change in properties. The role of gluten in baked products will be discussed under doughs and batters.

Lipids and Carbohydrates

The *lipids* of wheat are concentrated in the germ, so that although the wheat seed contains only about 2 per cent lipid, the germ contains close to 12 per cent. This does not mean, however, that all the lipids are held in the germ. Flour always contains 1 or 2 per cent lipid. The amount varies with the blend of flour and the variety of wheat used and the conditions of climate and soil under which it is grown. The lipids present include neutral glycerides as well as phospholipids and sterols. Wheat germ oil is a particularly rich source of the tocopherols, the vitamins E. Other small amounts of fat soluble vitamins and pigments separate with the ether extract, the "crude fat."

The principal *carbohydrate* of all cereal seeds is starch but there are always small quantities of others present. Dextrins are present in small amounts, particularly in flours since they are formed to a limited extent from the starch on milling. The water soluble carbohydrate is sucrose and it is more abundant in the germ than in the endosperm or bran. Some cellulose is present in all parts of the seeds since it is the chief component of the cell walls. There are also small amounts of lignins and pentosans including hemicelluloses and pectic substances mainly in the bran. Water extracts of either the whole grain or a flour yield gums. In wheat gum there are both pentosans and hexosans. A pentosan composed entirely of xylose has been isolated from durum wheat flour, and one composed of arabinose and xylose has been isolated from spring, winter, and durum wheats. (See p. 84.)

Vitamins

The *vitamin* content of cereal grains has been studied extensively. All the cereal grains contain important quantities of the vitamins of the B complex; and where the use of cereals is sizeable, they contribute significant quantities to the adequacy of the diet. Ascorbic acid is completely lacking in the seed or flour. Vitamins A and D are absent, but yellow corn derives its yellow color from carotenoids. The pigments in yellow corn are principally cryptoxanthin with small amounts of α- and β-carotenes, which are precursors of vitamin A. Wheat also contains carotenoids, principally xanthophyll which has no vitamin A activity in man or other animals. (See Carotenoids, p. 228). The germs of various cereal grains contain vitamins E, the tocopherols, which are pressed out or extracted with the lipid fraction. Wheat germ is particularly rich in the tocopherols.

Thiamin, riboflavin, niacin, and vitamin B_6 are present in fair amounts in all of the cereals while pantothenic acid and folic acid are in smaller amounts. The quantity of these vitamins present in any specific sample of seed or flour shows the same variation of values which is found in any plant product, differing with the conditions of climate and soil under which it was grown and with the variety and species. Thiamin varies from 3 μg per g of rough rice to 90 μg per g of oats. In 99 samples of wheats the thiamin level varied from 3.2 μg per g to 7.7 μg per g.

TABLE 9.2. DISTRIBUTION OF VITAMINS IN PRODUCTS OF WHEAT MILLING*

Product	Mill Yield (per cent)	Thiamin μg per g	Riboflavin μg per g	Niacin μg per g	Pantothenic μg per g	Pyridoxine μg per g
Patent Flour	63.0	0.68	0.34	12	5.7	2.2
First Clear Flour	7.0	3.0	0.62	26	9.6	3.9
Second Clear Flour	4.5	12.37	1.85	83	12.8	5.7
Red Dog Flour	4.0	29.66	3.80	120	—	—
Germ	0.2	22.93	—	68	15.3	9.6
Shorts	12.3	17.40	2.80	159	—	—
Bran	9.0	9.37	2.80	330	—	—
Wheat	100.0	5.03	1.00	70	13.3	4.6

*Part of a table in Geddes, W. F., "Cereal Grains," *loc. cit.*, p. 1100.

The riboflavin content of whole grain is lower than that of thiamin and does not show quite as wide variation for the different cereals. It is between 1.1 and 1.6 μg per g.

The niacin values for grains show some disagreement because the method influences the data obtained. With alkaline extraction, the values are higher. In wheats the quantity ranges from 47 to 106 μg per g. Other grains are in the same range, although rye and oats are lower.

The distribution of some of the B complex vitamins in different parts of the seed or in different flour streams has received considerable attention in recent years with the growth of interest in food fortification. In general the branny layers are richer in the B complex vitamins. Thus thiamin is distributed in approximately these amounts: bran, 61 per cent; endosperm, 24 per cent; germ, 15 per cent. While all flour contains small portions of germ and bran, highly milled flours with good baking quality are for the most part from the endosperm. Table 9.2 gives the amounts of B complex vitamins in various mill products.

Enrichment of Flour

It will be noticed from Table 9.2 that the amount of the B complex vitamins in flour is considerably lower than that in the whole wheat grain. In the diet of many individuals the consumption of white flour products represents about one-fourth of the total calories. If the rest of the diet is rich in B complex vitamins, the small amount in the white flour does not matter. But for many Americans this is not true. The diet is low in these very vitamins as well as in iron and calcium. Iron is likewise rather abundant in the branny layers of the wheat kernel but largely removed on milling. For many years nutritionists and doctors have recommended the use of

**TABLE 9.3. STANDARDS FOR ENRICHMENT OF FLOUR
EFFECTIVE OCTOBER 1, 1943**

	Content per lb	
	Min.	Max.
Required Ingredients:		
Thiamin, mg	2.0	2.5
Riboflavin, mg	1.2	1.5
Niacin or Niacin Amide, mg	16.0	20.0
Iron, mg	13.0	16.5
Optional Ingredients:		
Calcium, mg	500	625
Vitamin D, U.S.P. Units	250	1000
Wheat Germ	Not more than 5 per cent	

whole wheat. But the recommendation has been largely unheeded. Whole wheat does not produce as large or light a loaf of bread or as light a cake as does white flour. In recent years many have advocated the enrichment of white flour to its original level with thiamin, niacin, riboflavin, and iron. In 1938 a new Food, Drug, and Cosmetic Act, which became effective on October 1, 1943, prescribed the amounts of these compounds which must be incorporated in enriched flour and the amounts of calcium, vitamin D, and wheat germ which are allowed. See Table 9.3. Some nutritionists have opposed the addition of calcium and vitamin D, since they are not naturally present at those levels.

There has also been interest in the fortification of bread with lysine. The proteins of wheat flour contain little lysine; and when they are the sole source of amino acids in the diet, an animal soon develops deficiency symptoms. Low priced diets for man often contain bread or other wheat products in abundance since they supply sufficient calories cheaply. In these diets there are not large amounts of other proteins to make up for the lack of lysine in wheat and mild deficiencies can occur. Two methods are now used for this extra fortification: (1) addition of free lysine to the dough and (2) use of skimmed milk powder. Addition of lysine does not alter the dough structure, but incorporation of much skimmed milk powder produces a smaller, more compact loaf.

In many parts of the world the staple food for man is either rice or corn. Neither of these cereals have complete proteins. Rice is low in both lysine and threonine, while corn is low in lysine and tryptophan. Rosenberg[19] has shown that growth of white rats is improved by supplementation of a diet of cooked rice (90 per cent) with lysine up to 0.1 per cent but that improvement falls off at higher levels of lysine. On 90 per cent corn meal diet rats grow slowly, but this growth can be improved by the addition of up to 0.05 per cent lysine hydrochloride.

The *mineral* elements are present in the plant cells sometimes as salts and sometimes as part of organic molecules. We know, for example, that phosphates occur as hydrogen phosphate, $HPO_4^=$, and dihydrogen phosphate ions, $H_2PO_4^-$, but they also occur as the acid portion of many organic esters such as glucose-1-phosphate, lecithins, phosphopyruvic acid, and many others. The mineral elements are determined from the ash of natural products where the organic molecules have been destroyed and only the noncarbonaceous elements remain. Cereals contain a large number of elements, although many of them are present in trace amounts. Calcium, magnesium, potassium, sodium, iron, phosphorus, chlorine, sulfur are all present, and in the cereal seeds that contain hulls (covered grain) there are sizeable amounts of silicon. Traces of aluminum, arsenic, boron, bro-

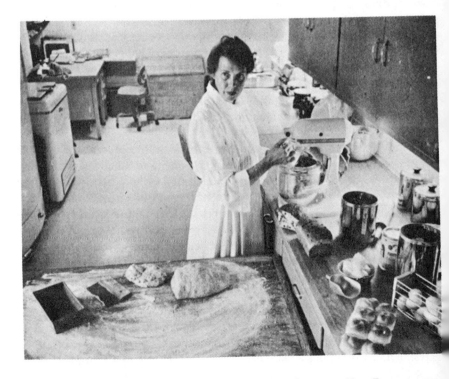

FIGURE 9.1. MRS. CATHERINE CLARK OF BROWNBERRY OVENS IN HER EXPERIMENTAL KITCHEN. A high protein bread supplemented with lysine is one of her specialties.

Courtesy of Brownberry Ovens.

mine, copper, cobalt, iodine, lithium, tin, selenium, titanium, and zinc are present. As a group the cereals are not important sources of the mineral elements; and since the bran is richer in most of these elements than the endosperm, highly milled white flour and products baked from it are poor sources of minerals. Calcium, the mineral most often poorly supplied in man's diet, occurs to the extent of only 0.15 mg per g in corn, 0.5 mg per g in wheat, and 1 mg per g in oats.

WHEATS AND THEIR BAKING QUALITIES

Wheats grown under different conditions of soil and climate and wheats of different varieties have different baking qualities. Some wheats yield a flour which will produce a loaf of bread of great height and good texture, others will form a small loaf of poor texture. Likewise in batter products the lightness and grain depend on the flour used.

Today in the United States technology of foods has progressed to the point where the consumer expects a uniform product and will not be satisfied with less. In order for the baker to turn out a loaf or cake exactly like the one produced last week, it is necessary to understand the factors which account for a good loaf or a poor, a good cake or a poor one. During the past fifty years much research has been done on flour products. But so complex is a loaf of bread or a cake in its chemical and physical-chemical relationships that all of the questions have not been answered. The problems group around three main areas: gas production, gas retention, and tenderness. In the yeast-leavened doughs of bread, rolls, coffee cake, and crackers the problem is a little different from that of the batters of cakes, cookies, muffins, and biscuits. Each will be briefly discussed under Doughs and then under Batters.

Doughs

Gas Production in Yeast-Leavened Doughs. The common agent for the production of gas in doughs is a selected strain of the yeast, *Saccharomyces cerevisiae*.[18] The yeast through its life process forms carbon dioxide as an excretory product and in so doing leavens the dough. The yeast also grows during the fermentation period and produces numerous offspring which likewise excrete carbon dioxide. The amount of gas formed depends on the strain of yeast and on the number of organisms present during fermentation. Some strains of yeast produce more gas in a given time than others. The strain of yeast is also important because carbon dioxide is only one of many compounds excreted, some of which have flavors.

"Starters" and Cultures. In ancient times and even today in countries where technology is not as advanced as in the United States, the yeast used

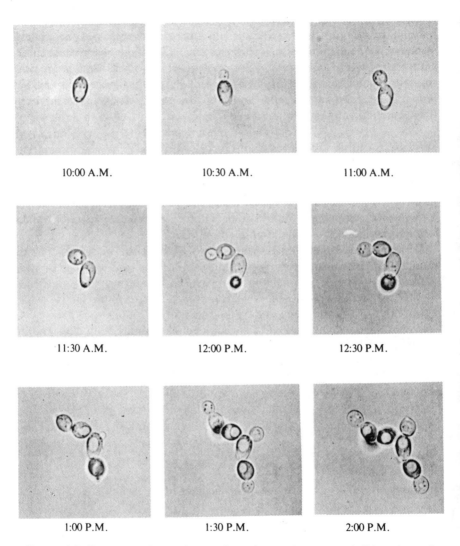

| 10:00 A.M. | 10:30 A.M. | 11:00 A.M. |

| 11:30 A.M. | 12:00 P.M. | 12:30 P.M. |

| 1:00 P.M. | 1:30 P.M. | 2:00 P.M. |

FIGURE 9.2. GROWTH OF YEAST CELLS. Photomicrographs taken at half-hour intervals. Magnification × 1000.   *Courtesy of Fleischmann Laboratories. Standard Brands, Inc.*

by bakers is a "starter," saved from the last baking. The "starter" is often a mixture not only of yeasts but also of bacteria and the rate of growth of the various organisms depends on the nutrients present as well as the temperature at which they are grown. Contamination with "wild" yeasts and bacteria readily occurs in kitchen or bakery. A baker might have a fine "starter" today but a different one next week, and the flavor of the loaf varies with the organisms.

It took many years for the companies which produce yeast to learn how to adequately protect their cultures and encourage the growth of selected strains of organisms as well as how to package them so that they reach the baker uncontaminated and highly viable. Much effort is devoted to maintaining a uniform strain. Today uniform yeast cultures are available either as compressed, dried, or powdered yeast.

In "sour dough" products bacteria which produce carbon dioxide and organic acids are used for either part or all of the leavening action. Although these organisms were formerly preserved as starters, most bakers today use cultures or some source of a uniform culture like cultured buttermilk instead of depending on an unreliable and variable "starter." Some of the organisms used are the acetobacters which form acetic acid, *Streptococcus lactis*, and the lactobaccilli, *L. acidophilus, L. bulgaricus,* and *L. casei*, which form lactic acid.

The Nutritional Needs of Yeast. Since a yeast cell is a living organism, it has numerous nutritional needs and it is only if these are met that it will grow vigorously and produce a large quantity of carbon dioxide. Some form of easily available carbohydrate and a utilizable source of nitrogen, calcium, and phosphate ions are important in the dough for rapid gas formation.

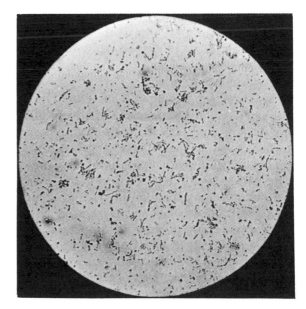

FIGURE 9.3. COMMER-CIAL TYPE BREAD SOURING ORGANISM STREPTOCOCCUS LACTIS. Magnification × 800.
Courtesy of Fleischmann Laboratories. Standard Brands, Inc.

Yeast cells use a number of mono- and disaccharides. These sugars are metabolized to ethyl alcohol and carbon dioxide. Glucose and fructose are the monosaccharides likely to be in yeast-leavened doughs, while sucrose and maltose are the disaccharides. Lactose may be present but is not used by the yeast cells. The sugar in flour is mostly sucrose with only small amounts of glucose and fructose. Most of the glucose and fructose is in the wheat seed in the germ; and since this is chiefly removed in the milling, the amount in the flour is low. Sucrose is present to the extent of 1 to 2 per cent in the flour, and a yeast, flour, and water mixture will ferment rapidly until this is used up. Too much sucrose, however, will slow down the rate of fermentation. If a very sweet dough is prepared by adding 10 per cent or more of sucrose at once, the growth of the yeast and the formation of carbon dioxide may be slow. Maltose is not present in flour to any extent; but when the dough is made, the amylases begin to hydrolyze starch and form maltose.

Amylase Content and Activity. The amylase (or diastase) content of the dough has an effect on the rate of gas production by the yeast because these enzymes form sugars from starch. Two types of amylases occur in nature, α-amylases which hydrolyze the internal 1–4 links in the amylose chain or the amylopectin brush, and β-amylases which hydrolyze the 1–4 links second from the end of the chain. Bernfeld[2] says that ungerminated cereals do not possess α-amylases but that they are formed during germination or malting. β-Amylases occur in both the ungerminated and the germinated cereals. α-Amylases through their hydrolytic action on internal 1–4 links rapidly produce dextrins of various sizes and then form smaller and smaller molecules. The final products are maltose and some glucose from amylose, but there may be larger molecules from amylopectin since the 1–6 links are not attacked. (See p. 76.)

β-Amylases hydrolyze the 1–4 link second from the end of the chain and consequently rapidly form maltose. On hydrolysis the configuration of carbon 1, the potential aldehyde, is changed from the alpha to beta configuration. Under optimum conditions β-amylase can catalyze the hydrolysis of the entire amylose chain to β-maltose; but with amylopectin, the 1–6 links are not attacked and a terminal dextrin is formed.

β-Amylase is found naturally in flour in small amounts and both α- and β-amylases in greater amounts in malt extract, malted barley, or malted wheat but not in yeast. Since the formation of maltose from starch is essential for the adequate growth of yeast in an unsweetened dough, the level of these enzymes in the dough is of paramount importance. They can be added to the dough or flour either in the form of malt extract or as an enzyme concentrate. Today most millers carefully regulate the amount of amylases in the flour so that the flour will have proper gas production qualities when it is inoculated with yeast. Some patent flours are low in amylase activity, and this is rectified by the addition of malted wheat flour or malted barley flour.

If amylase activity is great enough, the production of maltose will keep up with the demands of the speedily growing yeast and carbon dioxide formation will be rapid. Where amylase activity is low, the addition of dextrins and soluble starch will increase gas formation. Preparations composed of these compounds do not improve gas formation except when amylase activity is low. Overgrinding has a similar effect in the presence of low amylase activity by rupturing some of the starch granules and making them more readily hydrolyzed by that amylase which is present.

Amylase activity has an optimum pH of 7 and, if the dough is at some other pH, then amylase activity is lower. The addition of milk to the dough raises the pH because of the presence in the milk of buffer salts. Milk consequently retards amylase activity. However, in the presence of acid salts such as calcium hydrogen phosphate or the acetic acid of vinegar this retardation may be eliminated, and gas formation may even be increased by the milk through the improved nutrition of the yeast.

During baking the starch held in granules imbibes water, swells, and is gelatinized. The β-amylase from either malted barley or malted wheat is inactivated at a higher temperature (approximately 175°F) than the temperature for the gelatinization of wheat starch (approx. 150°F). Consequently, during the early stages of baking β-amylase affects the gelatinized starch. The baked loaf achieves an internal temperature close to 212°F and at this temperature enzymes are destroyed. See figure 9.4.

Malt, malted wheat, and malted barley are the sources of enzymes commonly used. These products contain not only α-amylase but proteinases as well. Both types of enzymes exert a pronounced effect on the dough, and the level at which they are added can be critical. In these products the level of enzyme activity is variable. Today purified enzyme extracts from fungi or bacteria are available and considerable investigation is being devoted to their use in doughs. With more sources of enzymes in which the amylase or proteinase levels are more accurately known, it will be possible to control dough performance more carefully.

"Dough conditioners" or "flour improvers" are on the market and contain calcium and phosphate ions. Both of these ions are necessary for yeast growth and gas formation but flour contains them in sufficient amounts for good leavening under normal conditions.

Pre-ferments. These have been used for many years in bread making, but recently their effectiveness in commercial baking has attracted attention. A

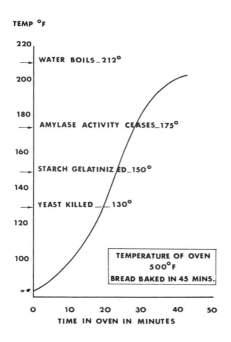

TEMP °F

pre-ferment is a more or less fluid mixture to which yeast is added and in which it grows rapidly. It differs from a sponge in the absence of or in a low level of flour. Potato water and yeast or a liquid mixture of mashed boiled potato and flour are examples of pre-ferments which have been used in the past. Johnson, Miller, Refai, and Miller[10] point out that Parisian Barm was brought from Paris to Scotland in 1865. It was a pre-ferment made of malt, salt, and flour inoculated with yeast and was used not only as a source of yeast but also to improve the flavor of bread. While the yeast is developing in the pre-ferment, it forms not only alcohol and carbon dioxide but other compounds such as organic acids which affect flavor.

Johnson, Miller, Refai, and Miller have studied six pre-ferments which differ in carbohydrate used (either glucose or sucrose) and the presence or absence of dry milk solids and flour. Their mixtures provide the nutrients demanded by yeast for rapid growth. In all mixtures containing sucrose, the carbohydrate disappears immediately when it is mixed with yeast. This rapid hydrolysis of sucrose by yeast had been known before. There are no marked differences in the six pre-ferments at the end of 5 or 6 hours. When milk solids are present, the rate of utilization of the carbohydrate is slightly slower and because of the buffering action of the milk salts, the pH does not fall as low. There is a slightly greater formation of carbon dioxide in the presence of milk solids. Production of organic acids is approximately the same in all pre-ferments, reaching a maximum in 3 hours and then falling.

Gas Retention in Doughs. When flour is mixed with water, a change takes place in the physical-chemical properties of the gluten. It has been customary to consider gluten the product of two protein fractions of wheat—gliadin and glutenin. (See p. 322.) It is in wheat flour and only in wheat flour that this elastic, rubbery protein develops when it is stirred with water. The ability of a dough to retain gas depends directly on the amount and quality of the gluten developed. The volume or lightness of a wheat loaf is therefore directly dependent on the factors which contribute to the development of gluten and which influence its elasticity. Doughs made of other cereals are not able to retain all of the gas generated and consequently are heavy and coarse textured. The factors of major importance for gluten development and strength in most doughs are: (1) mechanical action, (2) the amount of proteolytic enzymes, (3) the amount of oxidation of the flour either through aging or the addition of oxidizing agents, (4) the addition of milk, inorganic ions, etc., and (5) the addition of sucrose and fat—of importance in sweet rolls and coffee cake.

Mechanical Action. The mechanical action to which the dough is subjected either through mixing or knéading is important for the development of the gluten. Elastic strands of rubbery gluten form as the flour is mixed with water. Undermixing produces a dough in which there has not been sufficient gluten development to retain the gas well and the result is a loaf which has poor volume and is heavy. Overmixing or rough handling of the dough before panning has been reported to decrease the volume of the loaf. When flour and water are mixed, an increase in plasticity occurs as the gluten is developed. But if mixing is continued, a decline in plasticity occurs and the dough finally becomes slack and sticky.

Proteolytic Enzymes. These are a group of enzymes present in wheat and flour which catalyze the hydrolysis of proteins. They are also present in malted wheat flour, malted barley, malt extract, and yeast. Since the strength of the gluten depends on the intact protein, any reaction which hydrolyzes part of the protein reduces the amount of gluten. If too much of these proteolytic enzymes are present, too much hydrolysis occurs and the dough becomes sticky, difficult to machine in the mixers, and yields bread of poor volume. However, some protease activity is desirable since it improves the gluten. Doughs of low proteolytic activity are called "bucky" because the gluten is tough and inelastic. It does not machine well and it also produces loaves of poor volume because the dough will not stretch around the gas bubbles. The complexity of the problem, "what is gluten," is well demonstrated here. The quantity of proteolytic enzymes must be balanced so that enough hydrolysis occurs to produce an elastic gluten but not so much that the gluten is sticky. The addition of malt extract, malted wheat flour, or malted barley must be limited to avoid an excess of proteolytic activity.

Oxidation. Oxidation of flour likewise influences the buckiness or stickiness of the gluten developed and here, too, a nice balance must be achieved. It has been known for many years that storage of flour at moderate temperatures for several months lightens the creamy color and improves the baking quality. Eventually it was found that oxidizing agents can achieve this effect without the costly and hazardous period of storage. In the United States the Pure Food, Drug, and Cosmetic Act of 1938 allows the use of chlorine, chlorine dioxide, nitrogen oxides, nitrosyl chloride, and benzoyl peroxide. Nitrogen trichloride, "Agene," was used for many years but on August 1, 1949, it was removed from the list of permissible agents. See p. 352. It was found that dogs, cats, and a number of other species fed large amounts of flour bleached with Agene or bread made from that flour developed running fits and convulsions. Although no harmful effects were found in man, the compound is no longer used. Chlorine, chlorine dioxide, and nitrosyl chloride bleach and mature the flour, improving the properties of the gluten formed in working the dough. Nitrogen dioxide and benzoyl peroxide do not affect the baking properties of the flour at the levels used in milling but only exert a bleaching action.

Potassium bromate, $KBrO_3$, and potassium iodate, KIO_3, are often added as dough conditioners. They are oxidizing agents and are believed to have an effect similar to oxidizing agents added at the time of milling. The amount of potassium bromate which exerts this effect is surprisingly small, of the order of one or two thousandths of one per cent.

The effect of either oxidation or reduction on flour and dough has been studied for many years, but the exact nature of the reaction is still in doubt. If an oxidizing agent is added either to flour or to gluten, the strength of the gluten is increased. If relatively large amounts are added, the gluten becomes tough with little elasticity. Reducing agents have the opposite effect. They cause a decrease in the strength of the gluten, making it more extensible and sticky. When a reducing agent is added to gluten, a ball of it loses the ability to stand up and soon softens and flattens. Sullivan[21] points out that numerous factors influence the physical properties of gluten—fermentation, oxidation, the work of mixing, rounding, braking—and that all of these can be considered active in changing the bonding of the protein molecule. She reviews some of the evidence which indicates that oxidation affects the sulfhydryl groups and increases the number of sulfur-sulfur links.

$$2 -SH + (O) \rightarrow -S-S- + H_2O$$

If the sulfur-sulfur links are formed between polypeptide chains, they will hold the molecule together more firmly and increase its strength. Overoxidized flour can be used satisfactorily by increasing the working which it receives in the mixers or in braking. This can be considered an operation in which some of the less stable bonds between polypeptide chains are damaged and the molecule is rendered more extensible.

Whether the effect of oxidation is on sulfhydryls or some other group, it is an important reaction in practical baking. The amount of oxidation is critical and although the amount of oxidizing agent needed is small, sufficient must be present to yield gluten strong enough to retain the gas formed but not so tough that it will be unable to stretch around the gas bubbles.

Other Factors. Several other ingredients may have an effect on the strength of the gluten. Raw or pasteurized milk decreases the baking qualities of a flour unless the milk is first heated. It is believed that milk contains some substances which increase the activity of the proteolytic enzymes and consequently during the fermentation period foster the formation of a gluten which is too sticky. Heating the milk to 180° F for 30 minutes destroys these unknown substances and the milk, then, has no detrimental effect on gluten strength. In home baking it is always recommended that the milk be scalded before use in a dough. Pasteurization also provides a period of heating but for a much shorter length of time or at a lower temperature. The U. S. Public Health Service requires that pasteurized milk be held at 143° F for 30 minutes or at 160° F for 15 seconds.

When the fluid used in a dough is water, the softness or hardness of the water has an effect on the gluten. In general, calcium salts present in the hard water tend to increase the elasticity of the gluten. Sodium chloride likewise affects the gluten. Acids also alter gluten strength. Vinegar is sometimes added to dough, but too much must be avoided so that the gas retention is not diminished. Fats in small amounts increase the ability of the dough to retain gas.

The production of a uniform high quality loaf or roll is complicated by many factors. Much empirical work goes into the formulation of recipes and the control of the quality of the ingredients. The complex problem of the physical-chemical properties of a dough are still incompletely understood.

Batters

Gas Production in Batters. The satisfactory production of baked goods from batters involves the same type of problem as from doughs: adequate gas formation and gas retention. In batters the leavening agent is either air which is incorporated through beaten egg white or during creaming or it is a gas generated by chemical reaction in the batter from a baking powder or from ammonium bicarbonate. Often a combination of leavening agents is used. Thus in the production of some butter cakes, creaming sugar and fat introduces air and later in the mixing the addition of beaten egg white introduces still more air. Baking powder is added which in the moist batter and at the oven temperature forms carbon dioxide. Steam also acts as an important leavener.

Baking powders are composed of sodium bicarbonate, $NaHCO_3$, an acid salt, and starch. Occasionally some other ingredient such as a small amount of egg white is present as a drying agent. Starch helps keep the ingredients dry, prevents caking and standardizes the baking powder so that for a given volume of dry baking powder the amount of gas formed is equal to that of other baking powders. The chief difference is, therefore, in the compound or compounds which produce hydrogen ions (hydronium ions) to react with sodium bicarbonate in the wet batter and release carbon dioxide.

$$NaHCO_3 + H^+ \rightarrow Na^+ + H_2O + CO_2$$

Phosphate baking powders contain either primary calcium phosphate, $Ca(H_2PO_4)_2$ or disodium pyrophosphate, $Na_2H_2P_2O_7$. Those on the consumer market contain primary calcium phosphate, while some of the phosphate powders for bakers contain disodium pyrophosphate. *Tartrate* baking powders contain potassium acid tartrate (cream of tartar), $KHC_4H_4O_6$, as well as tartaric acid, $H_2C_4H_4O_6$, as a source of the hydrogen ion. S.A.S. baking powders contain sodium aluminum sulfate (alum), $NaAl(SO_4)_2$. The hydrated aluminum ion hydrolyzes on contact with water and forms a hydronium ion (hydrated hydrogen ion).

$$Al(H_2O)_6^{+++} + H_2O \rightleftarrows H_3O^+ + Al(H_2O)_5(OH)^{++}$$

The phosphate and tartrate baking powders react readily at room temperature to form carbon dioxide. S.A.S. baking powder forms little or no carbon dioxide at room temperature, and it is only at oven temperatures that leavening occurs. If much time elapses between mixing and baking, then the slow acting S.A.S. baking powder is much better. Leavening at room temperature is not effective unless baking follows immediately so that the gas formed will not escape from the batter. *Double action* baking powders that contain both sodium aluminum sulfate and primary calcium phosphate are available on the consumer market. An S.A.S. powder has the disadvantage of forming sodium sulfate which has a distinct bitter flavor and is therefore seldom used alone.

Ammonium bicarbonate or ammonium carbonate releases gas by decomposition.

$$NH_4HCO_3 \rightarrow NH_3 + H_2O + CO_2$$

They are used as a leavening agent by bakeries for cookies and are occasionally added to cream puffs and eclairs to improve their volume. The ammonia formed adds a faint flavor and raises the pH so that the color and spreading of the cookie are affected. In home baking ammonium bicarbonate or carbonate is seldom used because it decomposes readily and is difficult to store. Old cookie recipes sometimes call for "spirits of hart's horn," a solution of ammonium carbonate. Since ammonium bicarbonate reacts quickly, it is sometimes added to produce rapid leavening. In preparing base cookies which are topped with marshmallow, a soft, well leavened, rather thick cookie is required. In the oven the ammonium bicarbonate gives a quick spring before the cookie spreads excessively.

Air is a most important leavening agent in many types of batters. In pound cake and angel cake it is traditionally but not always actually the only leavening agent. The old pound cake recipe which used a pound each of butter, sugar, flour, and eggs is leavened by thorough creaming of the fat and sugar. As the mix becomes fluffy, small bubbles of air are adsorbed and in the oven the expansion of these bubbles raises the cake. Often today small amounts of other leavening agents are added in order to increase the volume and lightness of the cake. In angel cake the egg white foam holds the air, and mixing is controlled so that the maximum amount of air is incorporated and the minimum lost while the other ingredients are added. In sponge and butter cakes baking powders are used to produce carbon dioxide and achieve leavening, but air is also important and the amount incorporated in the egg white or whole egg influences the final volume of the cake.

Steam is the sole leavening agent in popovers and cream puffs. Commercially cream puffs and eclairs occasionally have small amounts of other leavening agents added to increase the volume produced. Steam is important in all batters since the vapor pressure of water increases rapidly with increased temperature. During the cooking of all batter products, whether doughnuts, griddle cakes, or oven heated cakes, the steam formed is one of the agents that gives a light porous structure.

Gas Retention in Batters. Batters are almost always extremely complex mixtures and the specific effect of many components of these systems is for the most part only partially understood at present. The production of a light, tender, high volume cake, doughnut or muffin requires the attainment of a delicate balance between elasticity and rigidity, cohesiveness and crumbliness, tenderness and toughness. Recipes for batters have developed through empirical methods where the cook tried various combinations and stumbled on some that were acceptable. Today enough research has gone into batter products so that some rules of thumb for balancing cake formulas, for instance, have been put together. Although the effect of each ingredient is known in a general way, the day when these effects can be explained by physical chemistry is still some way off. Those factors which make the batter elastic so that it can stretch around gas bubbles and not allow them to coalesce or escape from the top, are: (1) the gluten developed in the flour, and (2) the proteins of milk and eggs, particularly eggs. Those that decrease elasticity and consequently prevent a rubbery texture, thus increasing tenderness, are: (1) fats and (2) sugar.

Flour is the source of the gluten in the batter and consequently the type of flour and all its characteristics has a most important bearing on the qualities of the batter product. In bread flours a high quality gluten is desirable, but in batter products the amount of gluten development possible and the quality should be low. In general, flours produced from soft wheat yield better cakes than those from hard wheats. The gluten from soft wheats is much weaker than that from hard wheats and often, but not always, the total quantity of protein is lower. Flour formed in the shortest separation during milling, which contains the highest amount of starch and the lowest amount of protein, yields the best cakes. The granulation of the flour also has considerable effect. The finer the granulation, the higher the score of cakes baked with it. Cake flour is as finely granulated as is economically feasible.

The amount of mixing has an effect on the extent of gluten development and consequently is one of the factors which contributes to the strength of the batter. Lowe[13] and her students have found that with a standard cake recipe cake volume increases with mixing up to a maximum and then decreases. The batter of the overmixed cakes flows from the spoon in long ribbons, suggesting that gluten has been well developed. Methods of mixing, whether creaming, blending, or muffin method, yield cakes with different volumes and different scores (see p. 347).

The effect of egg protein on the structure is well demonstrated when muffins or one-egg cake is compared with cake made by conventional recipes. The grain tends to improve with the addition of egg up to a maximum and then, although the grain remains fine, the volume decreases and the texture becomes rubbery. It is also possible that the milk proteins are important in the final structure. The gluten and other proteins have two effects here—elasticity in the batter and rigidity after baking. In the batter it is essential that some elasticity occur to allow formation, retention and expansion of gas bubbles. It is probably the gluten which develops during mixing that is primarily responsible for elasticity in the batter. Although the ability of egg white and of whole egg to form elastic films is demonstrated when egg-white foams or whole-egg foams are prepared, the tenacity of these foams is not great. This can be observed when the foam is incorporated into the batter. Overmixing and rough handling must be avoided so that loss of gas does not occur. During baking the proteins are denatured and a rigid structure results which maintains itself around the gas bubbles even after the cake has cooled and the gas in the bubbles has contracted. Milk and egg proteins probably augment the effect of the flour proteins.

Fat. Fat has a two-fold influence on the retention of gas by the batter. Many fats available today, particularly the hydrogenated fats, contain emulsifying agents. These agents not only emulsify the fat so that it is better distributed through the batter, but probably also function to help retain the tiny bubbles of gas in the batter. Moreover the fat interferes with the development of gluten. This allows the protein particles to slide on one another and the product is tender when bitten or cut.

The emulsifying agents which can be used are fairly numerous. The mono- and diglycerides as well as other esters of fatty acids such as sorbityl stearate and the lecithins are useful. The addition of these emulsifying agents at 3 per cent of the total fat causes an increase in the volume of the cake as well as in the overall quality score. These agents also permit the addition of larger amounts of sugar and of moisture to the mixture, producing a sweeter, moister cake which is softer, which can retain moisture longer, and has better keeping qualities. Batters that are made with fats containing emulsifier do not show a tendency to separate or break into curds.

Study of the structure of the cake by staining in order to identify fat, starch, or protein shows that the fat tends to occur in tiny lakes surrounding the gas bubbles with some distribution through the crumb. The method of mixing, the temperature, and the kind of fat influence the distribution of it through the cake.

Sugar. In cakes and cookies sucrose is one of the principal ingredients which makes for tenderness. In the presence of sucrose development of gluten is retarded and restricted. Cakes in which the weight of sucrose is equal to or exceeds that of the flour (by measure, 1/2 cup sugar to one cup flour) are called high-sugar cakes and are very popular. Those in which the weight of sucrose is less than that of flour are called low-sugar cakes. High-sugar cakes are very sweet and have a fine texture with few tunnels. Cakes in which both the sugar and moisture are high (moisture includes the liquid

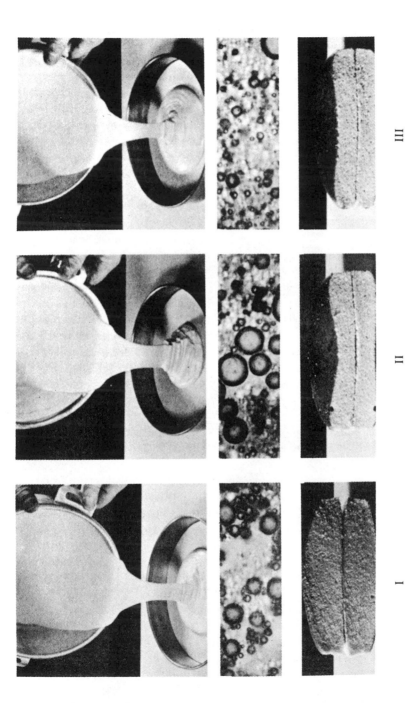

FIGURE 9.5. EFFECT OF EMULSIFIER ON BATTER AND CAKE STRUCTURE. Magnification of batter 100×'s. I. Excellent batter, rather thick, slightly curdled. The air bubbles are coated with fat and a few dark pools of reserve fat can be seen. Cake has good volume and fine texture. II. A type of emulsifier has been added which promotes the formation of large air bubbles with thick walls. Batter is smooth and viscous with no curdle. On baking, large bubbles finally burst and cake falls. III. Another type of emulsifier. Bubbles are small but with thick walls. Batter is thin and smooth. In oven bubbles break and cake has coarse grain and sunken center.

Courtesy of Distillation Products Industries.

added in milk and eggs) are high-ratio cakes and are possible today because of the emulsifying agents added to fats. (See p. 342.) Lowe and her students[13] have likewise demonstrated that high-sugar cakes require more mixing than low-sugar cakes in order to achieve optimum volume during baking and prevent falling when removed from the oven. Too much sugar causes a sticky crust and a gummy texture.

Sugar has a marked effect on the hydration of proteins and starch which occurs in any batter or dough and must through this competition for water affect the physical properties of the baked product. Proteins hold water on their surfaces and within the folded or coiled structure of the molecule by hydrogen bonding. A well-hydrated protein is probably a tender one in which the adherence of one protein molecule to another is weak and easily sundered. The starch hydrates and forms a gel in doughs and batters, particularly during baking. The extent of gel formation must also have a marked effect on the physical properties of the baked goods. Sucrose as a small molecule with many polar groups has an attraction for water and will compete with protein and starch molecules for it. Undoubtedly the differences in high-sugar and low-sugar cakes reflect the competition of sugar, protein, and starch molecules for water.

pH. The effect of pH on cake quality has been studied and the effect on texture observed, but there has been no direct study of the influence of pH

FIGURE 9.6. CAKE PREPARED WITHOUT SUGAR BUT WITH SUCARYL. It is difficult to produce good grain and volume without sugar. The recipe must be very carefully balanced.

Courtesy of Abbott Laboratories.

on gas retention. Stamberg and Bailey[20] reported that the optimum pH for white and yellow cakes could be narrowed to 7.22–7.35. These cakes were made with either phosphate or tartrate baking powders and those in which the pH was markedly different were inferior.

The ingredients which influence the pH of a white or yellow cake are the baking powder and the buffer salts which are present in milk. Disodium phosphate formed from a phosphate baking powder will give a pH a little above 7, depending on its concentration. Calcium phosphate is quite insoluble and has little effect. Sodium potassium tartrate from tartrate baking powders gives a pH above 7. Sodium sulfate, the product of an S.A.S. baking powder, gives a pH very close to 7. If bicarbonate of soda is added to any recipe, it raises the pH and if sodium carbonate is formed during baking from any soda which has not reacted with acid the pH is high. Acids are added to some cakes (but not usually to white or yellow) by use of sour milk, molasses, honey, or chocolate and cocoa. Since the amount of acid varies with the sample, it is very difficult to use these and achieve exact or close to exact neutralization of the soda. If the acids are present in insufficient amounts, the cake will be alkaline and have a high pH; if they are in excess, part of these acids will vaporize during baking but the pH of the cake will be lower than if they had not been added. Cream of tartar (potassium acid tartrate) is sometimes added to lower pH. Daniels and Heisig[6] measured the acidity of some of the syrups used in baking, many years ago, and found considerable variation. One cup of molasses varied from 12.8 to 22.6 ml of 0.1 N NaOH for neutralization.

Stamberg and Bailey[20] give the following pH values for commercial cakes:

White Cake	7.47
Yellow	7.59
Chocolate	8.48*
Angel Food	5.67

*Must have been made with soda.

Cathcart[4] gives the following pH ranges for commercial cakes:

Angel Food	5.2–5.6
Chocolate	7.2–7.6?
Devil's Food	7.5–8.0
Fruit Cake	4.4–5.0
Pound	6.6–7.1
Sponge	7.3–7.6
White	7.1–7.4
Yellow	6.7–7.1

The texture of a cake changes with changes in pH, tending to be finer and more compact at lower pH while it becomes coarser on the alkaline side. We would expect from our knowledge of the change in characteristics of proteins with change in pH that the elasticity and toughness of the proteins from flour, milk, and eggs are influenced by pH. This variation of texture with pH must surely reflect effects on protein.

In general the flavor and overall quality score for cakes is better when the pH is lower. The length of time for the development of staling is likewise prolonged with lower pH.

Chocolate and Devil's Food cakes have different meanings to different bakers and a wide range of pH values are found for commercial as well as home made cakes. (Hence the question mark which Cathcart has placed after the pH range given for chocolate cakes.) As the pH rises, the color changes through cinnamon brown to brown and then becomes more and more red. The flavor of the chocolate likewise changes. Many bakers add so much extra soda to Devil's Food cake that the pH is above 8, the color is very red, and the flavor very poor. The bitter soapy taste of sodium carbonate is readily detected and is most objectionable.

Baking Temperature and Time. Numerous studies have been made of the effect of baking temperature and time on the qualities of cakes. It is important that the batter temperature rise so that baking powder, particularly S.A.S. powders which react very slowly at room temperature, will form CO_2 and that the gases will expand. It is also essential that the walls set, so that the gas is not lost and the bubbles do not coalesce and form a coarse grain. In general it has been found that a relatively high temperature and consequently a short baking time starting with a preheated oven give the biggest volume and the highest scoring product. Charley[5] found that a cake baked in Japanned iron had a better volume when baked 24 minutes at 365° F than 27 minutes at 345° F. In tinned iron she found a temperature of 395° F preferable.

Moisture. The amount of moisture in a cake and the manner of its distribution are doubtless factors important to the quality of a batter product. The feeling in the mouth of moistness or of dryness adds a great contribution to the enjoyment of eating. Moisture is held in a batter or dough product as the solvent for some compounds, particularly sugar, and as a hydrate of protein and starch. During baking, proteins are denatured and the amount of water held in the molecule changes. The denatured protein can still hold water through hydrogen bonding. (See p. 122.) Starch swells and is gelatinized. The granules swell, water passes into the molecules, and is held there (see p. 104).

Staling

Staling is a process which occurs in all dough and batter products and renders them less desirable for eating. Attempts to understand and control the process have been made for many years; and although some progress has been made, there is still a lot to learn. Most of the work has been on bread, but the same process occurs in any flour mixture. In bread the crust loses its crispness and becomes leathery, while the crumb becomes hard and crumbly and dries out during the first two or three days after baking. The hard crumb first appears just under the crust, but with loss of moisture it progresses through the whole loaf. At the end of two or three weeks the entire loaf has become hard and dry. Early in the process the staling can be reversed by heating the loaf. Quantitative estimates of staling have been devised using (1) a penetrometer which gives a measure of the hardness of the crumb and (2) the amount of swelling which occurs when a crumb is placed in water. Fresh bread swells much more extensively than stale. For bread which is stored at 60° C or at very low temperature staling is very much delayed. But the flavor at 60° C suffers deterioration. Freezing bread and storing it well below the freezing point is a practical and much used method of maintaining baked products for considerable periods of time. The most rapid staling occurs at -2° C to -3° C.

Staling appears to be associated with changes in the starch. Chemically there is a loss in the amount of extractable starch which is obtained from bread as staling proceeds. Some workers believe that it is the relatively low molecular weight amylopectins which are involved while others believe it is the straight chain amyloses. Although the process is not completely analogous to the changes which occur when a simple starch-water paste or gel undergoes retrogradation, there are many indications that the process is similar. Today most investigators conclude that the basis of similarity between staling and retrogradation is that both are caused by a reorientation of the hydrogen bonds. It is thought that during both processes water molecules associated with starch molecules are lost. This leads in retrogradation to marked orientation of the starch molecules and the development of crystalinity. In staling the orientation does not appear to be as complete.

The addition of emulsifying agents such as mono- and diglycerides and other compounds similar to them lengthens the keeping quality of bread and delays staling. They are now extensively used in the United States. Fats up to 2 per cent of the flour improve bread and slightly retard the staling of the loaf. Addition of protein to the dough gives a softer crumb but a stronger structure to the loaf and produces a slight delay in staling. Part of the delay is apparent because of the initial softness of the crumb.

Staling retardants are appearing in the patent literature. In 1956 U. S. Patent 2,744,825 was issued to Thompson and Buddenmeyer for fatty acid lactylates. These compounds are patented for use in yeast-leavened doughs at concentration of 0.1 to 1 per cent of the weight of the flour. Their general structure is

$$RCO(-O\overset{\overset{\displaystyle H}{|}}{C}CO)_n-OX$$
$$\underset{CH_3}{|}$$

Where RCO is a fatty acid chain with 16-22 carbon atoms, n varies from 1 to 8 and X is a cation.

Scoring Baked Products

Many scoring forms have been developed by industries producing large amounts of baked products, by laboratories, and others interested in evaluating quality. Sometimes a rating of 1 to 10 is used for each attribute, and sometimes other numbers. Whatever the method a comparative numerical score is obtained that is useful in controlling the quality of the products or evaluating changes in methods and ingredients. Usually the qualities evaluated are: appearance, color of crust and crumb, aroma, texture, moistness, tenderness, flavor, symmetry and general eating quality. Occasionally odor and flavor may be combined since they are so intertwined and often reflect the same quality. Appearance is sometimes divided into general appearance including smoothness, greasiness, pebbling and sugary as well as the general contours such as humped, sunk, flat, etc. and into another evaluation for the volume of the product such as too large, good, too small. Texture refers to the grain, the thickness or thinness of the cell walls, the size of the gas bubbles, and occasionally includes a measure of tenderness.

Prevention of Mold

Prevention of mold on bakery products has been the object of much research. Mold growth, which causes considerable loss, develops in wrapped goods when humidity is high and temperature also fairly high. The baked goods are sterile when they leave the oven; but as they cool and are sliced and wrapped, they are contaminated by mold spores from the air. Air conditioning which brings in washed air and meticulous cleanliness in the cooling and wrapping rooms reduce the number of mold spores in the air and therefore the amount of contamination. Even so, some molding will occur.

FIGURE 9.7. MOLD GROWTH ON BREAD.
Courtesy of Fleischmann Laboratories. Standard Brands, Inc.

Numerous compounds can be added to the dough or batter in low enough concentration so that toxicity or flavor will not occur and yet in high enough concentration so that mold growth is sharply inhibited. Sodium and calcium propionate are now widely used in the baking industry

$$CH_3CH_2COONa \qquad\qquad CH_3CH{=}CHCH{=}CHCOOH$$

Sodium Propionate   Sorbic Acid

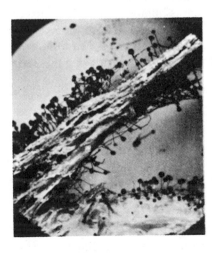

FIGURE 9.8. MOLD GROWTH ON BREAD.
Magnification × 25.
Courtesy of Fleischmann Laboratories.
Standard Brands, Inc.

TABLE 9.4. MOLD INHIBITION*

Supplement to the Batter		Mold Growth in Humidity Can after Days Indicated 90° F				
Substance	Concentration %	2	4	5	7	8
None	0.00	—	++++	++++	++++	++++
Sorbic Acid	0.05	—	++	+++	++++	++++
	0.10	—	—	—	+	+++
	0.20	—	—	—	—	—
Na Propionate	0.30	—	+	++	+++	++++
	0.60	—	—	+	++	++++
Na Benzoate	0.1	—	++	++++	++++	++++

*From Melnick, D., Vahlteich, H. W., and Hackett, A., "Sorbic Acid as a Fungistatic Agent for Foods," *Food Research*, **21**, 133–146 (1956).

to achieve this purpose. In more recent years sorbic acid and its salts have been studied and appear to be at least as effective as the propionates. Melnick, Vahlteich, and Hackett[14] added mixed mold culture to a variety of cake batters and measured the effectiveness of some compounds as inhibitors of mold growth in slices of cake stored in warm humid containers. Their results with white cake are shown in Table 9.4 and indicate good control at low concentration.

REFERENCES

1. ANDERSON, J. A., "Enzymes and Their Role in Wheat Technology," Interscience Publ., Inc., New York, N. Y., 1946.
2. BERNFELD, P., "Enzymes of Starch Degradation and Synthesis," *Advances in Enzymol.*, **12**, 380–428 (1951).
3. BLISH, M. J., "Wheat Gluten," *Advances in Protein Chem.*, **2**, 336–359 (1945).
4. CATHCART, W. E., "XXVI, Baking and Bakery Products," in Jacobs, M. B., ed., "The Chemistry and Technology of Foods and Food Products," Interscience Publ., Inc., New York, N. Y., 1951.
5. CHARLEY, H., "Characteristics of Shortened Cake, Baked in a Fast—and in a Slow—Pan at Different Oven Temperatures," *Food Res.*, **21**, 302–305 (1956).
6. DANIELS, A. L. and HEISIG, E. H., "The Acidity of Various Syrups Used in Cookery," *Home Econ.*, **11**, 193–199 (1919).
7. EMERSON, D. W., "Mold Control by Sorbic Acid," *Can. Food Inds.*, **27**, #7, 29 (1956).
8. GEDDES, W. F., "XXIV, Cereal Grains," in Jacobs, M. B., ed., "The Chemistry and Technology of Foods and Food Products," Interscience Publ., New York, N. Y., 1951.
9. HORDER, T. J., DODDS, C., and MORAN, T., "Bread," Constable, London, 1954.
10. JOHNSON, J. A., MILLER, B. S., REFAI, F. Y., and MILLER, D., "Dough Fermentation: Effect of Fermentation Time on Certain Chemical Constituents of Pre-ferments Used in Bread Making," *J. Agr. Food Chem.*, **4**, 82–84 (1956).

11. KENT-JONES, D. W., "Bread and Its Problems for the Chemist," *J. Royal Inst. Chem.,* **80,** 254–263 (1956).
12. KLIS, J. B., WITTER, L. D., and ORDAL, Z. J., "Effect of Several Antifungal Antibiotics on the Growth of Common Food Spoilage Fungi," *Food Technol.,* **13,** 124–127 (1959).
13. LOWE, B., "Experimental Cookery," 4th Ed., John Wiley & Sons, New York, N. Y., 1955.
14. MELNICK, D., VAHLTEICH, H. W., and HACKETT, A., "Sorbic acid as a fungistatic agent for foods," *Food Res.,* **21,** 133–146 (1956).
15. News Item, "Staling of Bread," *J. Am. Diet. Assoc.,* **30,** 598 (1954).
16. OSBORNE, T. B., "The Proteins of the Wheat Kernel," Carnegie Institute of Washington, Washington, D. C., publ. 84, 1947.
17. OSBORNE, T. B., "The Vegetable Proteins," Longmans, Green, New York, N. Y., 1912.
18. ROSE, A. H., "Yeasts," *Sci. Am.,* **202,** 136–146 (1960).
19. ROSENBERG, H. R., "Supplementation of Foods with Amino Acids," *J. Agr. Food Chem.,* **7,** 316–320 (1959).
20. STAMBERG, O. E. and BAILEY, C. H., "Hydrogen Ion Concentration of Cakes," *Cereal Chem.,* **16,** 419–423 (1939).
21. SULLIVAN, B., "Proteins in Flour. Review of the Physical Characteristics of Gluten and Reactive Groups Involved in Change in Oxidation," *J. Agr. Food Chem.,* **2,** 1231–1234 (1954).
22. THOMPSON, J. B. and BUDDEMEYER, B. D., "Staling of Bread," U. S. Pat. 2,744,825 (1956); *C.A.,* **50,** 12347g (1956).

CHAPTER TEN

Food Additives

THE MOST FREQUENTLY DISCUSSED food chemistry problem in the popular press today is that of food additives. A food additive is any substance not naturally present in a food but added during its preparation and remaining in the finished product; also in this category is any substance naturally present but with a concentration increased by fortification. However since salt, sugar, and vinegar have been used for centuries, they are not usually considered food additives. Today many compounds are added to foods and the list of substances that protect against spoilage, that enhance flavor, that improve nutritive value, or that give some new property to a food is increasing rapidly.

The market for these food additives is large and growing. Since in the United States the distance between producer and consumer of food can be great, the length of time between harvest and eating can present many problems in maintaining freshness. The economic status of the average consumer has improved until he is able to demand and pay for high grade products. There is a constant search for methods to improve quality: ways, for example, to prevent a fall-off in flavor as poultry or fish stands in the butcher's cold counter; to prevent the high loss of fresh fruits and vegetables between field and kitchen; and to prolong the shelf life of bakery products. The list of similar problems could be extended at length. Added to this is the highly competitive nature of the food industry where yesterday's methods and products do not necessarily mean success in today's market.

LEGAL SAFEGUARDS

As the production of new food additives increased, suspicion grew that safeguards against toxicity were not adequate. Considerable interest developed in the methods used by federal agencies to ensure the safety of new additives. The toxicity of pesticides for insects alarmed some consumers who believed that surely these compounds that kill insects might be equally lethal for themselves.

Agene, nitrogen trichloride, was used as a bleach for flour for many years. Bleaching improves the baking quality of flour in a short time and produces a flour comparable to the more expensive aged product. Agene was used in very low concentration but it was shown in the late 40's that flour bleached with this agent or bread baked from this flour caused running fits in dogs and convulsions in rabbits, ferrets, mink, and cats—a report which led to an almost hysterical distrust of all food additives in a small segment of our population. The animals adversely affected were fed diets composed principally of the flour or bread. The toxic compound is produced from the action of nitrogen trichloride on methionine to form methionine sulfoxamine. It has never been demonstrated that harmful effects are caused by this flour or bread in man although a number of experiments have been conducted to test the possible toxicity. Nevertheless, effective August 1, 1949, agene was removed from the list of permissible bleaching agents.

After much debate and a number of hearings and bills, the Miller Amendment was passed in 1954 and the Food Additives Amendment (Public Law 85-929) in 1958. The Miller Amendment covers spray residues and the use of pesticides; it attempts to establish procedures that will prevent any possible harm to consumers. The Food Additives Amendment sets up requirements for the demonstration of the safety of a substance before it can be used in foods; it went into effect on March 5, 1960, although a few items have had an extension beyond this date.

The Food Additives Amendment uses the following definition:—"'Food additives' include all substances not exempted by section 201 of the act, the intended use of which results or may reasonably be expected to result, directly or indirectly, either in their becoming a component of food or otherwise affecting the characteristics of food." The container is not considered a food additive but if some of the substances present in the container migrate into the food, then they become food additives and subject to the law.

A list of additives that are already in use and appear to have been adequately shown to be harmless has been prepared. No further testing is required with these compounds that are considered permissible. The list (see Appendix, p. 357) includes preservatives, buffers and neutralizers, nutrients, non-nutrient sweeteners, coloring agents, stabilizers, emulsifiers, and others. Although salt, sugar, and vinegar are not included, they are permitted. A list of permissible flavor substances is in preparation. Other additives cannot be used unless permission is granted by the Food and Drug Administration. A petition for use must include a thorough review of the substance, its stability under all conditions of possible use, and a complete report on animal tests that demonstrate its safety. Methods for the detection of the additive or its degradation products if they are formed in the food must also be given. The Food and Drug Administration guarantees protection of trade secrets. Only those facts about the substance that are required for its control will be revealed.

WHY USE FOOD ADDITIVES?

At a meeting of the Institute of Food Technologists in May, 1958, Harold Schultz[6] is reported to have said of food additives that "some are used to color foods, others to bleach them; some add flavor to food, others remove flavors; some make foods firmer, others soften them; some keep foods dry, others keep them moist; some thicken foods, others keep them from thickening; some produce foams, others prevent them; some make foods acid, others make them alkaline, still others suspend particles; some are added to increase the mineral ions, others to remove them; some are oxidizing agents, others are reducing agents; some hasten chemical changes, others retard them." This is a formidable list of functions although sometimes one food additive has more than one. It is a list that illustrates the needs of the food industry in its efforts to produce foods with sufficient shelf life to allow distribution over a wide area and with sufficient appeal to the eye and taste of the consumer to arouse and satisfy his appetite. These food products must nourish the eater well; but before they can nourish him, they must first attract his purchase. Justifiable uses of food additives have been described in some detail by the Food Protection Committee in Publication 398, "The Use of Chemical Additives in Food Processing." They are:

(1) "maintenance of nutritional quality, such as by the use of antioxidants."

(2) "enhancement of keeping quality or stability, with resulting reduction in food wastage, through the use of antioxidants, antimicrobial agents, inert gases, meat cures, etc."

(3) "enhancement of attractiveness of foods by means of coloring and flavoring agents, emulsifiers, stabilizers, thickeners, clarifiers, and bleaching agents. (That appeal to the eye as well as the palate is not a mere whim of the food producer, is recognized by all nutrition-

ists, since proteins, vitamins, and minerals serve no purpose unless eaten.)"

(4) "providing essential aids in food processing. Agents which function in this capacity include acids, alkalis, buffers, sequestrants, and various other types of chemicals."

METHODS OF DEMONSTRATING SAFETY

The current methods of demonstrating the safety of a chemical compound or crude extract to be used in food production are fairly well standardized. Those compounds which are not normally present in some foods are fed in graded amounts to rats for two years and to dogs for one year. The control animals are fed complete diets devoid of the additive. The level of feeding at which injury can be demonstrated is determined.

The success or failure of this method depends on recognition of "injury." A considerable amount of attention has been devoted to this effect.[7] The Food and Drug Administration uses the distinction between toxicity and hazard set forth by the National Research Council's Food and Nutrition Board: "Toxicity is the capacity of the substance to produce injury; hazard is the probability that injury will result from the use of the substance in the quantity and in the manner proposed." Since evaluation of injury depends on subjective judgment when changes are slight, the FDA requires a complete description of methods used. "No effect" must be supplemented by a description of the conditions under which there is "no effect." The decision of the safety, toxicity, or hazard of a chemical additive rests with this organization.

For some substances it is possible that use in small amounts proves safe but in very large amounts hazard exists. Even a substance such as sodium chloride, commonly considered completely safe to add to foods, can cause injury when the amount far exceeds that encountered in foods. With these compounds a limit is established by adding a margin of safety to their toxic level. This margin must include consideration of the difference in effect between man and test animals, individual variation, cumulative effect, and the possibility of the compound occurring in other items in the diet. Here again judgments rather than measurements are used and much debate is possible.

It is unlikely that most chemical additives will be chemically pure compounds. FDA requires that methods for the detection and estimation of impurities be developed and presented with the petition for permission to use. Although a pure substance might be safe, the impurities carried along might be injurious.

There is also a possibility that a compound might be safe as such but during the processing of a food, it could cause a change in some component to form a poisonous compound. The toxic factor in flour bleached with agene proved to be a derivative of methionine rather than the agene itself. It is therefore required that a compound not only be demonstrated safe, but also safe under the conditions of its use. Permissible chemical additives are restricted to those particular foods or processes which meet these requirements.

COLOR ADDITIVES LEGISLATION

The FDA is requesting legislation to modify the present regulations regarding color additives to foods (as well as to drugs and cosmetics.) At present color additives are either classed as "harmless" or not, with no regard for the quantities used. Since this is not very realistic, the FDA wishes controls set up so that the quantities in specific foods can be limited to safe levels. They also wish the law extended to include not only synthetic coloring substances but also natural pigments used as additives.

Regulations on food additives and explanations of the law and of attempts to put it into smooth operation are being published by the FDA. These publications are available free of charge to anyone who asks that his name be placed on the mailing list.

REFERENCES

1. Ed., "New Additives Regulations," *J. Agr. Food Chem.* **7**, 12–14 (1959).
2. Food Protection Committee, "The Use of Chemical Additives in Food Processing," Publ. 398, Natl. Acad. Sci. Natl. Res. Council, Washington, D. C., 1956.
3. HALL, R. L., "Flavoring Agents as Food Additives," *Food Technol.,* **13,** news sect. 14–19, (July, 1959).
4. HARRIS, R. S., "Fortification of Foods with Vitamins," *J. Agr. Food Chem.,* **7,** 88–102 (1959).
5. OSER, B. L., "The Functional Value of Food Additives," *Food Technol.,* **12,** 12–16 (June, 1958).
6. SCHULTZ, H., "Report of Meeting of Institute of Food Technologists," *Food Technol.,* **12,** 9 (1958).
7. Regulations on Food Additives are sent out frequently by the Food and Drug Administration, U. S. Department of Health, Education, and Welfare. They are available free of cost.

Appendix on Food Additives*

The U. S. Department of Health, Education, and Welfare, Food and Drug Administration is now issuing lists of permitted food additives. The following lists include those published in the Federal Register on November 20, 1959, and lists of January 19 and 26, 1960. It is expected that the Food and Drug Administration will continue to issue lists of permissible food additives.

CERTAIN FOOD ADDITIVES EXEMPTED FROM THE REQUIREMENT OF TOLERANCES

Substances That Are Generally Recognized As Safe

(a) It is impracticable to list all substances that are generally recognized as safe for their intended use. However, by way of illustration, the Commissioner regards such common food ingredients as salt, pepper, sugar, vinegar, baking powder, and monosodium glutamate as safe for their intended use. The lists in paragraph (d) of this section include additional substances that, when used for the purposes indicated, in accordance with good manufacturing practice, are regarded by the Commissioner as generally recognized as safe for such uses.

(b) For the purposes of this section, good manufacturing practice shall be defined to include the following restrictions:

(1) The quantity of a substance added to food does not exceed the amount reasonably required to accomplish its intended physical, nutritional, or other technical effect in food; and

(2) The quantity of a substance that becomes a component of food as a result of its use in the manufacturing, processing, or packaging of food, and which is not intended to accomplish any physical or other technical effect in the food itself, shall be reduced to the extent reasonably possible.

(3) The substance is of appropriate food grade and is prepared and handled as a food ingredient. Upon request the Commissioner will offer an opinion, based on specifications and intended use, as to whether or not a particular grade or lot of the substance is of suitable purity for use in food and would generally be regarded as safe for the purpose intended, by experts qualified to evaluate its safety.

---

*From the Federal Register, U. S. Department of Health, Education, and Welfare, Food and Drug Administration. Excerpts of November, 1959 and January, 1960.

(c) The inclusion of substances in the list of nutrients does not constitute a finding on the part of the Department that the substance is useful as a supplement to the diet for humans.

(d) Substances that are generally recognized as safe for their intended use within the meaning of section 409 of the act are as follows:

Chemical Preservatives

Ascorbic acid.
Ascorbyl palmitate.
Calcium ascorbate.
Calcium propionate.
Calcium sorbate.
Erythorbic acid.
Potassium sorbate.
Propionic acid.
Sodium ascorbate.
Sodium propionate.
Sodium sorbate.
Sorbic acid.
Tocopherols.

Buffers and Neutralizing Agents

Acetic acid.
Adipic acid.
Aluminum ammonium sulfate.
Aluminum sodium sulfate.
Aluminum potassium sulfate.
Ammonium bicarbonate.
Ammonium carbonate.
Ammonium hydroxide.
Ammonium phosphate (mono- and di-basic-).
Calcium carbonate.
Calcium chloride.
Calcium citrate.
Calcium gluconate.
Calcium hydroxide.
Calcium lactate.
Calcium oxide.
Calcium phosphate.
Citric acid.
Hydrochloric acid.
Lactic acid.
Magnesium carbonate.
Magnesium oxide.
Malic acid.
Phosphoric acid.
Potassium acid tartrate.
Potassium bicarbonate.
Potassium carbonate.
Potassium citrate.
Potassium hydroxide.
Sodium acetate.
Sodium acid pyrophosphate.
Sodium aluminum phosphate.
Sodium bicarbonate.

Sodium carbonate.
Sodium citrate.
Sodium hydroxide.
Sodium phosphate (mono-, di-, tri-).
Sodium potassium tartrate.
Sodium sesquicarbonate.
Succinic acid.
Sulfuric acid.
Tartaric acid.

Emulsifying Agents

Diacetyl tartaric acid esters of mono- and diglycerides from the glycerolysis of edible fats and oils.
Mono- and diglycerides from the glycerolysis of edible fats or oils.
Monosodium phosphate derivatives of mono- and diglycerides from the glycerolysis of edible fats or oils.
Propylene glycol.

Miscellaneous

Acetic acid.
Aluminum sodium sulfate.
Aluminum sulfate.
Ammonium bitartrate.
Ammonium sulfate.
Bees wax (yellow wax).
Bees wax bleached (white wax).
Bentonite.
Butane.
Calcium chloride.
Calcium phosphate, tribasic.
Caramel.
Carbon dioxide.
Carnauba wax.
Citric acid.
Dextrans (of average molecular weight below 100,000).
Glycerin.
Glycerol monostearate.
Helium.
Lecithin.
Magnesium carbonate.
Magnesium hydroxide.
Methyl cellulose (U.S.P. methylcellulose, except that the methoxy content shall not be less than 27.5 per cent and not more than 31.5 per cent on a dry-weight basis.)

Monoammonium glutamate.
Monopotassium glutamate.
Nitrogen.
Papain.
Phosphoric acid.
Potassium carbonate.
Potassium sulfate.
Propane.
Propylene glycol.
Rennet (rennin).
Sodium bicarbonate.
Sodium carbonate.
Sodium carboxymethylcellulose (the sodium salt of carboxymethylcellulose not less than 99.5 per cent on a dry-weight basis, with maximum substitution of 0.95 carboxymethyl groups per anhydroglucose unit, and with a minimum viscosity of 25 centipoises for 2 per cent by weight aqueous solution at 25° C.)
Sodium caseinate.
Sodium phosphate.
Sodium sesquicarbonate.
Sodium pectinate.
Sodium phosphate.
Sodium tripolyphosphate.
Torula yeast.
Triacetin (Glyceryl triacetate).
Tricalcium phosphate.

Nonnutritive Sweeteners

Ammonium Saccharin
Calcium cyclohexyl sulfamate.
Calcium saccharin.
Magnesium cyclohexyl sulfamate.
Potassium cyclohexyl sulfamate.
Saccharin.
Sodium cyclohexyl sulfamate.
Sodium saccharin.

Nutrients

Ascorbic acid.
Calcium carbonate.
Calcium oxide.
Calcium pantothenate.
Calcium phosphate (mono-, di-, tribasic).
Calcium sulfate.
Carotene.
Choline bitartrate.
Ferric phosphate.
Ferric pyrophosphate.
Ferric sodium pyrophosphate.
Ferrous sulfate.
Inositol.
Iron, reduced.
l-Lysine monohydrochloride.
Niacin.
Niacinamide.
D-Pantothenyl alcohol.
Potassium chloride.
Pyridoxine hydrochloride.

Riboflavin.
Riboflavin-5-phosphate.
Sodium pantothenate.
Sodium phosphate (mono-, di-, tribasic).
Thiamine hydrochloride.
Thiamine mononitrate.
Tocopherols.
α-Tocopherol acetate.
Vitamin A.
Vitamin A acetate.
Vitamin A palmitate.
Vitamin B_{12}.
Vitamin D_2.
Vitamin D_3.

Sequestrants

(For the purpose of this list, no attempt has been made to designate those sequestrants which may also function as chemical preservatives)
Calcium acetate.
Calcium chloride.
Calcium citrate.
Calcium diacetate.
Calcium gluconate.
Calcium hexametaphosphate.
Calcium phytate.
Citric acid.
Dipotassium phosphate.
Disodium phosphate.
Monocalcium acid phosphate.
Monoisopropyl citrate.
Potassium citrate.
Sodium acid phosphate.
Sodium citrate.
Sodium diacetate.
Sodium gluconate.
Sodium hexametaphosphate.
Sodium metaphosphate.
Sodium phosphate (mono-, di-, tribasic-).
Sodium potassium tartrate.
Sodium pyrophosphate.
Sodium tartrate.
Sodium tetrapyrophosphate.
Sodium tripolyphosphate.
Tartaric acid.

Stabilizers

Agar-agar.
Acacia (gum Arabic).
Ammonium alginate.
Calcium alginate.
Carob bean gum (locust bean gum).
Carragheenin.
Ghatti gum.
Potassium alginate.
Sodium alginate.
Sterculia gum (Karaya gum).
Tragacanth (gum tragacanth).
Guar gum.

FOOD CHEMISTRY

Substances Allowed If Restrictions Are Observed

Product	Tolerance	Specific uses or restrictions
ANTICAKING AGENTS		
Aluminum calcium silicate	2 per cent	In table salt.
Calcium silicate	5 per cent	In baking powder.
Calcium silicate	2 per cent	In table salt.
Magnesium silicate	do	Do.
Tricalcium silicate	do	Do.
CHEMICAL PRESERVATIVES		
Benzoic acid	0.1 per cent	
Butylated hydroxyanisole	Total content of antioxidants not over 0.02 per cent of fat or oil content, including essential (volatile) oil content, of food.	
Butylated hydroxytoluene	do	
Caprylic acid		In cheese wraps.
Dilauryl thiodipropionate	Total content of antioxidants not over 0.02 per cent of fat or oil content, including essential (volatile) oil content of the food.	
Gum guaiac	0.1 per cent (equivalent antioxidant activity 0.01 per cent).	In edible fats or oils.
Nordihydroguaiaretic acid	Total content of antioxidants not over 0.02 per cent of fat or oil content, including essential (volatile) oil content of the food.	
Potassium bisulfite		Not in meats or in food recognizable as a source of vitamin B_1
Potassium metabisulfite		Do.
Propyl gallate	Total content of antioxidants not over 0.02 per cent of fat or oil content, including essential (volatile) oil content of the food.	
Sodium benzoate	0.1 per cent	
Sodium bisulfite		Not in meats or in foods recognizable as a source of vitamin B_1.
Sodium metabisulfite		Do.
Sodium sulfite		Do.
Sulfur dioxide		Do.
Thiodipropionic acid	Total content of antioxidants not over 0.02 per cent of fat or oil content, including essential (volatile) oil content of the food.	
EMULSIFYING AGENTS		
Cholic acid	0.1 per cent	Dried egg whites.
Deoxycholic acid	do	Do.
Glycocholic acid	do	Do.
Ox bile extract	do	Do.
Taurocholic acid (or its sodium salt)	do	Do.
MISCELLANEOUS		
Caffeine	0.02 per cent	In cola type beverages.
Ethyl formate	0.0015 per cent	As fumigant for cashew nuts.
Magnesium stearate		As migratory substance from packaging materials when used as a stabilizer.
Sorbitol	7.0 per cent	In foods for special dietary use.
Triethyl citrate	0.25 per cent	Egg whites.
NUTRIENTS		
Copper gluconate	0.005 per cent	
Cuprous iodide	0.01 per cent	In table salt as a source of dietary iodine.
Potassium iodide	do	Do.
SEQUESTRANTS[1]		
Isopropyl citrate	0.02 per cent	
Sodium thiosulfate	0.1 per cent	In salt.
Stearyl citrate	0.15 per cent	

[1]For the purpose of this list no attempt has been made to designate those sequestrants which may also function as chemical preservatives.

360

SPICES, SEASONINGS, ESSENTIAL OILS, OLEORESINS, AND NATURAL EXTRACTIVES THAT ARE GENERALLY RECOGNIZED AS SAFE FOR THEIR INTENDED USE.

Spices and Other Natural Seasonings and Flavorings (Leaves, Roots, Barks, Berries, etc.)

Common name	Botanical name of plant source
Allspice	Pimenta officinalis Lindl.
Anise	Pimpinella anisum L.
Anise, star	Illicium verum Hook. f.
Basil, sweet	Ocimum basilicum L.
Basil, bush	Ocimum minimum L.
Bay	Laurus nobilis L.
Calendula	Calendula officinalis L.
Capers	Capparis spinosa L.
Capsicum	Capsicum frutescens L. or Capsicum annuum L.
Caraway	Carum carvi L.
Caraway, black (black cumin)	Nigella sativa L.
Cardamom (cardamon)	Elettaria cardamomum Maton.
Cassia, Chinese	Cinnamomum cassia Blume.
Cassia Padang or Batavia	Cinnamomum burmanni Blume.
Cassia, Saigon	Cinnamomum loureirii Nees.
Cayenne pepper	Capsicum frutescens L. or Capsicum annuum L.
Celery seed	Apium graveolens L.
Chives	Allium schoenoprasum L.
Cinnamon, Ceylon	Cinnamomum zeylanicum Nees.
Cinnamon, Chinese	Cinnamomum cassia Blume.
Cinnamon, Saigon	Cinnamomum loureirii Nees.
Clary (clary sage)	Salvia sclarea L.
Cloves	Eugenia caryophyllata Thunb.
Coriander	Coriandrum sativum L.
Cumin (cummin)	Cuminum cyminum L.
Cumin, black (black caraway)	Nigella sativa L.
Dill	Anethum graveolens L.
Fennel, common	Foeniculum vulgare Mill.
Fennel, sweet (finocchio, Florence fennel)	Foeniculum vulgare Mill. var. dulce (DC.) Alef.
Fenugreek	Trigonella foenum-graecum L.
Garlic	Allium sativum L.
Ginger	Zingiber cincinale Rosc.
Glycyrrhiza	Glycyrrhiza glabra L. and other spp. of Glycyrrhiza.
Grains of paradise	Amomum melegueta Rosc.
Horseradish	Armoracia lapathifolia Gilib.
Lavender	Lavandula officinalis Chaix.
Licorice	Glycyrrhiza glabra L. and other spp. of Glycyrrhiza.
Mace	Myristica fragrans Houtt.
Marigold, pot	Calendula officinalis L.
Marjoram, pot	Majorana onites (L.) Benth.
Marjoram, sweet	Majorana hortensis Moench.
Mustard, black or brown	Brassica nigra (L.) Koch.
Mustard, brown	Brassica juncea (L.) Coss.
Mustard, white or yellow	Brassica alba (L.) Boiss.
Nutmeg	Myristica fragrans Houtt.
Oregano (oreganum, Mexican oregano, Mexican sage, origan).	Lippia spp.

Paprika .... Capsicum annuum L.
Parsley.................................. Petroselinum crispum (Mill.) Mansf.
Pepper, black .... Piper nigrum L.
Pepper, cayenne .... Capsicum frutescens L. or Capsicum an-
nuum L.
Pepper, red .... Do.
Pepper, white .... Piper nigrum L.
Peppermint .... Mentha piperita L.
Poppy seed .... Papaver somniferum L.
Pot marigold .... Calendula officinalis L.
Pot marjoram .... Majorana onites (L.) Benth.
Rosemary................................ Rosmarinus officinalis L.
Rue..................................... Ruta graveolens L.
Saffron .... Crocus sativus L.
Sage .... Salvia officinalis L.
Savory, summer .... Satureia hortensis L. (Satureja).
Savory, winter........................... Satureia montana L. (Satureja).
Sesame .... Sesamum indicum L.
Spearmint............................... Mentha spicata L.
Star anise .... Illicium verum Hook. f.
Tarragon .... Artemisia dracunculus L.
Thyme.................................. Thymus vulgaris L.
Turmeric................................ Curcuma longa L.
Vanilla.................................. Vanilla planifolia Andr. or Vanilla tahitensis
J. W. Moore
Zedoary................................. Curcuma zedoaria Rosc.

Essential Oils, Oleoresins (Solvent-Free), and Natural Extractives (Including Distillates)

Common name	*Botanical name of plant source*
Allspice	Pimenta officinalis Lindl.
Almond, bitter (free from prussic acid)	Prunus amygdalus Batsch, Prunus armeniaca L., or Prunus persica (L.) Batsch.
Angelica root	Angelica archangelica L.
Angelica seed	Do.
Angelica stem	Do.
Angostura (cusparia bark)	Galipea officinalis Hancock.
Anise	Pimpinella anisum L.
Asafetida	Ferula assa-foetida L. and related spp. of Ferula.
Balsam of Peru	Myroxylon pereirae Klotzsch.
Basil	Ocimum basilicum L.
Bay leaves	Laurus nobilis L.
Bay (myrcia oil)	Pimenta racemosa (Mill.) J. W. Moore.
Bitter almond (free from prussic acid)........	Prunus amygdalus Batsch, Prunus armeniaca L., or Prunus persica (L.) Batsch.
Bois de rose	Aniba rosaeodora Ducke.
Cananga	Cananga odorata Hook. f. and Thoms.
Capsicum	Capsicum frutescens L. and Capsicum an-nuum L.
Caraway	Carum carvi L.
Cardamon seed (cardamon)	Elettaria cardamomum Maton.
Carob bean	Ceratonia siliqua L.
Cascarilla bark	Croton eluteria Benn.
Cassia bark, Chinese	Cinnamomum cassia Blume.
Cassia bark, Padang or Batavia	Cinnamomum burmanni Blume.
Cassia bark, Saigon	Cinnamomum loureirii Nees.
Celery seed	Apium graveolens L.
Chamomile flowers, Hungarian (camomile) ...	Matricaria chamomilla L.

Chamomile flowers, Roman or English (camomile)	Anthemis nobilis L.
Cherry, wild, bark	Prunus serotina Ehrh.
Chicory	Chicorium intybus L.
Cinnamon bark, Ceylon	Cinnamomum zeylanicum Nees.
Cinnamon bark, Chinese	Cinnamomum cassia Blume.
Cinnamon bark, Saigon	Cinnamomum loureirii Nees.
Cinnamon leaf, Ceylon	Cinnamomum zeylanicum Nees.
Cinnamon leaf, Chinese	Cinnamomum cassia Blume.
Cinnamon leaf, Saigon	Cinnamomum loureirii Nees.
Citronella	Cymbopogon nardus Rendle.
Citrus peels	Citrus spp.
Clary (clary sage)	Salvia sclarea L.
Clove bud	Eugenia caryophyllata Thunb.
Clove leaf	Do.
Clove stem	Eugenia caryophyllata Thunb.
Coca (decocainized)	Erythroxylum coca Lam. and other spp. of Erythroxylum.
Coffee	Coffea spp.
Cola nut	Cola acuminata Schott and Endl., and other spp. of Cola.
Coriander	Coriandrum sativum L.
Cumin (cummin)	Cuminum cyminum L.
Cusparia bark	Galipea officinalis Hancock.
Dill	Anethum graveolens L.
Estragole (esdragol, esdragon, tarragon)	Artemisia dracunculus L.
Fennel, sweet	Foeniculum vulgare Mill.
Fenugreek	Trigonella foenum-graecum L.
Garlic	Allium sativum L.
Geranium, East Indian	Cymbopogon martini Stapf.
Geranium, rose	Pelargonium graveolens L'Her.
Ginger	Zingiber officinale Rosc.
Glycyrrhiza	Glycyrrhiza glabra L. and other spp. of Glycyrrhiza.
Grapefruit	Citrus paradisi.
Guava	Psidium spp.
Hoarhound	Marrubium vulgare L.
Hops	Humulus lupulus L.
Jasmine	Jasminum officinale L. and other spp. of Jasminum.
Juniper (berries)	Juniperus communis L.
Kola nut	Cola acuminata Schott and Endl., and other spp. of Cola.
Laurel leaves	Laurus mobilis L.
Lavender	Lavandula officinalis Chaix.
Lavender, spike	Lavandula latifolia Vill.
Lavandin	Hybrids between Lavandula officinalis Chaix and Lavandula latifolia Vill.
Lemon	Citrus Limon (L.) Burm. f.
Lemon grass	Cymbopogon citratus DC. and Cymbopogon flexuosus Stapf.
Licorice	Glycyrrhiza glabra L. and other spp. of Glycyrrhiza.
Lime	Citrus aurantifolia Swingle.
Locust bean	Ceratonia siliqua L.
Mace	Myristica fragrans Houtt.
Mandarin	Citrus reticulata Blanco.
Marjoram, sweet	Majorana hortensis Moench.
Maté	Ilex paraguariensis St. Hil.
Mustard	Brassica spp.
Niaringin	Citrus paradisi Macf.

Neroli, bigarade	Citrus aurantium L.
Nutmeg	Myristica fragrans Houtt.
Onion	Allium cepa L.
Orange, bitter, flowers	Citrus aurantium L.
Orange leaf	Citrus sinensis (L.) Osbeck.
Orange, bitter, peel	Citrus aurantium L.
Orange, sweet	Citrus sinensis (L.) Osbeck.
Origanum	Origanum spp.
Palmarosa	Cymbopogon martini Stapf.
Paprika	Capsicum annuum L.
Parsley	Petroselinum crispum (Mill.) Mansf.
Pepper, black	Piper nigrum L.
Pepper, white	Piper nigrum L.
Peppermint	Mentha piperita L.
Peruvian balsam	Myroxylon pereirae Klotzsch.
Petitgrain	Citrus aurantium L.
Petitgrain lemon	Citrus limon (L.) Burm. f.
Petitgrain mandarin or tangerine	Citrus reticulata Blanco.
Pimenta	Pimenta officinalis Lindl.
Pimenta leaf	Do.
Pipsissewa leaves	Chimaphila umbellata Nutt.
Pomegranate	Punica granatum L.
Prickly ash bark	Xanthoxylum (or Zanthoxylum) Americanum Mill. or Xanthoxylum clava-herculis L.
Rose absolute	Rosa alba L., Rosa centifolia L., Rosa damascena Mill., Rosa gallica L., and vars. of these spp.
Rose (otto of roses, attar of roses)	Do.
Rose geranium	Pelargonium graveolens L'Her.
Rosemary	Rosmarinus officinalis L.
Rue	Ruta graveolens L.
Saffron	Crocus sativus L.
Sage	Salvia officinalis L.
Sage, Spanish	Salvia lavandulaefolia Vahl.
St. John's bread	Ceratonia siliqua L.
Schinus molle	Schinus molle L.
Spanish sage	Salvia lavandulaefolia Vahl.
Spearmint	Mentha spicata L.
Spike lavender	Lavandula latifolia Vill.
Tangerine	Citrus reticulata Blanco.
Tarragon	Artemisia dracunculus L.
Tea	Thea sinensis L.
Thyme	Thymus vulgaris L. and Thymus zygis var. gracilis Boiss.
Thyme, white	Do.
Tuberose	Polianthes tuberosa L.
Turmeric	Curcuma longa L.
Vanilla	Vanilla planifolia Andr. or Vanilla tahitensis J. W. Moore.
Violet leaves absolute	Viola odorata L.
Wild cherry bark	Prunus serotina Ehrh.
Ylang-ylang	Cananga odorata Hook. f. and Thoms.
Zedoary bark	Curcuma zedoaria Rosc.

Miscellaneous

Common name	Derivation
Civet (zibeth, zibet, zibetum)	Civet cats, Viverra civetta Schreber and Viverra zibetha Schreber.
Cognac oil, white and green	Ethyl oenanthate, so-called.
Musk (Tonquin musk)	Musk deer, Moschus moschiferus L.

Index

Acetaldehyde, 42
Acetone bodies in milk, 298
Acetyl methyl carbinol in butter, 52
Acetyl Value, 31
Achilles tendon, 178
Acidophilus milk, 305
Acids allowed in jellies, 94
Acids, in fruits and vegetables, 275–277
Acids, organic,
    changes in fruit, 286–288
    in foods, 276–277
Acids, volatile, 274–278
    and green vegetables, 275
Actin, 172
Active oxygen test, 39
Actomyosin, 172
Acuity of smell, 153, 154
Additives, food, 351–364
Adenosine triphosphate, 172
Adipose tissue, 14, 179–180
Adsorption of solvent and protein gels, 126
Adsorption spectra, fatty acid esters, 19
Adulteration of milk, 302–303
    of spices, 107
Aeration of fats, 34–38
Aftertaste, 148, 156, 158
Agar, 96
Agene, 336, 352
Aging of starch dispersion, 107
Air, intercellular in fruits and vegetables, 274
Air as leavening, 339
    in pound cake, 60–61
Albumin as antigen, 123
Albumin, egg, 129
Albumin in cereals, 322
Albumin in milk, determination, 301
Alcohol fermentation,
    methanol from pectins, 95
Aleurone layer in wheat, 317
Alga, amino acids in protein, 138–140
Algal xylan, 86
Alginic acid, 97
Alkali refining fats and oils, 46
Allicin, 278

Alliine, 278
Allylisothiocyanate, 279
Allyl propyl disulfide, 278
Almond oil,
    acid value, 33
    Crismer value, 28
    iodine number, 33
    saponification number, 33
    solidification point, 33
    specific gravity, 33
    Reichert Meissl, 33
Almond anthocyanin, 244
Amadori rearrangement, 109, 110
Amines, aromatic, antioxidants, 37
Amino acids, 114–115
    changes in cheese, 311
    determination, 133–135
    essential, 114
    in casein fractions, 142
    in plant proteins, 138–140
    in seed proteins, 141
    microbiological determination, 135
    separation by chromatography, 134–135
Ammonium bicarbonate in batters, 339
Ammonium carbonate in batters, 339
Amplitude in flavor, 158
Amphoterism of proteins, 122
Amylases,
    "dextrogenic", 76
    in carrot, 271
    in dough, 332–333
    in flour, 321
    "liquefying", 76
α- and β-Amylases,
    occurrence, 77
    activity, 77, 79
Amyloses,
    content in starches, 76
    in gels, 76
    micellar organization, 76
Anserine, 173, 174
Anthocyanidins, 241–243
    structure, 241–242
    occurrence, 244

Anthocyanins, 239–240, 241–247
  and browning, 256
  and metal ions, 245
  and pH, 242, 245
  and storage temperature, 246
  and tannins, 243
  factors affecting hue, 242–243
  in edible fruit, 244
  in cooking, 244–245
  properties, 246
  structure, 241–242
Anthoxanthins, 240, 247–250
  and cooking, 249
  and NaOH, 249
Anthrone and carbohydrate tests, 99
Antibodies for proteins, 123–125
Antigen, protein, 123
Antioxidants,
  and carotenoids, 234
  and dehydrated carrots, 234
  for fats, 32–38
Apigenin, 243, 248
Aponeuroses, 174
Apples,
  changes in hardness, 285
  cider and culinary, 252
  climacteric, 281
  firming canned, 266
  flavor compounds, 165
  flavor, synthetic, 165
  organic acids, 286, 287
  pectic substances, 262
  respiration, 281
  saucing quality, 267
  senescence, 281
  wax, 222–224
Arachidic acid, 19
Aroma, 152–154
Aromatic compounds in plant waxes, 224
Aronoff and Aronoff, 243
Ascorbic acid preventing browning, 257, 258, 260
Ash in wheat flour, 321
Asparagus cells, photographed, 223
Association of Official Agricultural Chemists, 2
Atheromas, 62
Atherosclerosis, 62
Atramalic acid in apples, 287
Available carbohydrate in fruit or vegetables, 219
Avenin, 322
Avidin, 143, 145
Avocado, lipid, 12

Babcock Method, 299
Bacteria, amino acids,
  in protein, 142–143
  in milk for margarine, 59
  in souring cream, 51–52
Bacterial invasion of fruit and epidermis, 222
Baked goods, rancidity, 38
  scoring, 347
Baking powders, 338–339
  double action, 338
  phosphate, 338
  S.A.S., 338
  tartrate, 338
Baking quality, 329–345
Bananas,
  changes in carbohydrates, 284
  pectic substances, 262
Barley gum, 86
Barley, lipid, 12
Barnell, 283
Batters, 337–345
  and gluten, 340
  egg protein, 340–341
  fat, 341
  gas production, 337–339
  gas retention, 339–342
  mixing, 340–341
  sugar, 341
Beans, canned with hard water, 266
Beef, nutritive value after heating, 137
Beef tallow, fatty acids, 19
Beer and tannins, 254
Beets,
  anthocyanins, 243–244
  cooking and color, 245
Benedict's reagent, 99
Benzoic acid in fruits and vegetables, 276
BHA, 37
Binding ions by protein, 122
Biscuit type shortenings, 56
Bitterness, 150, 151
Blair and Ayres, 238, 239
Bleaching fats and oils, 46
Bleaching oil, 54
Blends of flavors, 155–156
Blood cholesterol, 62
Blueberry pie color, 245
Bound water, 8
Bran in grain, 317
  composition, 318
Bread, fat in, 60
Breaks in milling, 320
Breed cow and milk composition, 296
Browning, 108–112, 254–259

control of, 257–258
in fruit and vegetables, 254–261
enzymatic, 255–259
nonenzymatic, 259–261
course, 259
factors, 259
and oxygen uptake, 260
prevention by patented compounds, 258
with sugar, 257
reactions, 108–112
Bryophyte, amino acids in protein, 139
Bucky dough, 335
Buffer salts in jelly, 94
Buffers allowed, 358
Bulgarlac, 305
Bulk in diet, 107
Butland, 253
Butter, 305–306
acid value, 33
composition, 306
consistency, 52
Crismer value, 28
fatty acids in, 19
foaming, 59
freezing, 306
iodine number, 33
Kirschner Value, 30
manufacturing, 51–52, 306
Reichert Meissl, 33
salting, 306
saponification value, 33
scoring, 306
solidification point, 33
specific gravity, 33
Butterfat determination in milk, 299
Buttermilk, 305
Butylated hydroxy anisole (BHA), 37
Butylated hydroxy toluene (BHT), 37
Butyric acid, 19

Cabbage,
color in cooking, 238
odor on cooking, 279
red, color change, 245
Cacao butter, Crismer value, 28
Caffeic acid, 251–252
formula, 240
oxidation, 256
Cake,
baking temperature, 345
fat in, 60
high sugar, 57, 343
leavening, 60, 61, 338–339

low sugar, 343
moisture, 345
pH, 343–345
pound, 60, 61, 339
sugar free, 343
Calcium in vegetables, 219
Calcium pectates, 266
and texture, 267
Cans, tin and discoloration, 245
Capillaries, fragile, 249
Caproic acid, 19
Capsanthin, 233
Carageenan, 97
Carbazole and carbohydrates, 104
Carbohydrates, 65–112
and cooking, 104–107
Benedict's reagent, 99
color reactions, 97
fermentation, 100–101
in dough, 332
in fruits and vegetables, 219
in wheat, 325
iodine for determination, 104
Molisch reaction, 98
osazones, 100
quantitative determination, 101–104
Seliwanoff's reagent, 98
separation by chromatography, 101
Tauber's test, 99
Carmelization, 109, 111
Carnauba wax, 222–224
Carnitine, 173, 174
Carnosine, 173, 174
Carotenes,
α, 230, 231
β, 230, 231
γ, 230, 231
κ, 233
conversion to vitamin A, 233
excretion, 233
fortification, margarine, 60
in American diet, 233
in carrot chromoplasts, 227
in food, 233
Carotenoids, 228–235
color, 228
distribution, 233
in fats, 17
occurrence, 228
oxidation, 234
solubility, 234
stability, 234
with chlorophyll, 228
Carré-Haynes method, 267

Carrots,
  carotenoids present, 233
  chromoplasts, 227
  dehydrated and color, 234
  pectic substances, 264, 266
  starch, 234, 271
Casein, 140-142
  α, β, and γ, 140
  fractions, amino acids, 142
  composition, 142
  determination in milk, 301
  heated, 137
Castor bean oil,
  acetyl value, 31
  acid value, 33
  iodine number, 33
  phytosterol, 16
  Reichert Meissl, 33
  saponification value, 33
  smoke, flash, fire points, 27
  solidification point, 33
  specific gravity, 33
Catalyst, hydrogenation of fats, 49
  poisoning, 49
Catechins, 240, 250
  structure, 250-251
Catechols, and browning, 255
  oxidation by enzymes, 256-257
Cathcart, 344, 345
Cauliflower,
  brown color, 249
  cooked, odor, 279
Cell turgor,
  and cooking, 226
  factors affecting, 226
Cell sap, 220
  anthocyanins, 243
Cell walls, 220
  and pectic substances, 261
Cells,
  chloroplasts, 220
  chlorophyll, 220
  chromoplasts, 220
  conducting, 222, 223
  leucoplasts, 220
  parenchyma, 221
  supporting, 222, 223
  vacuoles, 220
Cellobiose,
  from cellulose, 73
  structure, 69
Cellulose, 73-75
  changes on cooking, 269
  differs from hemicellulose, 84
  fibers, 74, 222

  in cell walls, 220
  in fruit and vegetables, 220
  in plants, 220
  molecular weights, 74
  structure, 74
  X-ray analysis, 74
Cephalins, formula, 14
Cereals, 316-349
  lipids, 12
  structure, 317-318
Cereal flours, antioxidants, 37
Cerotic acid, 224
Cetorhinene, 15
Cetyl lignocerate, 16
Charley, 345
Cheese, 306-314
  composition, 307
  food, 314
  milk for, 307
  process, 314
  ripened, 310-311
  ripening, 311
  spreads, 314
  unripened, 307-310
Cherry,
  cornelian, anthocyanin, 244
  sweet, anthocyanin, 244
Cherry, synthetic,
  flavor, 164
  flavor compounds, 164-165
Cherry pie color, 245
Cherries, discolored, 253
Chicken fat, acid value, 33
  iodine number, 33
  Reichert Meissl, 33
  saponification value, 33
  specific gravity, 33
  solidification point, 33
Chili, carotenoids present, 233
Chinese vegetable tallow, Crismer value, 28
Chitin, 80
Chlorogenic acid, 252
  formula, 240
  oxidation, 256
Chlorophyll,
  a formula, 235
  and pH, 237, 238
  and metallic ions, 239
  b formula, 235
  in chloroplasts, 220
  reactions, 235-239
  stability, 236
Chlorophylls in fats, 17
Chloroplasts, 228
  composition, 236

Chocolate ice cream discolored, 253
Chocolates, dipped,
    melting point, 23
Cholesterol formula, 15
Cholesterol, in blood, 62
    and diet, 62
Chondroitin sulfates A, B, C, 81–83
Chop in milling, 320
Chromatogram of methyl esters of fatty
    acids, 20
Chromatography, of amino acids, 134
    of carbohydrates, 101–102
Chromoplasts, 220
    in cell, 228
    photograph, 227
Cider, color, 252, 255
Cinnamic acid, formula, 240
Cis-trans isomers, fatty acids, 20, 49
β-citraurin, 233
Citraxanthin, 233
Citric acid,
    and browning prevention, 258
    in fruits and vegetables, 275–277
    in jelly, 94
Citrin, 249
Citrus fruit fermentation, methanol in, 95
Clear flour, 320
Climacteric in fruit, 281
Clotting milk with rennin, 141
Cloves, antioxidant, 38
Cocoa (Cacao) butter,
    acid value, 33
    interesterification, 54
    iodine number, 33
    phytosterol, 16
    Reichert Meissl, 33
    saponification value, 33
    solidification point, 33
    specific gravity, 33
    tocopherol, 18
Coconut oil,
    acid value, 33
    Crismer value, 28
    iodine number, 33
    Kirschner value, 30
    phytosterol, 16
    Reichert Meissl, 33
    saponification value, 33
    solidification point, 33
    specific gravity, 33
    tocopherol, 18
Coffee,
    and hard water, 252
    flavor composition, 167
    synthetic flavor, 165

Collagen, 175, 176–178
    from different phyla, 178
    hydroxy proline content, 177
    structure, 176–177
Color and vitamin A value, 234
Colostrum, 294
    composition, 294
    protein, 143
Compensation in taste, 152
Conalbumin, 143, 144
Connective tissue, 174–181
    ground substance, 179
Constant, protein, 130–131
Consumer acceptance in testing flavor, 160
Contrast in taste, 152
Cookies, fat in, 60
Cooking,
    and carbohydrates, 104–107
    and cellulose, 269
    and intercellular air, 274
    and pectic substances, 265–269
    and starch granules, 269–271
    oil, 54–55
    patterns, fats, 55
Copper reduction by carbohydrates, 103
Corn,
    canned, odor, 279
    carotenoids present, 233
    heated, 137
Corn cob hemicellulose, 86
Corn germ oil phytosterol, 16
Corn protein, amino acids, 115
Corn oil,
    and reversion, 41
    Crismer value, 28
    fatty acids, 19
    phosphatide, 14
    phytosterol, 16
    tocopherol, 18
    smoke, flash, fire point, 27
Cornflower,
    anthocyanin, 242
Cotton seed meal,
    amino acids in protein, 141
Cotton seed oil,
    Crismer value, 28
    fatty acids, 19
    influence of processing on tocopherol, 18
    in shortening, 55
    in margarine, 58
    phytosterol, 16
    tocopherol, 18
Coumarins, 240
Cowpea, carotenoids present, 233
Cracker type shortenings, 56

Cranberry, anthocyanin, 244
Cream for butter, 51
Crisco, phytosterol, 16
Crismer test, 26
    values for fats, 28
Crocetin, 232
Crocin, 232
Cruciferous family, peppery taste, 278
Crude fat, 13, 17
Crude fiber, 107
Crude protein, 129
Cryptoxanthin, 231, 233
p-Cumaric acid in anthocyanins, 242
Cutin, 222
    composition, 222-223
Cyanidin chloride, 241
    formula, 242
    properties, 246
Cyanin in fruit, 244
    reactions, 243
Cystein and browning prevention, 258

Daniels and Heisig, 344
Darkening of fats, 17
2, 4 Decadienal, 42
Deep fat frying oil, 54
Delphinidin chloride, 241
    formula, 242
    properties, 246
Delphinin in fruit, 244
Denaturation of proteins, 124-125
    agents, 124
    and enzymes, 125
    changes in structure, 125
    irreversible, 124
    reversible, 124
Denatured protein gels, 128
Depolymerase for pectic substances, 263
Desserts in diet, 62
Dextrinization of starch, 105
Dextrins in wheat, 325
Diacetyl in butter, 52
Diallyl disulfide, 278
Diallyl trisulfide, 278
Diet,
    American lipid, 12
    and milk composition, 297
    Eskimo's, 116
    fat, 62
    pattern with fruits and vegetables, 218
    primitive man, 116
    temperate zone, 116
    vegetable, 116
Difference tests for flavor, 157

Diglycerides, 56
    emulsifier, 57
    in frying fats, 57
    in margarine, 59
    preparation, 57
Dihydroxy benzenes, antioxidants, 37
o-dihydroxy phenols, oxidation, 256
Dilution test for flavor, 157-158
Di-n-propylketone, 42
Disaccharides, 68-70
Diurnal variation in milk composition, 297
Double action baking powder, 338
Dough, 329-337
    amylase, 332-333
    and hard water, 337
    and milk, 337
    bacteria, 331
    carbohydrate, 332
    conditioners, 333
    gas production, 329-335
    gas retention, 335-337
    proteolytic enzymes, 335-336
    yeast, 329-331
Doughnuts, fat in, 60
Dulcin, 69
Dumas nitrogen, 133

EDTA formula, 40
Egg,
    albumin, 129
    protein, 143
    protein and batters, 340-341
    yolk and vitamin A, 234
        carotenoids, 234
    yolk membrane, 146
    yolk proteins, 146
    white, trypsin inhibitor, 137
Eggplant anthocyanins, 244
Elastin, 178-179
    polar and nonpolar groups, 179
    structure, 178-179
Electrophoresis,
    paper, 120
    proteins, 117-120
Elder, red berried anthocyanin, 244
Elderberry, European anthocyanin, 244
Embryo in grain, 317
    composition, 318
Emulsifying agents,
    allowed, 358
    and batters, 341-342
    and staling, 346
    in margarine, 59
    restricted, 360

Endosperm in grain, 317
  composition, 318
Enrichment of flour, 326–329
  standards, 327
Enzyme inhibitors in protein, 136, 137
Enzymes,
  in denatured protein, 125
  in natural flavor, 168
  milk, 295–296
Epicatechins, 250
  structure, 250–251
Epidermis,
  cutin, 222
  stomata, 222
  suberin, 222
Epimysium, 172
Eryodictin, 250
Eskimo's diet, 116
Essential fatty acids, 63
Essential oils allowed, 362–365
Esters in plant waxes, 224
Ethers in plant waxes, 224
Ethyl esters of fatty acids, fractionation, 19
Ethylenediaminetetraacetic acid, 40
Evaporated milk,
  discolored, 253
  heated, 136
Euglobulin, of milk, 142
Extractives from muscle, 173
Extracts, flavoring, 161–162
  in flavors, 163
  preparation, 161
  synthetic, 164–165

Family patent flour, 320
Fat. See also Fats.
  and tenderness in cake, 60
  and texture in cake, 60
  animal, 14
  calories, 62
  crude, 13
  in diet, 62
  in leavening, 60
  in pastry, 61
  saponification, 13
  satiety value, 63
Fatigue for taste, 152
Fatigue of sense of smell, 154
Fats, 12–63
  acetyl value, 31
  added to food, 12
  alpha, 23
  and batters, 341

beta, 23
chemical properties, 27–33
composition, 14, 17
consumption, 12, 13
Crismer value, 26
crystalline forms, 23
darkening, 17
Fryer and Weston test, 26
gamma, 23
Hehner value, 30–31
identification, 23
in adipose tissue, 179
in cooking, 55
in wheat flour, 321
Iodine Number, 27, 31, 32
melting points, 23, 24
occurrence, 12
pigments, 17
plasticity, 23
Polenske Number, 27
polymorphism, 23
refractive index, 25
Reichert Meissl Number, 27–29, 32
Saponification Number, 27, 30, 32
shortening power, 61–62
shot melting point, 25
slipping point, 25
smoke, flash, fire points, 26
softening point, 24
solubility, 13
soluble vitamins, 63
specific gravity, 25
turbidity point, 26
Valenta Number, 26
Fatty acids, 18
  essentials, 63
  esters, fractionation, 19
  formulas, 22
  fractionation, 18
  free, 16, 45
  in common fats, 19
  in plant waxes, 224
  infra-red spectra, 20
  lead salts, 19
  natural, 21
  saturated, 19, 21
    melting points, 22
  steam distillable, 19
  unsaturated, 19, 21
    cis-trans, isomerism, 20–21
    determination, 20–21
    melting points, 22
  water soluble, 19
Fatty alcohols in plant waxes, 224
Feel factors in food, 154–155

Fehlings solution with anthocyanidins, 246
Fermentation, alcohol, methanol from pectins, 95
Ferricyanide reduction by carbohydrates, 103
Fiber, crude, 107, 222, 321
  determination, 107
  in plants, 222
  in wheat flour, 321
Fibrils of collagen, 176
Fibrin clot, 127–128
Fig, anthocyanin, 244
Fish oil in shortening, 55
Flash evaporation of fruit juices, 168
Flavanols, 247
Flavones, 247–250
  and cooking, 249
  and NaOH, 249
  basic ring structure, 239, 247
  physiological effect, 249
Flavonoids, 239–254
  and browning, 256
  classes, 240
  defined, 240
Flavor, 148–170
  control, 156–162
  extract, 161–162
  in butter, 52
  in cheese, 311–312
  in fats, 32
  intensifier, 160–161
  measurement, 157–160
  of milk, "oxidized," 298
    "sunlight," 298
    "cowy," 298
    "cooked," 298
  panel environment, 158–160
  profile, 158
  research, 162–168
  sensation, 148–156
Flavormatics, 163–165
Flavors,
  appropriate, 155
  imitation, 163–165
  synthetic, 163–165
Flesh, defined, 171
Flour, 319–329
  and dough, 321
  clear, 320
  composition, 321
  enrichment, 326–329
  gluten, 322–324
  improvers, 333
  milling of, 320
  minerals, 328

  patent, 320
  proteins, 321
  standards, 327
  straight, 320
  types, 320
Food additives, 351–364
  amendment, 352
  colors, 355
  defined, 352
  exempt from restrictions, 357–359
  need for, 353–354
  safeguards, 351–353, 354–355
Food, cheese, 314
Food chemistry development, 1–3
Food tannins, 240
Fortification bread, 327
  margarine, 60
Freeze dry and proteins, 130
Fructans, 79
Fructose, decomposition on heating, 8
  in fruit and vegetables, 219
  in potatoes, 68
  structure, 66
Fruit. *See also* fruits.
  bruised browning, 255
  concentrated, 167
  carbohydrates, 219, 282–286
  changes in organic acids, 286–288
  climacteric, 281
  defined, 218
  dried browning, 260
  gums, 97
  juice viscosity, 263
  nutritional importance, 218
  organic acids, 286–288
  pectic substances, 283–286
  pectin gels, 92–94
  pigment, 226–254
  post harvest changes, 280–288
  senescence, 281
  starches, 282, 283
  texture, 225–226
  water loss, 282
Fruits. *See also* fruit.
  and cooking, 265–280
  post harvest changes, 280–288
  structure, 220–225
  white, 247
Fryer and Weston Test, 26
  values, 28
Frying oil, 54–55
Fuller's earth, fat production, 17
Fungi, amino acids in protein, 138, 140
Furanose, 67

Furfural,
   and browning, 260–261
   and carbohydrate tests, 98

Gadusene, 15
Galactans, 79
Galactobiose, 71
Galactose, in anthocyanins, 241
   structure, 65
Gallic acid,
   formula, 240
   oxidation, 256
3-Galloyl epicatechin, 251
3-Galloyl gallocatechins, 251
Garlic odor, 278
Gas chromatography,
   in flavor research, 166
   of fatty acid esters, 19–20
   of meat vapors, 166
Gel,
   denatured protein, 128
Gel formation, 125
   factors affecting, 126
   in baked goods, 343
   theories, 126–128
Gel point, 126
Gelatin, 127–128
   and pH, 128
   gel, 128
   preparation, 128
Gelatinization of starch, 105–107
Gels, starch, 106–107
Germ in grain, 317
   composition, 318
"Ghosts" in flour, 321
Glands, lipid, 14
Gliadin, 115, 322
Globulins,
   in cereals, 322
   of egg white, 145
Glucamine, antioxidant, 37
D-Glucose,
   in anthocyanins, 241
   in fruit and vegetables, 219
   in muscles, 173
   in potatoes, 68
   structure, 65, 72
Glutamic acid, preparation, 160
Glutathione and browning prevention, 258
Gluten, 61, 322–324, 335, 336
   and fat, 61
   and oxidation, 336
   and proteolytic enzymes, 335
   composition, 322–323

   development, 323
   in batters, 340
   in cake, 61
   in pastry, 61
   hydration, 324
   reactions, 336
   strength, 324
   swelling, 323
Glycogen, 73, 183–184
Glycosides of anthocyanidins, 241
Gortner, 8
Gould, 252
Grading butter, 306
Grain, 317–319
   heating, 319
   moisture, 319
   molds, 319
   respiration, 318
   storage, 318–319
   structure, 317–318
Grapefruit, carotenoids present, 233
Grapes,
   anthocyanins, 244
   color in juice, 245
Green vegetables,
   cooking, 238
   pH, 238
Ground substance of connective tissue, 179
Guar, 86
Gum arabic, 95–96
Gums and mucilages, 95–97
Gums of fruit trees, 97

Hanus Iodine Number, 31
Hard water and dough, 337
Hard wheat and gluten, 324
Hazard in food additives, 354
   defined, 354
Heat treatment of proteins, 136–138
Heating of grain, 319
Hehner Value, 30–31
Helium in gas chromatography, 19–20
Hemicelluloses, 83–87, 268, 285
   and texture, 268
   composition, 85
   differences from cellulose, 84
   in fruit, 285
   structure, 84, 86
Hemoglobin, molecular weight, 121
   structure, 190–191
2-Heptenal, 42
Heptyl aldehyde, rancidity, 34
Hesperetin, 248

Hesperidin, 250
Hess, 322
Heteropolysaccharides, 73
Hexoses, 65–66
"High ratio" shortenings, 56, 57
Hirsutidin, 241
  properties, 246
Histidine, formula, 132
  in Kjeldahl determination, 132
Holocellulose, 83
Homopolysaccharides, 73
Hops, astringency and tannin, 254
Hulme, 282, 287
Hunger and fat, 63
Hyaluronic acid, 80–83
  distribution, 82
  occurrence, 80–81, 82
  structure, 80–81
Hyalurono sulfuric acid (mucoitin sulfuric acid), 82
Hydrates, 5, 7
Hydration of protein, 122–123
  extent, 123
Hydrocarbons, in natural fats, 15
  in plant waxes, 224
Hydrogen bond, 6–7
  disruption, 7
  in protein, 122–123
  in starch, 106
  with oxygen, 6
  with nitrogen, 7
Hydrogen for hydrogenation, 50
Hydrogen sulfide in vegetables, 279
Hydrogenated vegetable oils, 17
Hydrogenation of fats, 48
  catalyst, 49
  course of reaction, 48
  introduced in U. S., 55
Hydrolysis of carbohydrates on cooking, 104–105
Hydroperoxides, rancidity, 34, 36
Hydroquinone oxidation, 256
p-Hydroxy benzoic acid in anthocyanins, 242
p-Hydroxy cinnannic acid in anthocyanins, 242
Hydroxy proline in collagen, 177
Hydroxy-α-carotene, 233

Ice cream, chocolate, discolored, 252
L-Iduronic acid, 81
Imbibed water, 6
Imitation flavors, 163–165
Individual uniformity, 4
Individual variability, 3–4

Infra-red spectra of fatty acids, 20
Insects in grain, 319
  invasion and epidermis, 222
Interesterification of lard, 53
  catalyst, 53, 54
Invert sugar, in pectin gels, 93
  syrups, 69
  sweetness, 69
Iodine and carbohydrate determination, 100, 104
Iodine Number, 27, 31, 33
  Hanus, 31
  Wijs, 31
Ions, bound by protein, 122
Ions in vegetables, 219
Iron salts and anthocyanins, 246
Isoelectric point of proteins, 118
Isoflavone, 247
Isoleucine, formula, 114
Isolinoleate, 49
  and reversion, 40
Iso-oleate, 48
Isoprene, 230

Japan wax, 222–224
Jelly grade of pectin, 93
Jermyn and Isherwood, 285
Johnson, et al., 334
Josephson, 298
Joslyn and Ponting, 255
Juniperic acid, 224

Karrer and Jucker, 242
Kay and Graham test, 301
Ketones in plant waxes, 224
Kirchner, 276
Kirschner Value, 28, 29
Kjeldahl method, 131–133
  catalysts, 131
  digestion, 131
  measurement, ammonia, 132
  oxidizing agents, 132
  salts added, 132
Knowlton, 253
Koettstorfer Number. See Saponification Number.
Kreis and Shibstead test, 39
Krukovsky, 298
Kwaskiorkor, 116

Lactalbumin, 142
  heated, 137
Lactglobulin, 142

Lactic acid, in cream, 52
  in fruit and vegetables, 276
  in jelly, 94
Lactobacilli, 135
Lactobacillus acidophilus,
  in dough, 331
  in milk, 299, 305
Lactobacillus bulgaricus,
  in dough, 331
  in milk, 305
Lactobacillus casei,
  in dough, 331
Lactose,
  determination in milk, 300–301
  occurrence, 68
  structure, 68
Lactylates, 348
Lappin-Clark test, 39
Lard, 17, 52–54, 55
  composition, 53
  Crismer value, 28
  from various tissues, 180
  hydrogenated, 53
  interesterification, 53
  leaf, fatty acids, 19
Lard compound, 55
Lard oil,
  acid value, 33
  iodine number, 33
  Reichert Meissl, 33
  saponification value, 33
  solidification point, 33
  specific gravity, 33
Lea peroxide test, 39
Lead salts of fatty acids, 19
Leaf,
  cells, photographs, 223
  stomata, 222
  vegetables, 219
Leavening,
  in batters, 337–339
  in dough, 329–335
Lecithins, 12, 14, 38, 59
  added to food, 12
  antioxidants, 38
  formula, 14
  in margarine, 59
Lemon juice and browning prevention, 258
Leucine, formula, 114
Leucoanthocyanins, 240, 250, 251
Leucoplasts, 220
Ligaments, 174
Ligamentum nuchae, 179
Light scattering molecular weight determina-
    tion, 121

Lignins, 83, 222, 285
  development in supporting tissue, 222
  in fruit, 285
  low in phloem, 222
Lindet, 255
Linoleic acid, 19
Linoleate, rancidity, 36
Linolenate, and reversion, 40
  rancidity, 36
Linseed meal, amino acids in protein, 141
Linseed oil,
  acid value, 33
  Crismer value, 28
  iodine number, 33
  Reichert Meissl, 33
  saponification value, 33
  smoke, flash, fire points, 27
  solidification point, 33
  specific gravity, 33
Lipase in cheese, 310–311
Lipids, 12–63
  hydrolysis in cheese, 310
  in adipose tissue, 179
  in diet, 12
  in milk, 295, 296
  in muscle, 172
  occurrence, 12
  variation in composition, 179
Lipovitellin, 146
"Liquid sugar", 69
Livetin, 146
Locust bean gum, 97
Long patent flour, 320
Low ester pectins, 94–95
Low methoxyl pectins. *See* low ester pectins.
Lowe, 343
Lutein in egg yolk, 234
Lycopene, 230
  formula, 230
  occurrence, 230, 233
Lyophile and protein, 130
Lysine, formula, 114, 132
  fortification of bread, 327
  in Kjeldahl determination, 132
Lysozyme, 143, 144

Machining quality and gluten, 323
Mackinney and Weast, 236, 237
Macy, Kelly, and Sloan, 294
Maillard reaction, 259
Malic acid in fruit and vegetables, 275–277
Malonaldehyde, 39
Malonic acid, in anthocyanins, 242
Malt extract, 332, 333

Malted barley, 332, 333
Malted wheat, 332, 333
Maltose,
  in potatoes, 68
  structure, 69
Malvidin, 244
  properties, 246
Malvin in fruit, 244
Mann and Weier, 271
Mannans, 79
Manninotriose, 71
Mannose,
  in potatoes, 68
  structure, 66
Margaric acid, 58
Margarine, 57–60
  colored, 60
  consistency, 58
  emulsifying agents, 59
  fortified, 60
  history, 57
  leaking, 59
  plastic range, 58
  preservative, 60
  source, 58
  trans isomerism, 49
Mayonnaise, oil for, 54
Measurement of flavor, 157–160
Meat, 171–217
  cells, 171
  defined, 171
  fibers, 171
  structure, 171–172
  vapors, gas chromatogram, 166
Melanins, 110
Melanoidins, 109
Melibiose, 71
Melnick, et al., 349
Melting points, fats, 23, 24
  triglycerides, 24
Membrane, egg yolk, 146
Metal scavengers, 40
Methanol from pectins, 95
Methanolysis of fats, 20
Methionine, formula, 115
  in heated soy beans, 137
Methyl esters of fatty acids, fractionation, 19
Microorganisms, in milk, 299
  in souring cream, 51–52
  in souring milk for margarine, 59
Middle lamella pectic substances, 261, 262
Middlings, 320
Milk, 293–315
  acetone bodies, 298
  acidophilus, 305

albumin, 142
analysis, 299–303
butter fat, 299–303
certified, 303
checks for purity, 302
clotting with rennin, 141
composition, 294, 295–299
  enzymes, 295–296
  lipids, 295, 296
  variation, with breed, 296
    with diet, 297
    with lactational cycle, 297
    with season, 297
  vitamins, 295
defined, 293
euglobulin, 142
evaporated, 136
  discolored, 253
fermented, 305
filled, 305
flavor, 298
formation, 294, 295
freezing point, 302
homogenized, 303
in dough, 337
in margarine, 58–59
lactose determination, 300
microorganisms, 299
preservatives, 303
protein in, 140–143, 301
pseudoglobulins, 142
reconstituted, 305
regulations, 299
serum, 302
solids determination, 300
solids-not-fat, 302–303
special, 303–305
"standardized", 299
vitamin D, 304
whey, 302
Miller amendment, 352
Minerals, in vegetables, 219
Mixing batters, 340–341
Moisture determination, 8
  by drying, 8
  by distilling, 8
  by nuclear magnetic resonance, 9–11
  in baked goods, 345
  in food, 5–6
  in grain, 319
  in wheat flours, 321
  occurrence, 5
Molds,
  in cheese, 313
  in grain, 319

invasion of fruit, 222
prevention in baked goods, 347–349
Molecular weights of proteins, 117–122
Molisch reaction, 98
Monoglycerides, 56
    emulsifier, 57
    in frying fats, 57
    in margarine, 59
    preparation, 57
Monosaccharides and browning, 260
Monosodium glutamate, 160–161
    flavor intensifier, 160
    preparation, 160–161
"Mouth feel," 148
Mucoitin sulfuric acid, 82
Mucopolysaccharides, 82–83
Mucoproteins, 82–83
Mulberry, black, anthocyanin, 244
Muscle, 171–174
    composition, 172–174
    lipids, 14
    salts, 173
    smooth, 171
    striated, 171
    voluntary, 171
Mustard oil, 279
Mutton tallow, fatty acids, 19
Myogen, 172
Myosin, 172
Myricyl cerotate, 224
Myricyl isobehenate, 16
Myricyl lignocerate, 16
Myristic acid, 19

Native protein, 124
Neo-cryptoxanthin, 233
Nervous tissue, lipid, 14
Nessler's salt in Kjeldahl, 132–133
Neuraminic acid, 83
Neutralizing agents allowed, 358
Niger oil, Crismer value, 28
"Nitrogen bases" in muscle, 173
Nitrogen, Dumas, 133
    in protein, 130–135
    in wheat flours, 321
Nonsaponifiable, 13
Nuclear magnetic resonance, 9–11
Nucleus of plant cells, 220
Nutrients allowed, 359
    restricted, 360
Nuts, lipid, 12

Oatmeal, lipid, 12
Oats, amino acids in proteins, 141

Odor, 152–154
    fatigue, 154
    recall, 152
    sensation of, 153
    testing, 154
Off-flavor, 156
Official method of analysis of the Association of Official Agricultural Chemists, 2
Oils. See also Fats.
    green, 17
    vegetable, phytosterol, 16
Oleate, rancidity, 36
Olfactory nerves diagram, 153
Olfactory response, 153–154
Oleic acid, 19
Oleomargarine. See Margarine.
Oleo oil, 52
    in margarine, 56
Oleoresins allowed, 362–365
Oleostearin, 52, 55
Oligosaccharides, 71–72
Olive oil,
    acid value, 33
    Crismer value, 28
    fatty acids, 19
    iodine number, 33
    phytosterol, 16
    Reichert Meissl, 33
    saponification value, 33
    smoke, flash, fire point, 27
    solidification point, 33
    specific gravity, 33
    tocopherol, 18
Olive, ripe, lipid, 12
Onion cells, drawing, 221
    flavor, 167
    odor, 278
    photograph, 221
    yellow color, 249
Onslow, 255
Optical activity of carbohydrates, 102
Orange carotenoids present, 233
Orcinol, 104
    carbohydrate test, 98
Oregano, antioxidant, 38
Organic acids in foods, 276, 277
Organisms in cheese, 313
Oryzenin, 322
Osborne, 322
Osmotic pressure,
    and cell turgor, 225
    and molecular weight, 121
    and proteins, 121
Oven test, rancidity, 39

Overgrinding wheat, 333
Ovomucin, 144, 145
Ovomucoid, 137, 144, 145
Oxidases in browning, 255
Oxalic acid in fruit and vegetables, 275–277
Oxidizing agents in dough, 336
Oxygen and fats, 34–38

Palm oil,
  acid value, 33
  Crismer value, 28
  iodine number, 33
  Kirschner value, 30
  Reichert Meissl, 33
  saponification value, 33
  solidification point, 33
  specific gravity, 33
Palmitic acid, 19
Paper chromatography, amino acids, 134
Paper electrophoresis, 120
Paracasein, 141
Parenchyma cells, 220, 221
  living and dead, 229
  of squash, 229
  Parisian barm, 334
  Particle orientation and gel formation, 127–128
Pasteurization tests, 301
Pastry, function of fat, 61
Patent flour, 320
Peach,
  carotenoids present, 233
  flavor compounds, 165
  pectic substance, 262
Peanut flour, amino acids in proteins, 141
Peanut meal, heated, 137
Peanut oil,
  acid value, 33
  Crismer value, 28
  in margarine, 58
  in shortening, 56
  iodine number, 33
  phytosterols, 16
  Reichert Meissl, 33
  saponification value, 33
  solidification point, 33
  specific gravity, 33
  tocopherols, 18
Peanut skin color, 252
Pearl starch, 78
Pears,
  changes in carbohydrates, 284–286
  hemicellulose, 86
  pectic substances, 262–263

"sleepiness," 284
stone cells, 286
Pearson and Robertson, 281
Peas,
  canning and nutritive value, 137
  color in canned, 238
  seed coat, 268
  tall, anthocyanin, 244
  toughness, 239
Pecans, lipid, 12
Pectates, 262
Pectic acids,
  defined, 88
  structure, 88–90
Pectic substances, 87–95, 261–264
  and texture, 268
  changes in fruit, 283–286
  changes on cooking, 265–269
  defined, 88
  in cells, 261
  in fruits, 263
  jellies, 91–95
  occurrence, 87
  stained, 261
  structure, 87–88
Pectin esterase, 95
Pectin gels, 92–94
  factors influencing, 92
  jelly grade, 93
  setting time, 93
Pectin methyl esterase, 263
Pectinase, 263
Pectinic acids,
  defined, 88
  structure, 88–90
Pectins in alcoholic fermentation, 95
Pelargonidin,
  chloride, 241
  formula, 242
  properties, 246
Pelargonin in fruit, 244
Pentosans in wheat, 325
  in wheat flour, 321
Pentoses, 66–67
  in anthocyanins, 241
  in hemicelluloses, 84, 85, 86
Peonidin, 241
  properties, 246
Peonin in fruit, 244
Pepper, red,
  carotenoids present, 233
Perilla oil,
  acid value, 33
  Crismer value, 28
  iodine number, 33

Reichert Meissl, 33
saponification value, 33
smoke, flash, fire point, 27
solidification point, 33
specific gravity, 33
Perimysium, 172
Peroxidases, and browning, 255
test in milk, 301
Peroxide value, 39
Peroxides, and browning, 255
and rancidity, 34, 35
Petunidin, 241
pH,
and browning prevention, 258
and cake texture, 345
and chlorophyll, 237, 238
and pectin jels, 92
of cakes, 343–345
of syrups, 344
Phellonic acid, 224
Phenol oxidase, 256
Phenolase, 256
Phenols and browning, 255
Phenyl caffeate, 251
β-Phenyl ethyl isocyanate, 279
Phenyl thiocarbamide,
inheritance of taste for, 152
taste, 152
Phenylalanine, formula, 115
Phenylhydrazine and carbohydrates, 100
Pheophytin, 237, 238
Phlobaphene, 251
Phloem, 222
Phloroglucinol, oxidation, 256
Phosphatase, test in milk, 301
Phosphate baking powder, 338
Phosphate in wheat flour, 321
Phosphatide, 14
antioxidants, 38
Phospholipids, 13
Phytol, 236
Phytosterols in vegetable oil, 16
Phytylalcohol, 236
Picramic acid, 100
Picric acid and carbohydrate test, 100
Pie crust, fat in, 60
Pigments,
in fats, 17
in fruits and vegetables, 226–254
in plants, 228
Pineapple juice and browning prevention, 258
Plant protein, 138–140
amino acids, 140
Planteobiose, 71

Planteose, 71
Plasticity of margarine, 58
Plastids of cells, 220
Polenske Number, 27–28
Polygalacturonase, 263
Polyphenol oxidase, 256
Polyphenolase, 256
Polyphenols and browning, 111–112
Polysaccharides, 72–97
and anthocyanin color, 243
Pomegranate anthocyanin, 244
Poppy seed oil,
acid value, 33
iodine number, 33
Reichert Meissl, 33
saponification value, 33
solidification point, 33
specific gravity, 33
Porphyrins and chlorophyll, 235
Post harvest changes in fruit, 280–288
Potassium bromate in dough, 336
Potassium iodate in dough, 336
Potato chip browning, 112
Potatoes,
and starch granules, 270, 271, 272, 273
blackening, 258, 261
dehydrated mashed, 269
sugar content, 68
yellow color, 249
Pound cake, fat in, 60
volume, 61
Pre-ferments, 333–335
Preservatives allowed, 358
restricted, 360
sodium benzoate in margarine, 60
Pressing for oils, 43
Primitive fibers of collagen, 176
Pristane, 15
Process cheese, 314
emulsifying agents, 314
Prolamines in cereals, 322
Prooxidants, fat, 38
Propyl gallate, 37
Proteins, 114–146
amphoterism, 122
binding ions, 122
carbohydrate reaction, 136
constants, 130
crude, 129
denatured gel, 124, 128–129
determination, 129
determination in milk, 301
digestion, 136
and heat treatment, 136
heat treatment, 136–138

Proteins (*cont.*)
  hydration, 122–123
    in batters, 343
  hydrolysis in cheese, 311
  in colostrum, 142–143
  in diet, 115–116
  in fruit and vegetables, 218
  in egg white, 143–146
  in egg yolk, 146
  in milk, 140–143
  in muscle, 172
  in plants, 115, 138, 139, 140
  in wheat flour, 321
  in whey, 142
  molecular weights, 117–122
  native, 124
  nutritive value, 136
  properties, 116–129
Protocatechuic acid oxidation, 256
Protofibrils of collagen, 176
Protopectin, 88, 90–91, 262
  defined, 88
  isolation, 90–91
  structure, 91
Protoplasm in plant cells, 220
PTC. *See* Phenyl thiocarbamide.
Pteridophytes, amino acids in protein, 139
Pseudoglobulin of milk, 142
Puff-pastry, margarine for, 58
Pumpkin seed oil,
  acid value, 33
  iodine number, 33
  Reichert Meissl, 33
  saponification value, 33
  solidification point, 33
  specific gravity, 33
Pure Food and Drug Act, 2
Pyranose, 67
Pyrogallol oxidation by enzymes, 256

Quartering, 5
Quercetin, 248
Quévenne lactometer, 300
Quinic acid,
  change in apples, 287
  formula, 240
  in fruit and vegetables, 276

Raffinose,
  family, 71–72
  occurrence, 71
  structure, 71
Rancidity, 33–39
  course of reaction, 35

hydroperoxides, 34, 36
  in baked goods, 38
  in shortening, 56
  peroxides, 34, 36
  tests, 38–39
Ranking, test for flavor, 158
Rape seed oil,
  acid value, 33
  Crismer value, 28
  iodine number, 33
  Reichert Meissl, 33
  saponification value, 33
  solidification point, 33
  specific gravity, 33
Raspberry,
  pectic substances, 285
  synthetic flavor, 164
Recall of odors, 152
Reconstituted milk, 305
Red dog, 320
Red pepper, carotenoids present, 233
Reeve, 271, 274
Reeve and Leinbach, 267, 268, 274
Refined lard, 55
Refiners syrups, 71
Refining fats and oils, 44–47
Refractive index, fats, 25
Reichert Meissl Number, 27–29, 33
Reinecke salt, 135
Rendering, 42–43
Rennin (rennet), 141
Resorcinol and carbohydrate tests, 98
Respiration through stomata, 222
Respiration rate of,
  apples, 281
  bananas, 282
  pears, 281
Retrogradation of starch, 76, 94, 107
Reversion, 39–41
  factors affecting, 40
  in shortening, 56
  isolinoleate, 40
  linolenate, 40
  products, 42
  resistant, 41
  susceptible, 41
L-Rhamnose,
  in anthocyanins, 241
  structure, 67
D-Ribose,
  structure, 67
Rice,
  amino acids in protein, 141
  need for fortification, 327–328
  yellow color, 249

Ricinoleic acid, 31
Ripening cheese, 311
Rose anthocyanin, 242
Röse Gottlieb method, 300
Rosenberg, 327
Rotation of carbohydrates, 102
Rutin, 248, 250

Sabinic acid, 224
Saccharin, formula, 151
Saccharomyces bayanus, 101
Saccharomyces carlsbergensis, 101
Saccharomyces cerevisiae, 95, 100, 329, 330
Safflower oil, tocopherol in, 18
Saffron, 232
Sakaguchi test, 135
Salad oil, 54-55
    source, 54
    winterizing, 54
Saltiness, 150
Salting butter, 51
    margarine, 60
Salts in muscle, 173
Sampling methods, 4-5
Saponification, 13
Saponification Number, 27, 30, 32
Sarcolemma, 171
S.A.S. baking powder, 338
Saturated fatty acids, 19
Schardinger test, 301
Schiff base, 109
Scoring baked products, 347
Scoring butter, 306
Scott, 268
Season and milk composition, 297
Seaweed polysaccharides, 96
Sedimentation of protein, 120-122
Seliwanoff's reagent, 98
Sensation of smell, 153
Sense of smell, 152-154
Sensitivity to odors, 153
Sequestrants allowed, 359
    restricted, 360
Sesame oil,
    acid value, 33
    Crismer value, 28
    iodine number, 33
    phytosterol, 16
    Reichert Meissl, 33
    saponification value, 33
    solidification point, 33
    specific gravity, 33
    tocopherol, 18
Seshardi, 248

Seven Minute frosting, discolored, 253
Shigelle dysenteriae, 144
Shikimic acid, change in apples, 287
Short patent flour, 320
Shortening, 17, 55-57
    for crackers, 56
    source, 55
    value of fats, 60-62
Shortening power, 61-62
    determination, 61
    of fat, 61-62
    theories, 61-62
Shortometer, 61
Shot melting point, fat, 25
Sialic acid, 83
Simpson and Halliday, 279
Sitosterol formula, 15
Skimmed milk in margarine, 58-59
Skimmed milk powder in bread, 327
"Sleepiness" in pears, 284
Slipping point, fat, 25
Smell, 153
S-Methyl-cystein-S-Oxide, 278
Smoke, flash, and fire points, fats, 26, 27
Soda in cooking green vegetables, 238
Sodium benzoate, margarine, 60
    taste, 152
Sodium citrate buffer in jelly, 94
Sodium glutamate as flavor intensifier, 160
Sodium potassium tartrate buffer in jelly, 94
Sodium propionate and mold prevention,
    348-349
Sodium sulfide and browning prevention,
    258
Sodium sulfoacetate of glycerides, 60
Solvent extraction, oils, 43-44
Soft wheat, 324
Sorbic acid and mold prevention, 348
Sourness, 151
Soybean,
    carotenoids present, 233
    trypsin inhibitor, 137
Soybean meal, amino acids in protein, 141
Soybean oil,
    acid value, 33
    and reversion, 41
    Crismer value, 28
    in margarine, 58
    in shortening, 55
    iodine number, 33
    phosphatide, 14
    phytosterol, 16
    Reichert Meissl, 33
    saponification value, 33
    smoke, flash, fire points, 27

Soybean oil (*cont.*)
  solidification point, 33
  specific gravity, 33
  tocopherol, 18
Spattering of margarine, 59
Specific gravity, fats, 25
Specific rotation, 102
Spermatophytes, amino acids in protein, 140
Spices, adulteration, 107
  allowed, 361–362
  antioxidant, 38
Spinach, color on processing, 236, 238
"Spirits of hart's horn," 339
Squalene, formula, 15
Squash cells, 229
  carotenoids present, 233
Stability of fats, 35, 36
Stabilizers allowed, 359
Stachyose, 71, 72
Stadtman, 259, 260
Staling, 346–347
  and emulsifying agents, 346
  and starch, 346
Stamberg and Bailey, 344
Standard patent flour, 320
"Standardized" milk, 299
Starch, and enzymes, 76–77, 271
  crystal, 78
  changes in harvested fruit, 282
  changes on cooking, 269–271
  dispersions,
    aging, 107
    retrogradation, 107
  gelatinizations, 105–106
  gels, 106
    in baked goods, 345, 346
  hydrolysis on cooking, 105
  in carrots, 234
  in fruit and vegetables, 219
  lump, 78
  modified, 78
  powder, 78
  swelling of granules, 105–106
  technology, 77–79
  thin boiling, 78
Starches, 75–79
  amylose, 73, 76, 77
  amylopectins, 73
  and enzymes, 76–77
  gels, 76
  occurrence, 75
  retrogradation, 76, 346
"Starters" of yeast, 329–330
  of cheese, 310

Steam as leavening, 339
Steam deodorizing, fats and oils, 46
Steam refining, fats and oils, 46
Stearic acid, 19
Sterols, 13, 15
Stoll and Seebeck, 278
Stomata in leaves, 222
Stone cells of pears, 286
Strain, 239
Strawberry, anthocyanin, 244
  flavor, 164
  synthetic flavor, 164
Streptococcus citrovorus, 52
Streptococcus faecalis, 135
Streptococcus lactis, 52
  in cheese making, 310
  in dough, 331
Streptococcus paracitrovorus, 42
String beans, color on cooking, 236–238
Straight flour, 320
Suberic acid, 224
Suberin, 222, 225, 268
  composition, 223
Substances restricted, 360
Sucaryl sodium, formula, 151
Succinic acid in fruit and vegetables, 276
Sucrose, 68–69
  commercial products, 69
  in fruits and vegetables, 219
  in potatoes, 68
  -invert sugar solubility, 70
  occurrence, 68
  structure, 68
Sugar, and batters, 341–342
  antioxidant, 38
  liquid, 69
Sugars,
  in harvested fruit, 283, 284
  in wheat flour, 321
  sweetness, 70
Sulfides in onions, 167
Sulfonamides and browning prevention, 258
Sulfur compounds,
  in vegetables, 278–280
  origin, 280
Sulfurous acid and browning prevention, 258
Sullivan, 323, 336
Sunflower seed oil,
  acid value, 33
  Crismer value, 28
  iodine number, 33
  Reichert Meissl, 33
  saponification value, 33
  solidification point, 33
  specific gravity, 33

Sunflower meal, heated, 137
Superglycerinated shortenings, 56
Svedberg, The, 120
Sweet potatoes, canned discolored, 252
Sweetners, non-nutritive, allowed, 359
Sweetness of sugars, 70
Swelling of starch granules, 105-106
Synthetic flavors, 162, 163, 164
Syringidin, 241
    properties, 246
Syrups, refiners, 69
Syne and Wood, 278
Szent-Györgyi, 249

Tallow, beef,
    acid value, 33
    Crismer value, 28
    fatty acids, 19
    iodine number, 33
    Reichert Meissl, 33
    solidification point, 33
    specific gravity, 33
Tallow, Chinese vegetable, Crismer value, 28
Tallow, edible, 55
Tannic acid, formula, 240
Tannins, 240, 250-254
    and anthocyanins, 243
    and hard water, 251
    and iron salts, 252
    astringency, 253
    composition, 250
    use of term, 250
"Tannin red," 251
Tartrate baking powder, 338
Tallow, mutton,
    acid value, 33
    fatty acids, 19
    iodine number, 33
    Reichert Meissl, 33
    saponification, value, 33
    solidification point, 33
    specific gravity, 33
Taste, 148-152
    and chemical constitution, 150
    buds, 148-150
    compensation, 152
    contrast, 152
    fatigue, 152
    primary, 148
    threshold, 150
Tasters, 156, 157
TBA test, 39
Tea, 249, 251, 254
    and hard water, 251

and organoleptic quality, 254
color, 249
tannins, 251
Tempering wheat, 320
Tendons, 174
Testing flavor, 157-160
Testing sense of smell, 154
Texture in batter products, 341, 342, 343
Texture in fruits and vegetables, 225-226
Thallophytes, amino acids in proteins, 140
Thioamides, and browning prevention, 258
2-Thiobarbituric acid test, 39
Thiosemicarbazide, and browning prevention, 258
Thixotropic, 106
Thompson and Buddenmeyer, 348
Three dimensional network in gels, 126
Threonine, formula, 115
Tiselius, electrophoresis apparatus, 118, 119
Tissue, lipid, 14
Tobacco mosaic virus, 127
Tocopherols, 17, 37, 38
    antioxidants, 37, 38
    content of fats, 18
    formulas, 37
Tomatoes,
    canned, firming, 266
    carotenoids, 230
Toughness in peas, 239
Toxicity, defined, 354
Trans isomers of fatty acid in margarine, 49
Trans oleate, 49
Triangle tests, 157
Trigeminal nerve and food, 154
Triglycerides, 14
    melting points, 24
Trypsin inhibitor in soybean, 137
Tryptophan,
    formula, 115, 132
    in Kjeldahl determination, 132
Tung oil, Crismer value, 28
Turbidity points, fats, 26
Turgor cell, 225, 226
    and cooking, 226
    and osmosis, 225
    factors affecting, 226
Tyndall effect, 121
Tyrosinase, 255, 258
    in potato, 258

Ultracentrifuge, 120-121
Umbelliferone formula, 240
Uniformity, individual, 4
Unripened, cheese, 307-310
Unsaturated fatty acids, 19, 20,
    determination, 20

Unusual component in proteins, 121
Uronic acids,
  in hemicelluloses, 84, 85
  in heteropolysaccharides, 80, 81
Ursolic acid, 224
Urushiol, oxidation, 256

Vacuoles in cells, 220
Valenta test, 26
Valine, formula, 114
Vallate papilla, 149
Variability, individual, 3-4
Vegetable diet, 116
Vegetable oils, hydrogenated, 17
Vegetables,
  browning reaction, 255
  bulbs, roots, tubers, 219
  carbohydrates, 219
  cooking changes, 265-280
  defined, 218
  flowers, buds, 219
  fruits, 219
  general composition, 219-220
  green and volatile acids, 275
  leaf, 219
  nutritional importance, 218
  pigments, 226-254
  seeds, 219
  structures, 220-225
  texture, 225-226
Verbascose, 72
Violaxanthin, 233
Vitamins A, 232-233
  A₁ formula, 232
    from carotenes, 232
  A₂ formula, 233
  and American diet, 233
  value and color, 234
Vitamin D in milk, 304
Vitamin P, 249
Vitamins, fat soluble, 63
  in milk, 295
  in vegetables, 219
  in wheat, 325, 326
Vitellenin, 146
Vitellin, 146

Wackenroder, 230
Wallerstein, 258
Walnut oil,
  acid value, 33
  iodine number, 33
  Reichert Meissl, 33

saponification value, 33
solidification point, 33
specific gravity, 33
Walnut oil, black,
  acid value, 33
  iodine number, 33
  Reichert Meissl, 33
  solidification point, 33
  specific gravity, 33
Walnuts, English, lipid, 12
Water, imbibed, 6
Water in fruits and vegetables, 218
Water loss, harvested fruit, 282
Watering milk, 302
Wax acids, 224
  alcohols, 224
Waxes, 13, 16, 222-224
  apple, 222-224
  carnauba, 222-224
  Japan, 222-224
  plant, 224
Whale oil,
  acid value, 33
  in margarine, 58
  in shortening, 55
  iodine number, 33
  Reichert Meissl, 33
  saponification value, 33
  solidification point, 33
  specific gravity, 33
Weier, 271
Weier and Reeves, 234
Weier and Stocking, 234, 269, 274
Wheat, 324-325
  carbohydrates, 325
  germ carotenoids, 233
    phytosterols, 16
    tocopherols, 18
  gluten, heated, 137
  flour baking quality, 329-345
  hemicelluloses, 86
  lipids, 324-325
  minerals, 328
  protein, amino acids, 115
  vitamins, 325, 326
Whey proteins, 142
White fruits and vegetables, 247
Whortle berry, anthocyanin, 244
Wijs Iodine Number, 31
Wiley, Harvey, 2
Wilting in plants, 225
Wine,
  and tannins, 254
  clarification, 263
  color, 245

Winterized oils, 47–48, 54
Wort and tannins, 254

Xanthophylls, 231, 233
Xylem, 222
D-Xylose,
   in potatoes, 68
   structure, 67

Yogurt, 305
Yeast, 329–331
   nutrition, 331–332
   wild, 330

Zamene, 15, 16
Zeaxanthin, 233
   in egg yolk, 234
Zein, 115, 322
Zwitterions of proteins, 118

2